Second-Wave Enterprise Resource Planning Systems

The focus of this book is on the most important class of enterprise systems, namely Enterprise Resource Planning (ERP) systems. Organisations typically take the decision to employ ERP systems in an attempt to streamline existing processes. Once these systems are in place, however, their success depends on the effectiveness of the implementation, and on the additional benefits that can be obtained by further leveraging the technology. In this book, the editors have assembled some of the world's best research on ERP systems, with a view to providing a foundation for second wave improvements to enterprise systems. Written primarily for managers and consultants, this book is also an ideal reference for business schools and researchers.

Graeme Shanks is Associate Professor and Deputy Head of the Department of Information Systems at The University of Melbourne, Australia.

Peter B. Seddon is Associate Professor in the Department of Information Systems at The University of Melbourne, Australia.

Leslie P. Willcocks is Professor of Information Management and E-business at Warwick Business School, UK, Associate Fellow at Templeton College, Oxford, and is Visiting Professor at Erasmus University, University of Melbourne, and Distinguished Visitor at the Australian Graduate School of Management.

Second-Wave Enterprise Resource Planning Systems

Implementing for Effectiveness

Edited by

Graeme Shanks
Peter B. Seddon and
Leslie P. Willcocks

PUBLISHED BY THE PRESS SYNDICATE OF THE UNIVERSITY OF CAMBRIDGE
The Pitt Building, Trumpington Street, Cambridge, United Kingdom

CAMBRIDGE UNIVERSITY PRESS
The Edinburgh Building, Cambridge CB2 2RU, UK
40 West 20th Street, New York, NY 10011–4211, USA
477 Williamstown Road, Port Melbourne, VIC 3207, Australia
Ruiz de Alarcón 13, 28014 Madrid, Spain
Dock House, The Waterfront, Cape Town 8001, South Africa

http://www.cambridge.org

First published 2003

Printed in the United Kingdom at the University Press, Cambridge

Typefaces Minion 11/14 pt. and Helvetica Neue *System* LaTeX 2_ε [TB]

A catalogue record for this book is available from the British Library

Library of Congress Cataloguing in Publication data
Second-wave enterprise resource planning systems : implementing for effectiveness /
edited by Graeme Shanks, Peter B. Seddon, and Leslie P. Willcocks.
p. cm.
Includes bibliographical references and index.
ISBN 0 521 81902 4
1. Management information systems. 2. Information resources management.
3. Production planning. 4. Business planning. I. Shanks, Graeme, 1951– II. Seddon, Peter
B., 1947– III. Willcocks, Leslie.
T58.6 .S394 2003
658.4′012–dc21 2002073766

ISBN 0 521 81902 4 hardback

Contents

Part IV Cultural Aspects of Enterprise Systems

Part V Future Directions

Contributors

Frédéric Adam is a lecturer in the department of Accounting, Finance and Information Systems at University College Cork in Ireland and a senior researcher with the Executive Systems Research Centre (ESRC). He holds a Ph.D. from the National University of Ireland and Université Paris VI (France). His research has been published in the *Journal of Strategic Information Systems, Decision Support Systems*, the *Journal of Information Technology, Systèmes d'Information et Management*, and the *Journal of Decision Systems* for which he is Associate Editor. He is the co-author of the *Manager's Guide to Current Issues in IT* and *Postgraduate Research* (Blackhall Publishing, Dublin, Ireland).

Sheryl Axline earned her Ph.D. from the School of Behavioral and Organizational Sciences, Claremont Graduate University, in 2000. Her dissertation research addressed team learning, organizational memory, and information technology in the context of ERP implementations. In her current role as Program Director, Learning Services, for the Center for Excellence in Nonprofits, Sheryl oversees a learning community of non-profit executives in Silicon Valley.

Taizan Chan received his Ph.D. in Computer and Information Sciences from the National University of Singapore in 1998. He is currently a Lecturer in the Faculty of Information Technology in the Queensland University of Technology and was a visiting scholar in the Wharton School, University of Pennsylvania in 1995 and 2001. His current research interests include the economics of information systems maintenance, ERP management, as well as the efficiency of electronic market mechanisms. He has published in journals such as *IEEE Transactions of Software Engineering, Journal of Software Maintenance and Evolution, Journal of Software and Systems.*

Guy G. Gable is Professor of Information Systems and Director of the Information Systems Management Research Centre, Queensland University of Technology. He has published widely (for example *Management Science,*

Journal of Strategic Information Systems, European Journal of Information Systems, Journal of Software Maintenance and Evolution) and is on the editorial boards of eight journals. He is Chief Investigator on several collaborative grants with SAP, including 'Cooperative ERP Lifecycle Knowledge Management'. He has a particular interest in knowledge management practices of large consulting companies and in the client–consultant relationship. His 1991 doctoral thesis (Bradford) titled 'Consultant Engagement Success Factors' won the inaugural ICIS thesis award.

Bob Galliers has been Professor of Information Systems and Head of Research in the Information Systems Department at the London School of Economics. Prior to joining the LSE he was Professor of Information Management (and Dean for the period 1994–1998) at Warwick Business School. He is currently an Honorary Professor of the European Institute for Advanced Management Studies, Brussels and is Gemini Consulting Visiting Professor in Knowledge Management at the University of St Gallen. He was President of the Association of Information Systems (AIS) in 1999 and is Program Co-chair of ICIS 2002. He is editor-in-chief of the *Journal of Strategic Information Systems*. His books include *Information Analysis* (Addison-Wesley, 1987); *Towards Strategic Information Systems* (Abacus Press, 1987); *Information Systems Research* (Blackwell Scientific, 1992; Alfred Waller, 1995); *Information Technology and Organizational Transformation* (Wiley, 1997); *Strategic Information Management* (Butterworth-Heinemann, 1994, 1999), and *Rethinking Management Information Systems* (Oxford University Press, 1999). In late 2002 he moved to a chair at Bentley College, based in Boston.

Julia Galliers has been a lecturer in the Centre for HCI Design since January 1999. She has a Ph.D. in Artificial Intelligence from the Open University, an M.Sc. in Cognition, Computing and Psychology from Warwick University, and a B.Sc. Hons in Zoology from Nottingham University. Her current research interests include safety-critical user interface engineering and communication aid design for the disabled. Her background is in cooperative interaction, belief revision, and dialogue modelling.

Severin V. Grabski is Associate Professor of Accounting at Michigan State University and has over 25 publications in high-ranking journals and peer-reviewed books. He is on the editorial board of the *Journal of Information Systems, International Journal of Accounting Information Systems*, and *Accounting Horizons* and serves as a reviewer for numerous journals. He is a graduate of Arizona State University. He has been involved in various consultancy projects and was Visiting Professor at the University of Tasmania. Professor Grabski's research and teaching interests are enterprise and accounting information systems, their valuation, and the semantic modelling of accounting phenomena.

Jimmy C. Huang is Lecturer in the Information Technology and Strategy Group at the Nottingham University Business School, UK. Prior to joining Nottingham University he was the lecturer in the Department of Management Studies at the University of Aberdeen. He is also a member of the IKON research group based at the University of Warwick. His research focuses on the process of cross-functional knowledge integration underlying organization-wide project implementation. He has published articles on this subject in such journals as *Communications of the ACM*, *European Journal of Information Systems*, *Journal of Decision Systems*, and *Journal of Information Technology Management*.

Christopher P. Holland is Professor of Information Systems at Manchester Business School, University of Manchester. After working in the product-marketing department of P&P PLC he joined Manchester Business School as ICL Research Associate in IT. Christopher lectured at the University of Salford before returning to MBS as lecturer in Information Management. His research interests include IT strategy, legacy IS, e-commerce, ERP systems, systems integration, inter-organizational systems and EDI, banking, and implementation. He has published papers in these areas in *Strategic Management Journal*, *Sloan Management Review*, *Organization Science*, *Journal of Strategic Information Systems*, *IEEE Software and Communications of the AIS*. Christopher has consulted with a wide range of organizations in manufacturing, banking, IT, and retailing industries in the areas of IT strategy and implementation.

Marina Krumbholz is a Ph.D. student at the centre for HCI Design, at the School of Informatics at City University in London. She is currently writing up her thesis with the title 'The Impact of National and Organisational Culture on ERP Implementations'. She received a B.Sc. in Management and Systems at the City University Business School.

Jinyoul Lee is an Assistant Professor of Management Information Systems at State University of New York at Binghamton. He received his Ph.D. from the University of Nebraska, Lincoln. His research interests include virtual organization, knowledge management, and Enterprise Resource Planning implementation. Especially, his recent research focusses on enterprise integration and virtualization in the context of social structuration process.

Zoonky Lee is an Assistant Professor of Information Systems at the University of Nebraska, Lincoln. He holds a Ph.D. from the University of Southern California. He received his master's degree from University of Michigan in statistics and the Carnegie Mellon University in social and decision sciences, respectively. Prior to joining the University of Nebraska, he was a business consultant at Coopers & Lybrand. His current research interests include designing IT infrastructures for organizational learning, ERP implementation, and E-business (E-strategy development, sales channel conflicts, and pricing strategies).

Stewart A. Leech is Professor of Accounting and Business Information Systems at the University of Melbourne, Australia. He is currently the Chair of the Artificial Intelligence/Emerging Technologies Section of the American Accounting Association, having served previously as Chair of the Information Systems Section. Stewart is Associate Editor of the *International Journal of Accounting Information Systems*, *Journal of Financial Information Systems*, and *Accounting and Finance*. He is also a member of the editorial board of a number of other leading academic journals. His main research interests are in accounting information systems, intelligent decision aids, and ERP systems.

Ben Light is a member of the Information Systems Research Centre and Lecturer in Information Systems at the ISI, University of Salford, UK. Previously, he was Research Fellow at the Manchester Business School, UK. His work focuses on IS Strategy and Implementation and he is presently interested in the organizational adoption of standard software – particularly ERP and CRM systems. His work has been published in journals such as the *Journal of Software Maintenance and Evolution: Research and Practice*, *Data Base*, *IEEE Software and Electronic Markets.* He has also delivered executive seminars to companies such as Kodak, KPMG, and Arthur Andersen.

Bai Lu is assistant investment accountant at Colonial Mutual group, Australia. She holds an honours degree in commerce from the University of Tasmania. Her research interests are in accounting information systems and the complementary relationships between information systems and organizational factors.

Neil A.M. Maiden is a Reader and Head of the Centre for Human-Computer Interface Design, an independent research department in City University's School of Informatics. He received a Ph.D. in Computer Science from City University in 1992. He is and has been a principal and co-investigator of several EPSRC- and EU-funded research projects including SIMP, CREWS, and

BANKSEC. He is also founder and manager of City University's SAP R/3 Laboratory. His research interests include frameworks for requirements acquisition and negotiation, scenario-based systems development, component-based software engineering, ERP packages, requirements reuse, and more effective transfer of academic research results into software engineering practice. Neil has over 75 journal and conference publications. He is also co-founder and treasurer of the British Computer Society Requirements Engineering Specialist Group.

M. Lynne Markus is Professor (Chair) of Electronic Business at the City University of Hong Kong, on leave from the Peter F. Drucker Graduate School of Management, Claremont Graduate University. Her research on enterprise systems was funded by the National Science Foundation, the Financial Executives Research Foundation, SIM (the Society for Information Management) International, and Baan Research. Her current research addresses strategic systems integration and industry structure change associated with B2B e-commerce.

Sue Newell has been Professor of Innovation and Organizational Analysis in the School of Management at Royal Holloway, University of London, UK. Previously, she was Research Director at Nottingham Business School. She is a founding member of the ikon (Innovation, Knowledge and Organizational Networking) research group, based at the University of Warwick, where she was formerly employed. Her research focuses on exploring innovation as an interactive design-decision process, where networks are crucial for the sharing and integrating of knowledge. She has written extensively on this subject in journals including *Organization Studies*, *Organization*, *Human Relations*, *Information Systems Journal*, *Journal of Strategic Information Systems*, and *European Journal of Information Systems*. Recently she took up a chair at Bentley College, Boston.

Peter O'Doherty is a Business Analyst and Project Leader with Seabrook Research Limited, an Irish software company based in Cork and Dublin specializing in ERP software implementation. He has 14 years experience in the deployment of IT resources and has been helping companies carrying out successful ERP implementations for the last seven years. He is also an associate researcher of the Executive Systems Research Centre (ESRC) at University College Cork, Ireland and has special interest in developing project management methodologies better suited to ERP projects.

Anne Parr is a lecturer in the School of Business Systems at Monash University in Melbourne, Australia. She is currently completing a Ph.D. on Enterprise

Resource Planning (ERP) systems implementation, which has involved the study of a range of ERP implementation projects in Australia and in the USA. Active as in information systems consultant, she has developed a number of major PC-based business systems. Apart from ERP implementation, her research interests include health care systems and she is the author of an innovative piece of expert systems software currently being used at the Royal Childrens Hospital in Melbourne as a training tool for paediatric cardiologists. She also has a particular interest in IS research methods, especially case study research and its philosophical foundations. Anne has published widely in international conferences and journals.

David Petrie is a Ph.D. candidate at the School of Information Science, Claremont Graduate University and teaches at the University of Redlands. His dissertation research explores how companies deal with technological discontinuities posed by B2B e-commerce. Petrie has 20 years of practical IS experience with an emphasis on data warehousing and database marketing.

Michael Rosemann is Associate Professor in the School of Information Systems at the Queensland University of Technology, Brisbane. He received his MBA (1992) and his Ph.D. (1995) from the University of Münster, Germany. As an Associate Director of the Information Systems Management Research Center he is a core member of the team that manages the mySAP University Application Hosting Center. He is author of two books, editor of two books, author of more than 25 book chapters and more than 50 refereed journal and conference papers. His main areas of research are enterprise systems, knowledge management, conceptual modelling, ontologies, operations management, and distance education.

Jeanne W. Ross is Principal Research Scientist at MIT's Center for Information Systems Research where she lectures, conducts research, and directs executive education courses on IT management practices. Her research examines the relationship between technology investments and organizational process redesign to better understand how organizations derive value from their technology investments. Her work has appeared in academic and practitioner journals, including *Sloan Management Review*, and *MIS Quarterly*.

Judy E. Scott is Assistant Professor at the University of Colorado at Denver and was previously on the faculty of the University of Texas, Austin. She received an undergraduate degree in science from Sydney University, Australia and an MBA and Ph.D. from the University of California, Irvine. Her teaching and research interests are in the area of ERP, e-business, IT business value, organizational learning and organizational knowledge, focusing on cross-functional and inter-organizational integration issues. Her publications appear in journals including the *Journal of Management Information Systems*, *Journal of Computer-Mediated Communications*, *Information System Frontiers*, *Information and Management*, and *Communications of the ACM*.

Peter B. Seddon is Associate Professor in the Department of Information Systems at The University of Melbourne, Australia. His teaching and research interests focus on helping people and organizations gain greater benefits from the use of IT. His major publications have been in the areas of accounting information systems and evaluation of information systems success, including enterprise systems success. Peter is an Associate Editor for *Management Information Systems Quarterly*.

Shari Shang is a Research Fellow in the Department of Information Systems at the University of Melbourne, Australia. Her recently completed Ph.D. was on the benefits of enterprise systems. Before enrolling in her Ph.D., she was a consulting manager for the IBM Global Consulting Group, and KPMG Peat Marwick Consulting in Taiwan. She has led various process reengineering projects and helped implement numerous application packages.

Graeme Shanks is Associate Professor and Deputy Head of the Department of Information Systems at The University of Melbourne. He holds a Ph.D. in information systems from Monash University. His research interests include customer information quality, implementation and impact of enterprise systems, and conceptual modelling. His publications have appeared in the *Journal of Strategic Information Systems*, *Information Systems Journal*, *Requirements Engineering*, *Information and Management*, *Australian Computer Journal*, and *Australian Journal of Information Systems*.

Sia Siew Kien is the Associate Director of Information Management Research Centre at Nanyang Business School, NTU. His main research interests are in organizational control issues relating to business process redesign, enterprise systems, and IT-enabled organizational transformation. He has made various presentations in conferences like the ICIS and AoM. His papers have been published in international journals like *Communications of ACM*, *Journal of Management Information Systems*, and *Database*. His research and consulting experiences in the last eight years encompass both private and public organizations in the region.

Christina Soh is the Director of the Information Management Research Centre, and Head of the IT and Operations Management Division at the Nanyang Business School, NTU. For the past decade, she has been actively engaged in research and consulting with firms and government organizations in Singapore and Malaysia. She has published her work in the area of IT investment, national IT policy, ERP implementation, and EC strategy in journals like the *Communications of ACM*, *Journal of Strategic IS*, and *Database*. She has also made various presentations in highly reputed regional and international conferences and executive seminars.

Mary Sumner directs the undergraduate program in Management Information Systems (B.S. in MIS) at Southern Illinois University. She has published numerous texts and research papers in computer-supported collaborative work, the management of end-user computing, and electronic commerce. Her research has appeared in *Database*, *Journal of Systems Management*, and *Information and Management*. She has conducted numerous information systems design projects in industry and is currently serving as Assistant Dean for the School of Business. In that role, she organizes business/university partnerships, including the Technology and Commerce Roundtable and the e-business initiative. Her academic background includes a Bachelor's from Syracuse University, a Master's from the University of Chicago, a Master's from Columbia University, and a doctorate from Rutgers University.

E. Burton Swanson is Professor of Information Systems at the Anderson School at UCLA, where he also presently serves as Academic Unit Chair and Co-Director of the Information Systems Research Program. Professor Swanson was the founding Editor-in-Chief of the journal, *Information Systems Research*, 1987–1992. Earlier he was also a co-founder of the International Conference on Information Systems (ICIS), in 1980. His research examines the life cycles of systems in organizations, addressing issues of innovation, implementation, utilization, and maintenance. He has authored more than eighty publications.

Cornelis Tanis is an enterprise systems implementation and change management consultant with Coach & Commitments in The Netherlands. He was formerly Program Director for Implementation Research at Baan Research.

Iris Vessey is Professor of Information Systems at Indiana University's Kelley School of Business, Bloomington. Her research into evaluating emerging information technologies from both cognitive and analytical perspectives has been published in journals such as *Communications of the ACM*, *Information Systems Research*, *Journal of Management Information Systems*, and *MIS*

Quarterly. In recent years, her interests have focused on managerial issues associated with the management and implementation of enterprise systems with papers, either published or accepted, in *Communications of the ACM*, *International Conference on Information Systems*, and *Information Systems Frontiers*. She currently serves as Secretary of the Association for Information Systems and of the International Conference on Information Systems and is an AIS Fellow. She received her M.Sc., MBA, and Ph.D. in MIS from the University of Queensland, Australia.

Michael R. Vitale is the Dean and Director of the Australian Graduate School of Management in Sydney. He was formerly a Professor in the Centre for Management of IT at the Melbourne Business School, and the Foundation Professor of Information Systems and Head of the Information Systems Department at the University of Melbourne. Prior to coming to Australia, he was a Fellow at the Ernst & Young Center for Business Innovation in Boston, USA. Dr Vitale was also an Associate Professor of Business Administration at Harvard Business School, where he wrote more than 50 case studies on the use of information technology in organizations. His industry experience includes four years as Vice President of Technology and Corporate Services at the Prudential Insurance Company of America.

Leslie P. Willcocks is Professor of Information Management and e-business at Warwick Business School, UK, Associate Fellow at Templeton College, Oxford, and is Visiting Professor at Erasmus University, University of Melbourne, and Distinguished Visitor at the Australian Graduate School of Management. He holds a doctorate in information systems from the University of Cambridge, and has been for the last 12 years Editor-in-Chief of the *Journal of Information Technology*. He is co-author of 22 books and 140 refereed papers, is a regular keynote speaker at academic and practitioner conferences, is retained as advisor by major corporations and has provided expert witness at several international government enquiries into IT. In February 2001 he won the PriceWaterhouseCoopers/Michael Corbett Associates World Outsourcing Achievement Award for his contribution to this field.

Introduction: ERP – The Quiet Revolution?

Peter Seddon, Graeme Shanks and Leslie Willcocks

Organizations invest in enterprise system software from vendors such as SAP, Oracle, PeopleSoft, Siebel, and i2 Corporation to gain access to powerful computer-based information systems more cheaply than through custom-built software development. What they acquire is a highly flexible software product, containing solutions to the needs of many of the vendor's existing customers, that impounds deep and detailed knowledge of many good and less good ways of conducting a wide range processes in a broad range of industries. The difficulty for the licensing organization is to identify the right combination of processes for its own changing needs and to implement those processes in its own organization.

Enterprise systems and the internet were probably the two most important information technologies to emerge into widespread use in the 1990s. According to Technology Evaluation.com, the total revenue from the enterprise system software and services market was US$18.3 billion in 1999 and US$19.9 billion in 2000 (Gilbert, 2000; Jakovljevic, 2001). Enterprise system implementation costs are often reported to be five to ten times the cost of software licenses (Davenport, 2000; Scheer and Habermann, 2000). If so, organizations worldwide spent something like US$100 billion per annum on enterprise systems in both 1999 and 2000. In short, organizations around the world have made huge investments in enterprise systems in the past decade.

Enterprise systems (ES) are large-scale organizational systems,[1] built around packaged enterprise system software. Enterprise system software (ESS):

1 is a set of packaged application software modules, with an integrated architecture, that can be used by organizations as their primary engine for integrating data, processes, and information technology, in real time, across internal and external value chains;
2 impound deep knowledge of business practices that vendors have accumulated from implementations in a wide range of client organizations, that can

[1] Systems composed of people, processes, and information technology.

exert considerable influence on the design of processes within new client organizations;

3 is a generic 'semi-finished' product with tables and parameters that client organizations and their implementation partners must configure, customize, and integrate with other computer-based information systems to meet their business needs.

Enterprise system software includes enterprise resource planning (ERP), customer relationship management (CRM), supply chain management (SCM), product life cycle management (PLM), enterprise application integration (EAI), data warehousing and decision support, intelligent presentation layer, and eProcurement/eMarketplace/electronic exchange software. In June 2000, AMR Research forecast that by 2004, sales of CRM, SCM, and eMarketplace software would *each* match the current US$20 billion per annum sales of ERP software. By 2002, sales forecasts were not as rosy, but as vendors rushed to 'internet enable' their products and allow 'best of breed' solutions to cooperate, the enterprise software market was still undergoing a period of tumultuous change.

The focus of this book is on the most important class of enterprise system, namely ERP systems. Today's ERP systems have evolved from packaged software for supporting material requirements planning (MRP) and manufacturing resource planning (MRP II), hence the strange name: enterprise resource planning. To those unfamiliar with the capabilities of MRP software, the combination of words 'enterprise', 'resource', and 'planning' conveys very little information about the capabilities and purpose of ERP software, but the label has stuck; the acronym 'ERP' has effectively become a new three-syllable word in the English language. There is no universally accepted definition of 'ERP' software (Klaus, Rosemann, and Gable, 2000), but Deloitte Consulting's ERP's Second Wave report provides a useful starting point. According to Deloitte:

An Enterprise Resource Planning system is a packaged business software system that allows a company to:

- Automate and integrate the majority of its business processes
- Share common data and practices across the entire enterprise
- Produce and access information in a real-time environment

(Deloitte Consulting, New York, NY, 1999)

This definition, which treats an 'ERP system' as a piece of software (not an organizational system), is similar in meaning to the first point of our definition of ESS above. What it does not say is that ERP is essentially 'back office' software: organizations use ERP software to integrate enterprise-wide information and processes for their financial, human resources, manufacturing and logistics, and sales and marketing functions. In addition, the Deloitte definition does not address points 2 and 3 in our definition above.

The consequence of the second point in our definition of enterprise system is eloquently expressed in Davenport's (1998) much-quoted line:

> An enterprise system imposes its own logic on a company's strategy, culture and organization. (Davenport, 1998: 122)

Essentially Davenport is saying that unlike custom-built software, which is by definition tailored to the precise needs and nuances of the organization, packaged enterprise system software is unlikely to be a perfect fit with every client's needs and this can cause problems. Although highly configurable to different situations, ERP systems' data structures, program code, and in-built assumptions about processes can impose patterns of behaviour on organizations that some find very difficult to accept. If the client organization is not careful, it may find the enterprise system in the driver's seat, defining how the business will be run.

The problem for the client organization is to understand the various business process options supported by the software and to make the right choices about which process variants to implement. The knowledge impounded in the various process variants has been accumulated by vendors from implementations of their software in a wide range of their clients' organizations. Furthermore, the capabilities of the software grow day by day as vendors respond to existing clients' requests for new features. Understanding what the software is capable of doing, then either customizing the software or changing processes in the organization to fit the software, is probably the central challenge in enterprise system implementation.

This brings us to the third point in the definition above. Enterprise system software is a generic 'semi-finished' product that user organizations must tailor to their needs. Tailoring the software using parameter settings provided by the vendor is usually called *configuration* (Bancroft, Seip, and Spregel, 1998: Chapter 5; Davenport, 2000: 150–153). For example, financial software must be configured so that it 'knows' which companies exist, which companies are subsidiaries of which other companies, the base currency of account for each subsidiary, the sales tax regimes for each subsidiary, and so on. Adding non-standard features to the software is usually called *customization*. Customization involves writing program code. It can range from relatively simple changes, such as developing very special new reports, through to major changes, such as changing the actual program code in the vendor's software (Brehm, Heinzle, and Markus, 2001). An implementation that involves configuring but not customizing the software is called a 'Vanilla' implementation.

Vendors usually recommend against customization, both because of the software-development risk and because new releases of the software may need re-customization. One way that organizations can avoid customization is by changing their processes to match those supported by the software. The risk in this option is that changing organizational processes is hard. Furthermore, the processes embedded in the system software may not be appropriate for the needs of the organization.

Apart from all the normal problems of information system project management, the novel difficulty for teams implementing ESS is to decide which mix of configuration, customization, and process change is best for the organization. This difficulty is compounded by uncertainty about which processes will be best for the organization (both now and in the future) and, frequently, by lack of understanding of the capabilities of the ES software. After user managers have learnt more about what the software can do, for example, after a year or so, they are often in a much better position to specify what they want to achieve with their software.

ERP vendors experienced rapid growth throughout the 1990s and by 2002 most Fortune 500 companies had them installed. Revenues flattened out in 1999, both because year 2000-related implementations were being completed, and because the market was saturating: many large organizations had already converted to ERP systems. By 2002 five major vendors dominated the ERP market. Market shares in 2000 were SAP 32%, Oracle 14.5%, PeopleSoft 9%, JD Edwards 5%, Baan 3%, and 'Other' 41% (Jakovljevic, 2001). By 2002, there was also a sizeable and growing market in upgrading ERP software to accommodate new software releases, particularly as ERP vendors internet-enable their product offerings.

Because they affect so many parts of an organization, ERP systems can provide a huge range of benefits and problems, often with different benefits in different organizations. Many organizations around the world, having collectively spent hundreds of billions of dollars implementing enterprise systems, are now asking, 'How can we gain greater benefits from our investment in enterprise systems?'. In fact, according to Markus and Tanis (2000), 'the key questions about enterprise systems*from the perspective of an adopting organization's executive leadership* are questions about success. For example: Will our investment pay off? Did our investment pay off?' These questions are echoed by Davenport (2000), Deloitte Consulting (1998), Markus and Tanis

(2000), and Ross and Vitale (2000) who note that business benefits from ES use are multidimensional, ranging from operational improvements through decision-making enhancement to support for strategic goals.

One of the best attempts, to date, to answer these questions about benefits from enterprise systems and how to achieve them is the benchmarking partners study published by Deloitte Consulting (1999). That report, based on 'in-depth interviews with 164 individuals at 62 Fortune 500 companies' concludes:

> Until now, conventional wisdom saw going live as the end. In sharp contrast to this view, our study uncovers at least two distinct waves of ERP-enabled enterprise transformation. The First Wave refers to the changes to an organization that include and accompany going live with ERP. The Second Wave, on the other hand, refers to the actions that are taken after going live that help organizations achieve the full capabilities and benefits of ERP-enabled processes. (Deloitte Consulting, 1999)

It is our belief, as we assemble this book, that many organizations have now begun to focus on the second wave, in terms of maximizing benefits, making continuous improvements, and taking advantage of new, including web-based, technologies and new ways of configuring systems in a journey to establish the integrated, extended business enterprise. Hence the title of this book: *Second-Wave Enterprise Resource Planning Systems: Implementing for Effectiveness*. In this book, we have assembled some of the world's best research on ERP systems, with a view to providing a foundation for second wave improvements to enterprise systems.

One recent study of ERP systems concludes that the way to maximize benefits from enterprise systems over the long term is to set up process improvement teams, led by highly motivated process owners, whose job is to understand both the evolving capabilities of the software and needs of the organization, and to strive to maintain a reasonable on-going level of fit between the software and changing organizational needs (Shang, 2001). Achieving such on-going fit is hard. Enterprise system software is so complex that, no matter how good the initial implementation, it is unlikely that it will be a good fit for the organization. Over time, a growing understanding throughout the organization of the potential benefits of enterprise system use, and changes in leadership, strategy, competitors' behaviour, and organizational structure, will lead to changes in perceived needs. In addition, the enterprise system software will change as vendors issue new releases, extend the scope of their software, implement support for new technologies (for example, the internet), and as the vendor firms themselves merge or fail. These pressures result in an on-going need for

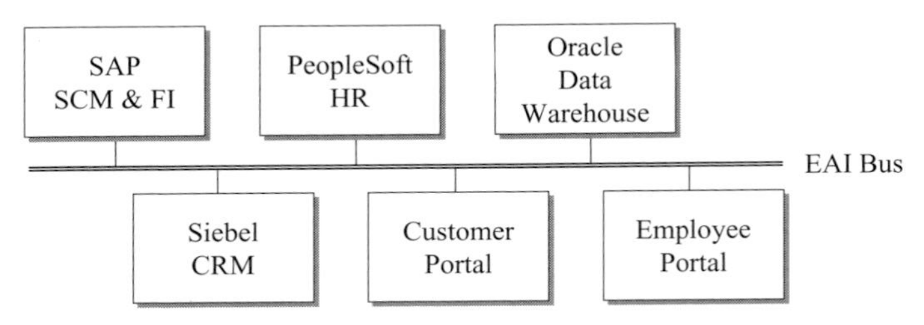

Figure 1 Integration of application software through enterprise application integration (EAI)

enterprise system improvement initiatives. In the organizations Shang studied, the best people to lead such initiatives were the process owners. These managers had both the necessary deep understanding of organizational needs and the authority to implement necessary changes in the organization. However, they also had to be prepared to devote considerable energy to understanding the capabilities of their enterprise system software.

The future for ERP software and its brothers and sisters in the enterprise system software family is, at time of writing, unclear, though the final chapter of this book suggests two plausible scenarios. Three issues commanding attention in 2002 were:

- application software integration (for example, through EAI and intelligent portals);
- inter-organizational process integration through exchanges;
- use of shared services and application service provision.

With respect to application software integration, many organizations are presently asking themselves, 'How do we integrate our application software, including our various "legacy" or custom-built systems?'. The most-favoured solution at present is to use so-called enterprise application integration (EAI) software from vendors such as WebMethods, BEA, IBM (MQ Series, WebSphere), and Tibco to create a middleware or EAI communication channel or bus that links the various applications. This situation is depicted in Figure 1.

Enterprise Application Integration (EAI)

The second issue for the future is inter-organizational process integration through exchanges, hubs, and eMarketplaces (see also Chapter 18). Whereas ERP systems support back office functions, such as financial, human resources, manufacturing, and logistics *within* an organization, its younger siblings, such

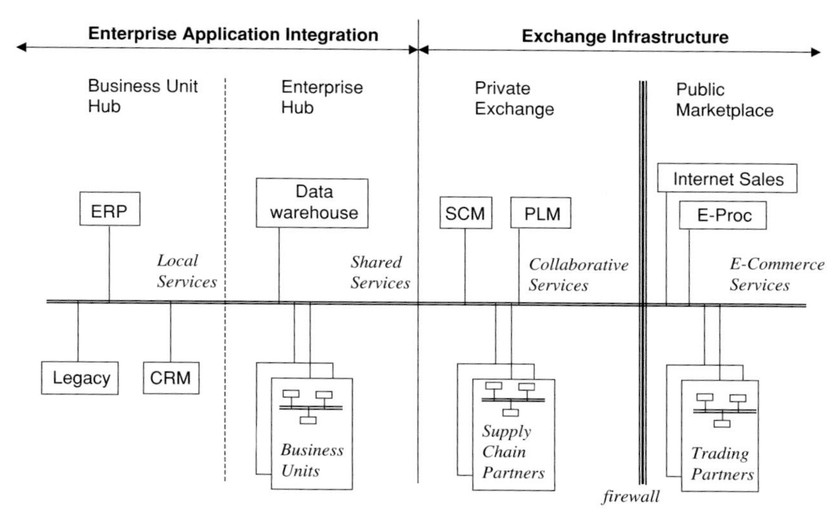

Figure 2 Possible future enterprise system hub and exchange architecture (*Source:* Based on H. Kagermann, Sapphire Conference Presentation, Nov. 2001.)

as CRM and SCM, are more concerned with collaborative links to partner organizations up and down the supply chain. As noted earlier, developments in the area of customer-facing CRM systems, and supplier-facing SCM systems and exchanges are where major future growth in the use of packaged application software is expected. ERP systems provide a core facilitating infrastructure for many of these new applications. One possible future hub and exchange architecture that integrates the various components of a firm's enterprise system architecture is depicted in Figure 2. Moving from left to right, from deep inside the organization to the view from the outside, the figure depicts:

- local application services, not necessarily from the same vendor, being provided from each business unit hub;
- enterprise-wide shared services, possibly including financial accounting, human resources, data warehousing, and analytic services being provided from the enterprise hub;
- links to business partners being provided through collaborative SCM and PLM services provided by a private exchange hub inside the corporate firewall;
- e-commerce services, such as internet sales and e-procurement, being provided to trading partners via a public marketplace exchange.

In this vision of the future, the enterprise application integration (EAI) bus from Figure 1 links services within the organization at both the business unit hub and enterprise hub (the two left-hand columns in Figure 2), and collaborative links to business partners might be provided using message-based technologies delivered over the internet, such as XML, and managed by exchange and marketplace software probably based around so-called 'web services'.

The third issue for the future is the use of shared services and application service provision. Wide area networks are now so fast and cheap that it really does not matter where in the world the application and database servers for enterprise software are located. For instance, in teaching with SAP R/3 at The University of Melbourne, we issue presentation layer software (SAP GUI) to students to install on their home computers and use an application service provider 2000km away to provide all ERP services. Students access the service from their homes through their own internet service providers. In return for a fixed sum per semester, Queensland University of Technology manages the entire technical infrastructure (R/3 version 4.6c running on Sun hardware with an Oracle database). The system works faultlessly. ERP services are delivered to the screen through the internet just as mains power is delivered through the power plug in the wall. It does not matter where in the world the application and database servers are located.

Various organizations have moved to cut the cost of back office services by centralizing accounts payable services in a single city where labour is cheap, rather than distributing the accounts payable function around the world. This can work on an internal ASP/hub model of shared services, as introduced globally, for example, by the oil conglomerate Shell during 2001, or as Hagel and Brown (2001) posit may involve a sourcing model incorporating in-house supply with a single third-party supplier, or multiple third-party suppliers, of applications, services, and infrastructure. The relevance of external sourcing models is that, as this book will make clear, ERP systems can be difficult and expensive to implement. From 2000, the ASP model has been offered on the promise of relatively pain free ERP outsourcing. Pain free in this context means cutting costs and offering a fixed fee per month, removing the complexity of owning and operating these systems in house, and bridging the ERP skills gap. However, users have found that they need to find out what level of support they are getting for their money. In fact by 2002 a broad range of services was on offer, ranging from basic hosting – in which the ERP application runs on the ASP's computers and is delivered over the Internet or leased lines – to managed services – in which the ASP maintains and upgrades the ERP system – to complete installation, hosting, and maintenance (Kern, Lacity, and Willcocks, 2002).

Hosted and managed ERP services were originally intended to appeal particularly to mid-sized companies previously excluded by the cost and complexity of ERP. For leading ERP vendors, the ASP model has been seen, amongst other things, as a route to the smaller companies market. However, the ASP model depends on a one-to-many service. Thus US-based Host Logic founded itself as a one-size-fits-all managed service provider. Its 2000 product, Smart-Enterprise, was offered at an all-inclusive, fixed, monthly price, and included all the hardware, services, implementation, and software needed to install and run SAP R/3 in a mid-size company. It was delivered partly by Host Logic, and partly by partners dealing with, for example, servers, network services, load balancing, and encryption. However, such 'out-of-the-box' applications provide, at best, only 65–75% fit to enterprise requirements. Indeed the ASP model may well heighten customers' typical complaints about ERP: lack of flexibity, lack of customizability, and difficulty and cost of implementation. One likely further development then will be a continuum of ERP offerings based on the ASP, net-enabled delivery model, ranging from off-the-shelf, standardized systems to premium services supporting fully customized ERP systems (Moran, 2000; Kern, Lacity, and Willcocks, 2002). The issues surrounding this model's application to ERP implementation for effectiveness are discussed in detail in Chapter 4 (see also Chapter 18).

Having defined the ERP world, and given an overview of the major recent developments therein, we now move on to detail the precise focus, organization, and contents of this book.

Organization of the Book

We have organized the book into five parts, each dealing with a critical issue in the effective implementation of ERP systems. The first part provides an overview of implementation for effectiveness of ERP systems, and includes discussions about process models for ERP implementation, the reasons organizations have moved to ERP systems, problems encountered and success achieved during ERP implementation, and how to assess and manage benefits from ERP implementation. The second part focuses on critical success factors and risk management during ERP implementation and maintenance, and highlights the areas in which management should focus their attention during ERP implementation. The third part addresses organizational learning and knowledge management issues in ERP implementation. Lessons learned from ERP implementations in both large and small to medium firms are presented, together with discussions about how knowledge about ERP systems may be represented, integrated, shared, and transferred. The fourth part provides an analysis of cultural issues in ERP implementation, and explains implementation problems using organizational and national culture characteristics that differ from values embedded within the ERP system. The final part discusses two possible scenarios that highlight very different futures for ERP systems.

Part I Implementation and Effectiveness: Overview

We selected four papers to provide an overview of implementation for effectiveness of ERP systems. In the first paper, 'Learning from Adopters' Experience with ERP: Problems Encountered and Success Achieved', Lynne Markus, Sheryl Axline, David Petrie, and Cornelius Tanis note that ERP packages touch many aspects of a company's internal and external operations. Consequently, successful deployment and use of ERP systems are critical to organizational performance and survival. They present the results of a study of problems and outcomes in ERP projects, conducted under the sponsorship of an ERP system vendor. Two basic research questions are addressed: First, how successful are companies at different points in time in their ERP experiences, and how are different measures of success related (that is, can early success be followed by failure and vice versa)? Second, what problems do ERP adopters encounter as they implement and deploy ERP, and how are these problems related to outcomes? Findings show that the success of ERP systems depends on when it is measured, and success at one point in time may be only loosely related to success at another point. Companies experience problems at all phases of the ERP system life cycle, and many of the problems experienced in later phases originated earlier but remained unnoticed or uncorrected. These findings suggest that researchers and companies do well to adopt broad definitions and multiple measures of success and pay particular attention to the early identification and correction of problems.

In his paper, 'Innovating with Packaged Business Software: Toward an Assessment', E. Burton Swanson notes that in the 1990s many firms turned to software packages when they replaced the older systems in their application portfolios. While packaged business software had already long been in use, it was in the 1990s with the advent of ERP that it began to dominate enterprise decisions throughout the world. This 'package transition', which continued into 2002, represents a broad business change of significant importance. It entails both organizational and inter-organizational innovation on a vast scale. Swanson reports findings from a survey aimed at studying and assessing this phenomenon. It focuses on both the know-why and the know-how associated with successfully undertaking the transition.

In the third paper, 'A Comprehensive Framework for Assessing and Managing the Benefits of Enterprise Systems: The Business Manager's Perspective', Shari Shang and Peter Seddon present a framework for assessing and managing the business benefits of enterprise systems. After analysing the features of enterprise systems, the literature on information technology evaluation, data from 233 enterprise system vendor success stories published on the web, and interviewing managers in 34 enterprise system using organizations, they present a consolidated enterprise system benefits framework with five dimensions. The five dimensions are: operational, managerial, strategic, information technology infrastructure, and organizational benefits. Finally, they show how the framework has been used to assess benefits in four case study organizations.

Jeanne Ross, Michael Vitale, and Leslie Willcocks, in their paper 'The Continuing ERP Revolution: Sustainable Lessons, New Modes of Delivery', present findings from a research project that examined how firms are generating business value from their investments in ERP systems. The research describes the stages of ERP system implementation, the obstacles that firms encountered in generating benefits from the systems, and some critical success factors for getting business value from the implementation of an ERP system. The chapter considers these lessons and then evaluates the extent to which delivering ERP over the net, including by third-party service providers, will be adopted, the likely pitfalls, and whether the CSFs and lessons will still be applicable.

Part II From Risks to Critical Success Factors

A further five papers were selected to provide a comprehensive analysis of risks and critical success factors in implementation for effectiveness of ERP systems. In the first paper, 'Enterprise System Implementation Risks and Controls', Severin Grabski, Stewart Leech, and Bai Lu develop a theoretically grounded framework that identifies the major factors and associated procedures related to the successful implementation of ERP systems in organizations. The framework, which is based on the theory of complementarity and supermodularity, indicates that ERP system implementation success is dependent upon the performance of complementary factors. A field study of a recent ERP system implementation project provides a limited test of the framework. The results of the study provide strong support for the existence of complementary relationships among the factors identified in both the literature and the case study as vital for the success of an ERP system implementation project. For the research community, the results provide a much-needed foundation for further research into the implementation of ERP systems. For the practice community, the results provide advice to management, consultants, and auditors on how best to utilize their limited resources to effect a successful ERP system implementation.

Mary Sumner, in her paper 'Risk Factors in Enterprise-Wide/ERP Projects', identifies the risk factors in implementing traditional MIS projects, describes the risk factors associated with enterprise-wide/ERP projects, and identifies the risk factors in ERP projects which are unique to these projects. Some of the unique challenges in managing enterprise-wide projects include: the challenge of reengineering business processes to 'fit' the process which the ERP software supports, the investment in recruiting and reskilling technology professionals, the challenge of using external consultants and integrating their application-specific knowledge and technical expertise with existing teams, the risk of technological bottlenecks through client–server implementation, and the challenge of recruiting and retaining business analysts who combine technology and business skills.

In the third paper, 'A Framework for Understanding Success and Failure in Enterprise Resource Planning System Implementation', Chris Holland and Ben Light identify the scale and strategic importance of ERP systems and define the problem of ERP implementation through a theoretical framework of critical success factors. They offer a framework for managers which provides the basis for developing an ERP implementation strategy. Analysis of two case studies identifies different approaches to ERP implementation, highlights the critical role of legacy systems in influencing the implementation process, and identifies the importance of business process change and software configuration in addition to factors already cited in the literature, such as top management support and communication.

Anne Parr and Graeme Shanks, in their paper 'Critical Success Factors Revisited: A Model for ERP Project Implementation', present a project phase model of ERP implementation projects that is a synthesis of existing ERP implementation process models and focuses on the implementation project. Two case studies of ERP implementation within the same organization, one unsuccessful and a later one successful, are reported and analysed to determine which critical success factors are necessary within each phase of the project phase model. The project phase model is used as a 'lens' for understanding ERP implementation projects, by highlighting the differences between the two cases. They then offer an explanation for these differences, focusing particularly on the successful case. First, the organizational learning that occurred during the unsuccessful project and the associated early appointment of an experienced 'champion' with clearly defined responsibilities were critical to the successful project. Second, organizations implementing ERP systems should partition large projects into several smaller, simpler projects identified here as 'vanilla' implementations. The project phase model, together with associated critical success factors, provides guidance for practitioners when planning ERP implementation projects, and also provides researchers with a foundation for further empirical research.

In their paper, 'Offsetting ERP Risk through Maintaining Standardized Application Software', Guy Gable, Taizan Chan, and Wui-Gee Tan note that the standardization of business applications has changed the patterns of development and use of software, as well as many aspects of their maintenance. They propose a framework for future research on large application packaged software maintenance that accounts for some of these changes, thereby revealing important new research questions. The commodification of software development and services, emerging alongside the pervasiveness of standard business software, has amplified issues of maintenance economics and related business strategy. The user–organization-centric research framework presented commences from the simple observations that maintenance generates benefits as well as costs, and that maintenance strategy fundamentals are impacted by a range of factors, including (a) software source, (b) support source, (c) organizational contexts, and (d) environmental contexts. Within the new distributed maintenance arrangement, four key stakeholders participate: the user organization, the software vendor, third-party service providers, and society (national economy). Important economic and business strategy issues arise from the fact that various software and related support sourcing alternatives have substantial maintenance incidence implications (incidence of costs, benefits, and responsibilities). In terms of organizational knowledge strategy, the complexity of standard business software and the concomitant commercial arrangements necessitate that maintenance knowledge sourcing decisions are made in light of lifecycle-wide maintenance knowledge requirements. To optimize maintenance it appears necessary that all stakeholders have a lifecycle-wide view of maintenance costs and benefits, considering the four key factors, and underpinned by an understanding of the other stakeholders' perspectives.

Part III From Learning to Knowledge

Six papers were selected to address organizational learning and knowledge management issues in ERP implementation. In the first paper, 'Implementing Enterprise Resource Planning Systems: The Role of Learning from Failure', Judy Scott and Iris Vessey adapt Sitkin's theory of intelligent failure to ERP implementations resulting in a theory that they call 'learning from failure'. They then examine from the viewpoint of this theory the details of two SAP R/3 implementations, one of which failed while the other succeeded. Although it is impossible to state, unequivocally, that the implementation that failed did so because it did not use the approach that was derived from the theory, the analysis reveals that the company that followed many of the tenets of the theory succeeded while the other did not.

Frédéric Adam and Peter O'Doherty, in their paper 'ERP Projects – Good or Bad for SMEs?' studied 14 ERP implementation projects in Irish organizations and focused on the key relationships between organizations that attempt to implement ERP systems and their implementing partner. They found that the ERP implementations in Ireland are different to the projects that have been reported elsewhere in two key respects. First, organizations interested in ERP software are, on average, far smaller than the case studies reported in the literature and the majority of the cases they reviewed were small and medium enterprises (SMEs). Second, the durations of implementation were far shorter than reported elsewhere. These results are not surprising considering the smaller average size of Irish organizations, but they indicate that the ERP movement is truly ready for an extension towards the SME market. The results also indicate that the duration of the implementation of ERP software may be related to the size and complexity of the client organization and that SMEs can expect to have an easier time implementing ERP systems than the current literature suggests. They also found that software implementers play a key role not only in technical terms, but also in managerial and political terms because they can help their clients to correct their expectations and perceptions of ERP systems and implementations.

In their paper, 'The Role of the CIO and IT Function in ERP', Leslie Willcocks and Richard Sykes report that disappointment is frequently due to the inappropriate roles or capabilities of the CIO and the IT function more generally. It points to three ways in which this occurs, then develops from the successes analysed a detailed picture of how CIOs and IT specialists can be effectively involved in ERP implementation.

Michael Rosemann, in his paper 'Enterprise Systems Management with Reference Process Models', notes that reference models describe on a conceptual level the structure and functionality of enterprise systems. They allow easy access to the functionality of enterprise systems from a process-oriented viewpoint. However, these models focus on depicting executable processes and do not take into account tasks related to business engineering, system selection, implementation or change. This chapter discusses how reference process models can be extended to support the entire enterprise system lifecycle.

In their paper 'An ERP Implementation Case Study from a Knowledge Transfer Perspective', Zoonky Lee and Jinyoul Lee note that ERP implementation entails transferring the business knowledge incorporated in the basic architecture of the software package into the adopting organization. They propose a new approach to analysing ERP implementations from a knowledge transfer perspective. They begin by identifying the types of knowledge transferred during an ERP implementation and the factors affecting this transfer. They then investigate how conflicts between the business knowledge transferred from the ERP package and the existing organizational knowledge are resolved. During the investigation, they used in-depth interviews, process analysis, and documentation analysis to analyse an early implementation stage of ERP. The results indicate that business processes, incorporated in an ERP package, are transferred into an organization along with the business rules inherent in the processes due to process automation, the limited flexibility of such packages, and the cross-functional nature of an ERP package. The results also suggest that an organization's adaptive capability concerning role and responsibility redistribution, development of new types of required knowledge, and the introduction of a different knowledge structure influence an organization's ability to internalize these standardized processes into business routines that provide a competitive advantage.

Jimmy Huang, Sue Newell, and Robert Galliers, in their paper 'Knowledge Integration Processes within the Context of Enterprise Resource Planning System Implementation', note that an increasing number of multinational enterprises have adopted ERP systems in the hope of increasing productivity and efficiency as a means of leveraging organizational competitiveness. While some are starting to harvest the benefits from their initial investments, others are still struggling to release the promised potential of their ERP systems. This can be seen as an illustration of the 'productivity paradox', that is, that firms face significant problems in both measuring and achieving the return on their

IT investments. While many accounts have examined the adoption of technology, including ERP, few of these accounts have considered such adoption from the perspective of cross-functional knowledge integration. Against this backdrop, a case study was conducted as a means of exploring and theorizing the dynamics of knowledge integration underlying the process of ERP implementation. In this paper they focus not only on presenting the theoretical framework but also on describing the stages that were undertaken to derive this framework.

Part IV Contextual and Cultural Aspects of Enterprise Systems

Two papers were selected to address contextual and cultural aspects of enterprise systems. In the first paper, 'An Exploratory Analysis of the Sources and Nature of Misfits in ERP Implementations', Siew-Kien Sia and Christina Soh note that ERP vendors often market their products as incorporating 'industry best practices'. But are these embedded business models of 'industry best practices' universal or context specific? How are the functional gaps handled? To what extent do they jeopardize the integration benefits of ERP? Through a systematic analysis of ERP misfits and their resolution in a study of three hospitals in Singapore, they provide a framework for identifying misfits by the sources of specificity and the nature of the misalignment.

Marina Krumbholtz, Julia Galliers, and Neil Maiden, in their paper 'Implementing Enterprise Resource Packages? Consider Different Organizational and National Cultures!', note that packages are implemented in companies with different organizational and national cultures, and there is growing evidence that failure to adapt ERP packages to fit these cultures leads to projects that are expensive and overdue. Their paper describes research that synthesizes social science theories of culture in order to model and predict the impact of culture on ERP package implementation. They describe a knowledge meta-schema for modelling the surface and the deeper manifestations of culture and predictions of ERP implementation problems based on national culture differences. They report an empirical study of the implementation of SAP's R/3 sales and distribution (SD) module in a large pharmaceuticals organization in Scandinavia, based in Sweden. Results provide evidence for an association between organizational and national culture and ERP implementation problems. Furthermore, results demonstrate that these diverse implementation problems can be caused by a mismatch between a small set of core values indicative of a customer's organizational culture.

Part V Future Directions

Finally, one paper was selected to address future directions of ERP systems. Lynne Markus, David Petrie, and Sheryl Axline, in their paper 'Continuity Versus Discontinuity: Weighing The Future of ERP Packages', argue that, with so much going for them, ERP packages seem likely to remain popular for some time to come. In this context, it makes sense to try to anticipate how ERP packages will evolve in the future. One view of the future can be constructed from an assessment of the issues faced by today's ERP package adopters and the responses under development by ERP package vendors. In this view, the functionality of ERP packages will expand, and the architecture of ERP packages will evolve, in ways that address many of today's business opportunities and technical challenges. This view of the future assumes a high degree of continuity with today's in-house IT management regimes; in particular, it involves a continuation of the current division of labour between ERP package adopting organizations and ERP package vendors.

But experts in strategic planning contend that it is often not wise to rely solely on views of the future that are extensions of the past. Instead, they argue, there is value in visualizing alternative future scenarios that incorporate discontinuities. Doing so can enable planners to identify and invest in options that preserve one's flexibility if unexpected situations unfold. In the scenario-planning tradition, this paper offers a plausible discontinuity view of the future of ERP packages. The claim is not that this scenario *will* transpire, but only that it *could*. However, if the scenario did occur, it would have major implications for both ERP package adopters and ERP package vendors. Companies that had developed the knowledge, capabilities, and skills suited to this alternative future would prosper relative to those that had not. Therefore, companies would be well advised to assess their ERP strategies in light of the discontinuity view.

After the general IT experiences in the developed economies throughout 2000–2002, there would seem, to us, every reason for pursuing the path of developing alternative scenarios – about the IT services market, the use of IT, ERP developments, and the nature of the business climate – for at least the 2002–2007 period. Justifiably, Moschella (1997) dubbed what can be roughly considered as the 1990–2010 period the net- or network-centric era. While the central technological thrust towards ever more networking can be easily discerned, what has always been less clear is the organizational, social, and political and economic factors that massively influence what actually happens – and does not happen. This argues for increased use of scenarios, but also learning from what has already been experienced, before over-optimistic technological scenarios are accepted. What is true for the network-centric era is also true for one of its sub-sets. In the realm of ERP, this book provides a rich set of lessons and scenarios about how to implement ERP for effectiveness.

REFERENCES

AMR Research, Inc. (2000) AMR Research Predicts Enterprise Applications Market Will Reach $78 Billion by 2004. AMR Research, 12 June. http://www.amrresearch.com/pressroom/files/00613.asp (accessed 18 June 2000).

Bancroft, N. H., Seip, H., and Spregel, A. (1998) *Implementing SAP R/3.*, Greenwich, CT: Manning Publications Inc.

Brehm, L. Heinzl, A., and Markus, M. L. (2001) Tailoring ERP Systems: A Spectrum of Choices and Their Implications. Proceedings of the 34th Annual Hawaii International Conference on Information Systems, Hawaii, January.

Davenport, T. H. (1998) Putting the Enterprise into the Enterprise System. *Harvard Business Review*, July–August, 121–131.

Davenport, T. H. (2000) *Mission Critical, Realizing the Promise of Enterprise Systems.* Boston, MA: Harvard Business School Press.

Deloitte Consulting (1999) *ERP's Second Wave – Maximizing the Value of ERP-Enabled Processes.* New York: Deloitte Consulting.

Gilbert, A. (2000) ERP Vendors Look for Rebound after Slowdown. *Information Week*, 14 February. http://www.informationweek.com/773/vaerp.htm (accessed 18 June 2000).

Hagel 111, J., and Brown, J. (2001) Your Next IT Strategy. *Harvard Business Review*, October, 105–113.

Hurst, Q. and Nowak, D. (2000) *Configuring SAP R/3 FI/CO.*, Alameda, CA: Sybex.

Jakovljevic, P. J. (2001) The ERP Market 2001 and Beyond – Aging Gracefully with the 'New Kids on the Block. Technology Evaluation.com, 3 October. http://www.technologyevaluation.com/research/researchhighlights/erp/2001/10/research_notes/mn_er_pj_10_03_01_1.asp

Kern., T., Lacity, M., and Willcocks, L. (2002) *Netsourcing: Renting Your Business Applications and Services over Networks.* New York: Prentice Hall.

Klaus, H. Rosemann, M., and Gable, G.G. (2000) What is ERP? *Information Systems Frontiers*, **2**(2), 141–162.

Markus, L. M. and Tanis, C. (2000) The Enterprise Systems Experience – From Adoption to Success. In *Framing the Domains of IT Research: Glimpsing the Future Through the Past*, Zmud, R. W. (eds), Cincinnati, OH: Pinnaflex Educational Resources.

Moschella, D. (1997) *Waves of Power: Dynamics of Global Technology Leadership.* New York: AMACOM.

Moran, N. (2000) ASP Outsourcing. Special Report: ERP and Beyond. *Financial Times*, London, p. 6.

Richardson, B. (2000) When Worlds Collide: Trillion Dollar Maybes, AMR Research Spring Executive Conference Presentations, 'Deadline 2001: Accelerating

End-to-End E-Business'. The Phoenician Scottsdale, AZ, accessed 28 June. http://www.amrresearch.com/Events/presentations.asp.

Ross, J. and Vitale, M. R. (2000) The ERP Revolution: Surviving Versus Thriving. *Information Systems Frontiers*, **2** (2), 233–241 (http://www.wkap.nl/oasis.htm/277131).

Scheer, A-W. and Habermann, F. (2000) Making ERP a Success. *Communications of the ACM*, **41**(4), 57–61.

Shang, S. (2001) Maximizing Benefits from Enterprise Systems. Unpublished Ph.D. Dissertation, The University of Melbourne, Australia.

Part I

Implementation and Effectiveness: Overview

1 Learning from Experiences with ERP: Problems Encountered and Success Achieved

M. Lynne Markus, Sheryl Axline, David Petrie, and Cornelis Tanis

1.1 Introduction

One of the most enduring research topics in the field of information systems is that of systems success (Ballantine et al., 1996; DeLone and McLean, 1992; Lyytinen and Hirschheim, 1987). Prior research has addressed the measurement of success, the antecedents of success, and the explanations of success or failure. Yet, with each new type of information technology or application, the question of success comes up again. In the case of enterprise resource planning (ERP) systems, success takes on a special urgency since the costs and risks of these massive technology investments rival their potential payoffs. Failures of ERP system implementation projects have been known to lead to organizational bankruptcy (Bulkeley, 1996; Davenport, 1998; Markus and Tanis, 2000).

Briefly, ERP are commercial software packages that enable the integration of transactions-oriented data and business processes throughout an organization. From a base in manufacturing and financial systems, ERP systems may eventually allow for integration of inter-organizational supply chains (Davenport, 1998; Markus and Tanis, 2000). Because these systems touch so many aspects of a company's internal and external operations, their successful deployment and use are critical to organizational performance and survival.

This chapter describes the results of a study of problems and outcomes in ERP projects. The study was conducted under the sponsorship of an ERP vendor who was interested in helping its customers be more successful in ERP implementation. Two basic research questions are addressed: first, how successful are companies at different points in time in their ERP experiences, and how are different measures of success related (that is, can early success be followed by failure and vice versa)? Second, what problems do ERP adopters encounter as they implement and deploy ERP, and how are these problems related to outcomes?

1.2 Success with ERP and How it Happens

The definition and measurement of success are thorny matters. First, success depends on *the point of view* from which you measure it. Early in our research, it became clear that people often meant different things when talking about ERP success. For example, people whose job was to *implement* ERP systems (for example, project managers and implementation consultants) often defined success in terms of completing the project plan on time and within budget. But people whose job was to *adopt* ERP systems and use them to achieve business results tended to emphasize having a smooth transition to stable operations with the new system, achieving intended business improvements like inventory reductions, and gaining improved decision support capabilities.

In this paper we adopt an inclusive perspective that focuses on the organizations that adopt ERP systems and the individuals within these organizations (rather than on ERP vendors and external implementation consultants). We recognize that our 'etic'[1] perspective may not correspond with that of any particular actor(s) in the organizations we study, but it allows us to include in our assessment of success many different dimensions, among them:

- success viewed in technical terms;
- success viewed in economic, financial, or strategic business terms;
- success viewed in terms of the smooth running of business operations;
- success as viewed by the ERP-adopting organization's managers and employees;
- success as viewed by the ERP-adopting organization's customers, suppliers, and investors.

A second important issue in the measurement of success concerns *when* one measures it. Some years ago, Peters and Waterman (1982) attracted much attention with their study of 'excellent companies'. A few years later, a sizeable number of their excellent companies were no longer star performers. Project managers and implementers can afford to declare success in the short run, but executives and investors are in it for the long haul. The organizations that adopt ERP systems need to be concerned with success not just at the point of adoption, but also farther down the road. The importance of considering ERP success at multiple points in time was made clear in a case study by

[1] Anthropologists use the term etic (from phonetics) to refer to an external, researcher-oriented perspective and the term emic (from phonemics) to refer to the perspective of the research subjects.

Larsen and Myers (1997), in which a successfully installed ERP system was later terminated when the company merged with another.

In this study, we are concerned with the assessment of success at three different points in time during the adopting organization's experience with an ERP system. In the 'ERP Experience Cycle' (Markus and Tanis, 2000), we can conceptually differentiate three distinct phases: (1) the *Project phase*, during which ERP software is configured and rolled out to the organization, (2) the *Shakedown phase*, during which the company makes the transition from 'go live' to 'normal operations', and (3) the *Onward and Upward phase*, during which the company captures the majority of business benefits (if any) from the ERP system and plans the next steps for technology implementation and business improvement. For each of these phases, a number of success metrics can be defined.

1 Success in the Project phase:
   - project cost relative to budget;
   - project completion time relative to schedule;
   - completed and installed system functionality relative to original project scope.

2 Success in the Shakedown phase:
   - short-term changes, occurring after system 'go-live', in key business performance indicators such as operating labor costs;
   - length of time before key performance indicators achieve 'normal' or expected levels;
   - short-term impacts on the organization's adopters, suppliers, and customers such as average time on hold when placing a telephone order.

3 Success in the Onward and Upward phase:
   - achievement of business results expected for the ERP project, such as reduced IT operating costs and reduced inventory carrying costs;
   - ongoing improvements in business results, *after* the expected results have been achieved;
   - ease in adopting new ERP releases, other new information technologies, improved business practices, improved decision making, etc., *after* the ERP system has achieved stable operations.

These success metrics include indicators of *human and organizational learning*. How well the ERP system itself performs (for example, accuracy, reliability, and response time) is important, but also how well people in the organization are able to use, maintain, and upgrade the ERP system and how well the business improves its performance with the ERP system.

An unresolved question is the relationship among the measures of success at different points in time. Larson and Myers (1997) found that an ERP experience could be an early success and a later failure. But can an ERP experience be an early failure yet a later success? How important is it for organizations to be successful at all three phases of the ERP Experience cycle? And how often do organizations overcome initial failure to achieve an ultimate measure of success? These are empirical questions.

A third important issue in the measurement of success is the *yardstick* or *criterion* against which to compare an actual level of achievement. It is quite common in systems evaluation, technology assessment, and impact studies to use the adopters' *objectives*, *expectations*, and *perceptions* as the standard for defining and measuring success. Naturally, these subjective judgments of success can be quite important in understanding how organizations behave. If a company stops using an ERP system because corporate objectives have not been met, it does not matter that an outside observer might have assessed the implementation project and system operation as successful.

However, there are serious disadvantages of using perceptions, objectives, and expectations as the sole measures of success. In the first place, it is hard to normalize them across individuals and organizations, making comparisons difficult. Second, their relationship with so-called 'objective' measures of success (such as whether or not a project is terminated prior to completion, cf. Sauer, 1993) is unclear. People's objectives and expectations for ERP systems may be overly ambitious, so that they are unrealizable, no matter what people do. Or they may be insufficiently ambitious, so that people do not take full advantage of capabilities 'in' the technology available for them to use (Markus and Tanis, 2000).

If one wants to compare the outcomes achieved by the organizations that have adopted ERP systems, it is useful to have an external yardstick of success, in addition to internal perceptual measures and local objectives and expectations. For this purpose, we propose using as our critierion *optimal success*, defined by Markus and Tanis (2000) as follows:

> Optimal success refers to the best outcomes the organization *could possibly* achieve with enterprise systems, given its business situation, measured against a portfolio of project, early operational, and longer term business results metrics. Optimal success can be *far more or less* than the organization's goals for an enterprise system. Further, optimal success can be *dynamic*; what is possible for an organization to achieve may *change over time* as business conditions change. (Markus and Tanis, 2000)

Naturally, the concept of optimal success, so defined, is difficult to operationalize. However, the advantage of attempting to assess outcomes by using such an external, non-interpretive yardstick is that it helps us compare the results achieved in different organizations and explore the interesting relationships between 'objective' outcomes and people's perceptions of results.

The phrase *optimal* success suggests that most organizations experience outcomes that fall somewhat short of what a 'best in class' organization might achieve. This observation directs attention to the *problems* companies experience when they adopt, deploy, and use ERP systems, and how they *respond* when problems arise. This is not a focus on 'success factors' per se, but on aspects of the 'lived experience' of organizations' ERP journeys. One wants to know *how* (the process by which) some companies realize better or worse outcomes than other companies, and what they *do* that makes the difference. Put differently, one wants to know whether all companies experience the same types of problems with ERP systems, whether they respond similarly to the problems, and whether the problems and responses are related to the outcomes they experience.

It should be clear that we believe the outcomes companies achieve with ERP systems (varying degrees of sub-optimality, relative to what they could achieve, if all went perfectly well) are non-deterministic. Problems, such as a turnover of personnel or lack of resources, can arise in each phase of the ERP experience cycle. They may or may not be perceived as problems by the people in the organization. And, even if people perceive the problems as problems, they may or may not take appropriate actions to resolve them. As a result, the outcomes in a particular phase may be optimal or less than optimal, and the problems may or may not remain unresolved, affecting outcomes later (see Markus and Tanis, 2000, for a fuller treatment of this 'theory' of ERP success).

In practice, it can be extremely difficult to differentiate between problems, symptoms of problems, and outcomes (that is, the consequences of problems). Nevertheless, the importance and complexity of the ERP experience suggests the need to try. Therefore, this study addresses two related questions about the ERP experience. First, how successful are companies at different points in time in their ERP experiences, and how are different measures of success related (that is, can early success be followed by failure and vice versa)? Second, what problems do ERP adopters encounter as they implement and deploy ERP, and how are these problems related to outcomes?

Table 1.1 helps to frame the findings of this research by providing a more complete description of the issues involved in assessing the success of ERP

Table 1.1 Assessing achieved success and problems in the ERP experience

Major activities of phase	Implications of phase activities for success in phase and later	Common problems	Phase success measures
Project phase			
Project team formation and training. Develop enterprise model for configuration; develop and validate kernel in multiple implementations. Configure ERP software to reflect either current operations or planned new business processes. Design and execute changes (if any are planned) in the organization's business processes and related organizational elements (organization structure, jobs, compensation, etc.). Implement add-ons, modifications, and interfaces with other enterprise systems and legacy systems. Acquire IT infrastructure resources and integrate ERP system with infrastructure and legacy systems (if any). Document configuration	The majority of ERP expenditure plans are made during this phase. Few benefits are experienced during this phase unless the organization pursues a 'quick wins' or 'low hanging fruit' strategy of identifying and implementing business process improvements while ERP planning and configuration is underway. The longer the project phase, the lower the overall financial benefits from the system on a discounted cash flow basis. If the project goes very badly, decision makers may terminate it. When the schedule gets tight, team may decide to cut scope, so that strategically essential processes are not supported.	Inability to acquire/retain employees and external advisors with requisite expertise in ERP, project management, and supporting technologies. Turnover of project sponsor or project managers. Excessive turnover (and/or stress-related health problems) on project team. Unwillingness of business managers and key users to make time for project activities. Major changes in project scope after start of project. Poor quality software, documentation, training materials. Modifications that do not work; delays in development of modifications and interfaces. Conflicts with implementation consultants over project plans and management.	Project cost relative to budget. Project completion time relative to schedule. Completed system functionality relative to original project scope.

Table 1.1 (*cont.*)

Major activities of phase	Implications of phase activities for success in phase and later	Common problems	Phase success measures
decisions and rationale. Decide how to satisfy decision support/reporting needs. Communication and perform change management. Clean up data and convert data to new ERP system. Test the new system. Train users.		Cutting testing and/or training when schedule gets tight. Pressure to terminate project if cost and schedule overruns occur.	
Shakedown phase			
Make the transition to 'normal operation' of the new system and the new business processes. 'Rework' (mistake correcting) activities may include: changing configuration settings, upgrading IT infrastructure, revising business practices and procedures, retraining users.	The organization may not be able to realize planned improvements in IT costs and/or business process efficiency until (1) the new system stabilizes, (2) the old systems are interfaced or turned off and older IT resources are removed from maintenance agreements, and (3) users achieve full proficiency with the new system. In addition, the organization may have to make significant new expenditures for temporary and overtime labour, consulting help, and additional IT resources	Extremely poor system performance. Excessive stress and/or turnover of key users and/or key system support personnel. Excessive dependence on 'key users' (project team members) and/or IT specialists. Maintenance of old procedures or manual workarounds in lieu of learning the relevant system capabilities. Data input errors. Inability to diagnose and remedy system and/or business process performance problems.	Short-term deterioration in key (business) performance indicators (KPIs, e.g., process cycle times, inventory levels, operating labour costs). Length of time before KPIs and business impacts return to normal. Short-term negative impacts on organization's suppliers and customers (e.g., increased average time on hold, lost calls, lost sales, drop in customer satisfaction levels).

(*cont.*)

Table 1.1 *(cont.)*

Major activities of phase	Implications of phase activities for success in phase and later	Common problems	Phase success measures
	to complete the transition from 'go live' to 'normal operations'. The longer the Shakedown phase, the lower the overall business benefits on a discounted cash flow basis. If Shakedown phase goes very badly, the system may be removed or the organization may become unwilling to undertake future system improvements (e.g., upgrades).	Extremely negative reactions from customers and suppliers (e.g., large losses of business). Absence of sharp and fast improvements during shakedown. Absence of satisfactory management information/analysis and reporting. Pressure to de-install system.	
Onward and Upward phase			
Ongoing operation and use of system and business process after the Shakedown phase. Planning for upgrades and migration to later releases/version of hardware and ERP software. Adoption of additional modules/packages and integration with ERP. Business decision making based on data provided by the ERP system. Continuous improvement of users' IT skills.	The majority of business benefits (if any) are achieved after shakedown. Many desired business benefits may not be possible with the current release, but may require the organization to undertake a series of upgrades (e.g., reductions in an organization's IT personnel expenditures may not be realizable during the initial ERP experience cycle; achieving business process 'visibility' across sites occurs after	'Normal operation' never materializes. Non-improvement in users' ERP skill levels (e.g., many potential users remain untrained; users routinely rely on project team members and technical support personnel to perform 'normal' job activities). Failure to retain people who understand the implementation and use of ERP systems.	Achievement of planned business results (e.g., IT operating costs, inventory carrying costs, business process cost and cycle time). Use of data and decision analyses produced by the system. Ongoing improvements in business results (after planned results have been achieved). Ease in developing/ adopting/ implementing additional innovations in technology, business practices and managerial decision making.

Table 1.1 (*cont.*)

Major activities of phase	Implications of phase activities for success in phase and later	Common problems	Phase success measures
Continuous business process improvement to achieve better business results. Reconfiguration of current release/version.	all sites have been implemented). Many benefits cannot occur until: (1) users have learned how to use the system well, (2) managers have used the data collected by the system to make business decisions and to plan improvements in business processes, and (3) additional changes are made in business processes, practices, software configuration, etc.	No documentation of rationale for business rules and configuration decisions. Difficulty in optimizing system performance and in reconfiguring the system to support business innovations. Unwillingness of organization to adopt additional changes in business processes, system configurations, or IT infrastructure. Pressure to de-install system.	Original decision to implement ERP still makes sense in light of subsequent business decisions and events (e.g., mergers and acquisitions). (Over time) decreases in length of Project Planning and Shakedown phases for subsequent ERP implementations.

projects. The table outlines (1) the activities that characterize each phase of the ERP lifecycle; (2) how and why the activities in each phase may affect the outcomes an adopter achieves, not just in the phase, but also downstream; (3) some of the problems an ERP adopter may experience during the phase (which can also be understood as indicators that the experience may be heading for sub-optimal success); and (4) the success measures relevant to the phase.

1.3 Approach

This research study combined several methods: (1) reviews of published and in-process research studies and teaching cases of ERP implementations; (2) in-depth case studies of the ERP experience in five ERP adopter organizations, following procedures prescribed by Yin (1994); (3) interviews with 11 additional ERP-adopting organizations; and (4) approximately 20 interviews with ERP implementation consultants and members of the ERP vendor company sponsoring this study. Table 1.2 describes each of the 16 ERP adopter

Table 1.2 Overview of companies in the study

Company identifier	Company description	ERP implementation description – status at time of data collection	Data sources
Company A	US company with $250m. annual revenues. Company had grown through past mergers that were never fully integrated. Single manufacturing location, functionally organized. Assemble to order manufacturing plus some custom business processes. Industry details omitted at company request.	Project was justified and approved by company board. Single site implementation with approximately 400 users. All major business functions included in project, including manufacturing, finance, and distribution (but excluding HR). System went live June 1997. Total project cost roughly $17m. including hardware and consulting services.	Four researchers, approximately 12 hours of interviews with one key informant, plus three hour interview with CIO (one researcher). Data collected in the USA, March through September 1998.
Company B	Global manufacturer based in the USA with 7000 employees and 170 worldwide sites; $1.2bn. annual revenues. 5th or 6th in their industry. Make-to-stock, assemble-to-order, and a small proportion of engineer-to-order processes. Company formed through 1997 merger of two companies roughly equal in size and previously competed in distinct, industrial equipment niches; little integration of the companies has occurred.	Company was having difficulty deciding which software release to implement. Approximately 500 concurrent users world wide were expected. Project budget approximately $133.5m.; planned schedule 30 months. The ERP project was cancelled in 1999 after an expenditure of $70m.	Three researchers, approximately 24 hours of interviews with 16 members of implementation team, including CIO and consultants. Reviewed internal documentation and videotapes of ERP project introduction at leadership meeting January 1998. Data collected in USA in June 1998.

Table 1.2 *(cont.)*

Company identifier	Company description	ERP implementation description – status at time of data collection	Data sources
Company C	European subsidiary of US multinational apparel company. Make to stock manufacturing. Facing declining market demand for core product.	Multi-site, pan-European rollout in process (eight sites over four years). HQ went live with finance modules December 1997 (on time) and Raw Materials Management June 1998 (six months delay). First affiliate and sales office live July 1998 (six months delay). Multi-site configuration: eight (logistical and financial) companies were set up Distributed architecture – separate server at each site. $5.5m. worth of modifications and interfaces (includes consulting).	Two interviews with five key informants, Summer 1998. Two researchers, approximately 27 hours of on-site interviews with 19 informants, September 1998. Reviewed key internal documentation Data collected in Belgium and Netherlands.
Company D	Arm of global energy and engineering firm, located in Scandinavia; $330m. annual revenue. 1500 employees. Company is the result of a 1996 merger of three companies, preserved in 1998 as three divisions. Manufactures components and systems that serve the entire supply chain of the electrical power industry.	Three divisions, each with separate single site instance of same ERP package. First division went live January 1996, on time and within budget. Second division live went May 1998, six months behind schedule, 25% over budget, due to buggy software and more customizations than planned. Third division was in the process of implemention.	Two researchers, approximately 30 hours of on-site interviews with 17 informants from operations, finance, and IS and one external IT consultant, September 1998. Reviewed key internal documentation. Data collected in Scandinavia.
Company E	Multinational conglomerate based in the UK.	Implementation began Fall 1997.	Three researchers, approximately 20 hours of on-site interviews with 13 informants, October 1998;

(cont.)

Table 1.2 *(cont.)*

Company identifier	Company description	ERP implementation description – status at time of data collection	Data sources
			plus three hours of interviews prior to site visit.
		30 ERP projects planned over five year period.	On-site data collected in the UK.
Company N	Small private US company manufactures health care equipment and supplies.	Single site implementation. Initial implementation in 1993, re-implementation in 1997.	One hour telephone interview, company President, March 1998.
Company O	Electronics company based in the USA.; $1bn. annual revenues. Manufacturing locations in five countries. 6000 employees.	Multiple implementations in different international sites. First implementation of early ERP package in Germany, then company-wide rollout. In process of re-implementation and worldwide rollout, starting with manufacturing.	One hour telephone interview with member of worldwide rollout team, March 1998.
Company P	Japanese-owned automotive supplier, operating in the USA. 1000 employees in the USA. Regional offices in five sites. Experiencing rapid growth.	Planned live dates May 1998. Canadian operations, August 1998 main office. All modules to go live at once.	One hour telephone interview, Manager-Corporate Services, Corporate Production Planning, March 1998.
Company Q	Canadian electronics semi-conductor manufacturer; $40m. annual revenues. Growing at 20+% per year. High-tech company with strong IT experience. Company culture tolerates change reasonably well.	Beta test site for early ERP package. In process of re-implementing ERP with new software.	One hour telephone interview, CIO/IT project manager, April 1998.

Table 1.2 *(cont.)*

Company identifier	Company description	ERP implementation description – status at time of data collection	Data sources
Company R	Small private company with plants in the USA, Canada, and UK. Combination of assemble-to-order and repetitive manufacturing.	Used ERP since 1993. Currently implementing ERP upgrade.	One hour telephone interview, CIO, April 1998.
Company S	Growing industrial equipment manufacturer based in the USA with 5200 employees worldwide; $1bn. annual revenues. 50 years old, a wholly owned subsidiary of a major industrial equipment company.	Enterprise rollout planned for 1999. Description of 'Unit A', the first division to go live with ERP. Single-site implementation with ten licenses. Went live November 1997. Project budget: $1.2m. First release implemented was 'very buggy', now re-implementing later release. Still have three major interfaces to homegrown systems. Plan to replace with ERP in the future.	One hour telephone interview, Manufacturing Systems Manager, Unit A, April 1998.
Company T	Small US based producer of security systems. Approximately 600 direct labour manufacturing employees. 15–20% annual growth rate past five years.	$1.5m. project. Server upgraded numerous times due to undersizing by ERP vendor. Due to company's performance problems, vendor performed special modifications to enable special software release.	Electronic interview, April 1998.
Company U	Small European electronics assembler. 25–30% annual growth rate. 500 employees.	First ERP implementation 1990, logistics. Re-implemented current release.	One hour telephone interview, June 1998.

(cont.)

Table 1.2 *(cont.)*

Company identifier	Company description	ERP implementation description – status at time of data collection	Data sources
Company V	Manufacturer based in the USA with $175m. annual revenue. Products sold in 100+ countries through distributors and dealers. 900 employees, most of which are at HQ/manufacturing facilities. Many of the products are engineered to order.	Went live on ERP early 1997. Elapsed time from initial project concept to go-live approximately 2.5 years. 150+ users of the system.	1.5 hour telephone interview, IS Director, March 1998.
Company X	Large manufacturing firm. Details omitted at company request.	Rollout completed December 1997.	One hour telephone interview, June 1998.
Company Y	Large US manufacturing firm in the aerospace industry.	Two US locations, three phase rollout. First phase went live May 1997 Other two planned to be complete by September 1998.	One hour telephone interview, June 1998. Reviewed internal documentation.

organizations that participated directly in this research. At the same time, the analysis and interpretation of results presented in this report reflect the experiences of a much larger number of companies (approximately 40 in total), including those described in teaching cases, other research reports, and the trade press.

The 11 ERP adopter interviews were conducted by phone or in person: one or more members of the research team discussed the ERP experience with one or more members of the adopter organization. The interviews ranged in length from one to three hours. The case studies involved a much more significant level of effort. Two to four members of the research team visited the case site for two to four days, interviewing 12 to 25 people. Documents describing the company and the implementation effort were collected and analysed. Notes were transcribed and reviewed by project team members, and summaries were written. The detail and thoroughness of the case study method mean that it is not necessary to examine a large number of cases to gain the benefits of this research

Table 1.3 Companies studied by stage of ERP experience cycle

Stage of experience cycle reached at time of data collection	Company identifier
Project phase	Company B
	Company O
	Company P
Shakedown phase	Company C
	Company E
	Company X
	Company Y
Onward and Upward phase	Company A
	Company D
	Company N
	Company Q
	Company R
	Company S
	Company T
	Company U
	Company V

strategy for analysing 'how and why' research questions (Yin, 1994). For such scientific purposes, four to 12 case studies are considered perfectly adequate.

Several criteria were used to select the companies for this study. First, we selected companies that were interested in learning from the research about how to improve their ERP experiences. These companies were recruited at public presentations where we described our research project.

Second, we studied companies *at different stages* of the ERP experience. Studying projects in process provides useful knowledge about how the ERP experience unfolds over time. This is particularly useful for identifying why companies act the way they do. After the project is over, people forget many details and reconstruct the past to be consistent with known outcomes. Studying completed projects allows researchers to identify the key causal factors in success or failure. Thus, we aimed for a mix of both completed and in-process projects. Table 1.3 shows the stage of completion reached by each company at the time we collected data. Because some of the cases we studied were in process at the time of data collection, we do not have complete outcome data for all companies.

Third, we went out of our way to select projects that had experienced problems, rather than projects that were unqualified successes. A major goal of the study was to understand the problems adopters experience with ERP systems, why these problems occur, and what could be done about them. Therefore, we skewed our sample toward companies with problems and sub-optimal success. This means that the companies examined in this study *may not be a representative sample* of all companies using ERP systems. It would *not* be valid to draw conclusions from this study about *how frequently* ERP adopters experience certain problems or how frequently they achieve success (or lack of it) on different measures.

Two additional factors may limit the potential *statistical* generalizability (Yin, 1994) of the results. First, all 16 of the adopter companies we studied were based in North America or Europe. Second, all 16 companies used the ERP products of a single software vendor. However, we do not believe that these factors materially affect our findings *about the kinds of problems and outcomes* companies experience with ERP systems. Our findings closely track reports by other academics and by journalists. Further, these factors are not likely to affect the analytical generalizability (Yin, 1994) of our results: although the current study design does not provide reliable data about frequencies, it can provide reliable insights into *how and why problems and outcomes occur, when they do occur.*

1.4 Findings

Table 1.4 presents a summary of the problems and outcomes reported by the companies participating in this research. Immediately below, we present some interesting generalizations about the nature of success across the ERP lifecycle. In a subsequent section, we discuss the problems companies experienced.

Findings about Adopters' Achieved Success with ERP Systems

First, *none* of the ERP adopters we studied was an unqualified success at all of the stages of the experience cycle completed at the time of our data collection. This is to be expected given the nature of our sampling (overselection of companies that had experienced or were experiencing difficulties), and it may not be a representative finding. We do believe that *some* companies are successful on all three categories of success measures.

However, Ross and Vitale (2000) have found that a performance dip after initial implementation of an ERP system is very common. And many of our companies similarly experienced moderate to severe business disruption when their ERP systems 'went live'. They had difficulty diagnosing problems

Table 1.4 Problems and outcomes experienced by phase

Companies by phase of experience cycle at time of study	Project phase problems and outcomes	Shakedown phase problems and outcomes	Onward and Upward phase problems and outcomes
Companies in Project phase			
Company B	In process ERP project halted and rechartered when company merged. ERP system faced huge organizational integration issues. The ERP project was cancelled in 1999 after an expenditure of $70m.	N/A	N/A
Company O	Had to modify ERP software for essential functionality despite policy against it. First implementation partner replaced for lack of relevant experience. ERP project combined with company-wide standardization and reengineering; big change management issues.	N/A	N/A
Company P	Had to modify ERP software for essential functionality despite policy against it. Project team had difficulty getting involvement from local sites.	N/A	N/A
Companies in Shakedown phase			
Company C	Project team communicated well with management and sites. Management centralized formerly decentralized IS units for ERP implementation; local sites resisted. Experienced cost and schedule overruns	Experienced system performance problems. KPIs deteriorated in short run. Customers and suppliers experienced negative effects. Managers not happy with reporting capabilities.	N/A

(*cont.*)

Table 1.4 *(cont.)*

Companies by phase of experience cycle at time of study	Project phase problems and outcomes	Shakedown phase problems and outcomes	Onward and Upward phase problems and outcomes
Company E	On time and within budget. Expected scope achieved. Project team did excellent job of building consensus around need for common systems in this decentralized company.	KPIs deteriorated in short-run.	N/A
Company X	Not discussed.	Experienced difficulties with data conversion. Experienced system performance problems. KPIs deteriorated in short run.	N/A
Company Y	Budget and schedule overruns. Experienced software bugs.	Experienced system performance problems. KPIs deteriorated in short run. Customers and suppliers experienced negative effects.	N/A
Companies in Onward and Upward phase			
Company A	On schedule and within budget. Scope cuts.	KPIs deteriorated in short run.	Turnover of experienced user and support personnel. User skill with system is low. Some improvements in key business measures. Other planned improvements not achieved (owing to scope cuts). System may be partially de-installed (owing to unanticipated merger).

Table 1.4 (*cont.*)

Companies by phase of experience cycle at time of study	Project phase problems and outcomes	Shakedown phase problems and outcomes	Onward and Upward phase problems and outcomes
Company D	One division on time and within budget. A second division experienced schedule delays due to software modifications.	Experienced heavy data entry errors by users. Had to increase staff to cope with errors. KPIs deteriorated in short run.	User skill with system remains low; errors remain high. System not used in managerial decision making. Insufficient plans for ongoing system support and business improvement.
Company N	Acceptable project outcomes despite heavy customizations.	Disastrous performance problems. Severe business disruption. Severe negative impact on customers and suppliers.	Never achieved normal operations. Experienced permanent loss of business.
Company Q	Acceptable despite entirely in-house implementation with no prior experience.	Users challenged by conversion to client–server environment. Severe system performance problems.	Business improvements were not sought as part of ERP implementation.
Company R	Experienced poor consulting advice. Made excessive software modifications.	Experienced many software errors due to lack of integrated system testing. Experienced difficulty recovering from errors.	Business improvements were not sought as part of ERP implementation.
Company S	Within budget. Greatly behind schedule. Greatly reduced scope.	Not discussed.	Planned business benefits achieved.
Company T	Within budget. Schedule overruns. Scope cuts.	Not discussed.	Headcount increased instead of decreased as planned. Improved reporting and data access. Some planned business improvements not achieved.

(*cont.*)

Table 1.4 *(cont.)*

Companies by phase of experience cycle at time of study	Project phase problems and outcomes	Shakedown phase problems and outcomes	Onward and Upward phase problems and outcomes
Company U	Project team took excessively functional approach to implementation. Management support gained quite late in process.	Extensive reconfiguration occurred as a result of learning about integrated operations. KPIs improved after reconfiguration.	Poor data quality continues to hamper business. Company has gained improved awareness of benefits of cross-functional integration. Management is extremely satisfied with system. Upgrade to new release is planned.
Company V	Ignored advice of experienced implementation consultants. Did not follow conventional implementation methodology. Excessive project manager turnover (5 times).	Extreme business disruption owing to configuration errors, system integration problems, and incomplete/inaccurate data. KPIs deteriorated in short-run Customers and suppliers experienced negative effects.	Permanent loss of customers. Business processes streamlined. Company is now ready to undertake process reengineering.

(which had many possible causes), and they had difficulty recovering from them. They sometimes achieved 'normal' operations only by permanently increasing staffing levels and reducing expectations about labour efficiency. In general, ERP adopters seemed unprepared both physically and psychologically for Shakedown phase difficulties.

Further, extreme difficulties in the Shakedown phase appeared to have strong negative influences on companies' willingness to continue with the ERP experience. Several companies with Shakedown phase problems reported strong pressure to de-install their ERP system. Even when the ERP system was retained, there was great unwillingness to upgrade to 'enhanced' versions of the software. In essence, these companies have implemented 'legacy' ERP systems.

Second, mixed 'success' results were observed even with a single phase. For example, a number of companies achieved their budget and schedule targets, but had to cut scope, often substantially (companies S, A, and T). In the case of company T, these scope reductions led to failure later on: the company did not achieve the business results it had hoped for. But company S did achieve its desired business results, despite a massive cut in scope. While company S implemented only 15% of the ERP functionality it had originally planned to implement, the company claimed to have achieved substantial inventory reductions, as intended. This result shows that it is possible for 'failed' projects to achieve eventual business success.

We found that companies differed substantially in how they defined success in the Project phase, because they differed in their definitions of the project itself. Some companies defined the project as 'implementing ERP as quickly and cheaply as possible'. Others defined the project as 'adopting best practices enabled by ERP' (which entails business process reengineering). Still another defined the project as 'achieving commonality of systems and business practices in a decentralized organization' (which entails a process of organizational development and consensus building). In general, the larger the organization's definition of the project, the more willing the organization was to expand the project's budget and schedule. These companies were less likely to judge the overall ERP experience as unsuccessful when project budget and schedule were not met.

We found that larger organizations tended to define the ERP experience in much more expansive terms than smaller ones. They often demanded business results from 'IT' projects. In many cases, these organizations were planning for multiple (perhaps dozens of) ERP installations, and realized the importance of learning how to implement and upgrade ERP systems better each time. They were more likely than smaller organizations to start planning for the Onward and Upward phase during the Project phase.

Third, as Larsen and Myers (1997) have observed, some companies that achieved 'success' in the Project phase could be classified as failures later on. Either they experienced substantial difficulties during Shakedown (companies E and N), or they reported a lack of business benefits during the Onward and Upward phase (company Q). Similarly, one of the companies studied by Dolmetsch et al. (1998) successfully implemented SAP R/3 within four months but was later disappointed not to have achieved business performance improvements, because it had not reengineered its processes.

We were surprised that several companies in the Onward and Upward phase could not say with any confidence whether they had achieved business benefits from using ERP. They gave a variety of related reasons for their inability to assess their results:

- The ERP system had been adopted for technical reasons (for example, Y2K, cost or lack of capacity in current system), not for business reasons.
- No business goals for the ERP project had been set.
- The company does not manage by metrics.
- Existing systems did not allow the company to measure where it was on key business metrics prior to the implementation of the ERP system.
- The company did not perform a post-implementation audit of the ERP project to assess whether projected benefits were achieved.

In general, companies that do not deliberately set out to achieve measurable business results do not get them (or do not know that they have achieved them). Further, the inability to document measured benefits from an ERP implementation appears to discourage organizations from undertaking future upgrades and/or migrations.

In conclusion, success in the ERP experience is multidimensional and often hard to measure. Early success (or success on project measures) is not closely linked with later success (success on business measures). And early failure (failure on project measures) is not tightly linked with later failure (failure of business measures). Clearly, then, success in an ERP experience is not predetermined by a set of success factors in place at the start of the project and continuing unchanged throughout. Either conditions change over the course of the experience or different types of actions are required at different phases, and the ways a company responds to conditions at each phase influences the subsequent progress and ultimate success of the ERP experience. This observation suggests one obvious normative recommendation: companies should be concerned with success in all phases of the ERP experience and should not concern themselves exclusively with what happens during the Project phase. And this observation suggests one obvious research issue: to understand the success of an ERP experience, one needs to look at what goes on (for example, problems experienced and attempts at problem resolution) at each phase of the experience cycle. In the next section, we focus more deeply on why the companies we studied achieved sub-optimal success.

Findings about Adopters' Problems with ERP

We asked adopters what problems they experienced with ERP systems, how they had dealt successfully with these problems (if they had), and what they had learned as a result of their experience. We also formed our own impressions of their experiences based on what we observed when we visited their companies. We came away with a deep respect for the challenges they faced. If what they were trying to do was easy, more of them would have been successful on all measures. But many of the problems they experienced were 'wicked', that is, hard to recognize and diagnose owing to multiple interacting causes and varying symptoms and effects. In this section, we describe what we believe to be the most difficult problems adopters experienced and why the problems occurred.

Project phase problems

The most challenging Project phase problems reported by our respondents involved software modifications, system integration, product and implementation consultants, and turnover of project personnel.

Software modifications Almost every analyst of the ERP experience strongly advises companies to avoid modifying the software. Companies are advised to live with existing ERP functionality and to change their procedures to adapt to it. We found, however:

- Many adopters could not avoid some degree of ERP software modification. In some cases, ERP packages are selected on a centralized basis to fit the majority of corporate needs. Often, there are a few sites that cannot operate effectively with the software's functionality, even if people there are willing in principle to modify their business processes. For example, one company reported having an order of magnitude more entities (for example, sales reps) than were allowed by the relevant field size in the software package. Other companies explained that the software simply did not fit business rules around commissions and royalties, and that these rules could not be changed without serious negative business implications.
- Many adopters had difficulty getting modifications to work well. They complained about implementation consultants who did not deliver well-tested and working modifications in a timely manner.
- Most distressingly, several adopters reported that, after wrestling with modifications (and sometimes failing to make them work well), they eventually learned that their modifications were unnecessary after all. They had usually made plans for software modifications early in the Project phase when they did not understand the software thoroughly (particularly the integrations across modules). Later on when they understood the software better, they discovered ways to implement the needed capabilities without modifications.

For a more general treatment of the issues involved in tailoring ERP software to a company's specific needs see Brehm, Heinzl, and Markus (2000).

Problems with system integration ERP systems are sold as 'integrated packages', implying that they contain everything one needs and that ERP software configuration (plus tailoring) is the major activity of the Project phase. But there are a number of respects in which this is not so.

- First, an ERP system needs to be integrated with the computing platform on which it will run. We found that companies had great difficulty integrating their enterprise software with a package of hardware, operating systems, database management systems software, and telecommunications' systems suited to their particular organization size, structure, and geographic dispersion. They reported having difficulty finding experts who could advise them on the precise operating requirements of their ERP configuration. They described having made unplanned upgrades of processors and memory to support their systems. One company reported making several changes of database management system before finding one that 'worked'.
- Second, even though ERP systems are said to be comprehensive packages that cover every organizational function, most of the companies we studied (large and small) reported needing to retain some legacy systems that performed specialized functions not available in ERP packages. (Alternatively, they acquired specialized software from third parties.) These systems needed to be interfaced with ERP systems – a process both challenging and expensive.
- A particular area in which many organizations found ERP systems deficient was that of data reporting. ERP systems are essentially transaction processing systems that do not (without expensive add-ons) solve companies' needs for decision support. For descriptions of the measures companies must often take to solve their ERP-related data reporting problems, see the cases of Microsoft (Bashein, Markus, and Finley, 1997) and MSC Software (Bashein and Markus, 2000).

Problems with product and implementation consultants ERP implementations are socially complex activities. As many as a dozen or more external companies – including the ERP vendor, vendors of ERP product extensions, vendors of supporting hardware, software, and telecommunications, capabilities, implementation consultants, and so forth – may be involved in different aspects of an organization's ERP experience. Coordinating the efforts of all these firms is, to put it mildly, a challenge. We found that:

- Few IT product and service firms were willing to take end-to-end responsibility for coordinating all parties; and adopters were often rightly reluctant to cede authority for project management to an outside party, even when they were willing to pay the steep fees for outside assistance.

- IT product and service firms generally seem to resent taking subordinate roles to other such firms. They do not to cooperate well. There is much finger pointing when problems occur.
- Despite representations during the sales cycle, there is widespread lack of knowledge about the details of ERP products, particularly where integrations, tools, and interfaces with 'partner' products are concerned.
- Because IT product and service firms are growing rapidly, they find it difficult to provide continuity in personnel assigned to adopter projects. And adopters strongly value continuity in personnel.
- Several adopters reported having had conflicts (sometimes severe) with IT product and service vendors over contractual provisions (for example, pricing and billing arrangements) and over project direction (for example, project management).

Turnover of project personnel An all too common complaint was the frequency with which adopters lose key personnel experienced with ERP or supporting technologies. As already noted, external service providers themselves are unable to maintain continuity of customer support personnel. In addition, adopters frequently reported:

- Losing key IT specialists and user representatives working on the project while the project was going on, often despite handsome retention bonuses.
- Losing experienced people after the project is complete. Many IT specialists thrive on project work and view assignment to a 'competence center' (support unit) as unpleasant maintenance work.

In short, the Project phase of the ERP lifecycle posed severe challenges for the adopters we studied, and not all companies resolved these problems well. In some cases, unresolved issues 'left over' from the Project phase became the source of problematic outcomes later in the Shakedown phase.

Unfortunately, it was also the case that companies experienced during the Shakedown phase problems that had originated in the Project phase, but were not perceived as problems or rectified at that time. Although these problems are more rightly classified by their origins as Project phase problems, we list them below as Shakedown phase problems, because that is where their symptoms show up.

Shakedown phase problems

As mentioned earlier during the discussion of 'success', many of our companies experienced negative outcomes during the Shakedown phase. Among the outcomes experienced were the following:

- performance problems with the ERP system (and underlying IT infrastructure);
- slowdown in business processes;
- errors made by users entering data into the system;
- increased staffing required to cope with slowdowns and errors;
- drop in the company's key performance indicators;
- negative impacts on customers and suppliers from an inability to answer their queries and from delayed shipments and payments;
- need for manual procedures to address lack of functionality in ERP software;
- data quality problems;
- inadequate management reporting.

This list is an uncomfortable mélange of symptoms of leftover problems (for example, performance problems with the system, slowdown in processes), attempts to resolve problems (for example, manual processes, workarounds, and increased staffing) that create new problems in their turn, and true outcomes – consequences of problems (for example, negative impacts on customers). These elements are difficult to disentangle analytically. But, after detailed examination, we concluded that many Shakedown phase difficulties were caused by problems that occurred during the Project phase but were not recognized as problems or successfully resolved at the time they occurred. Among these problems, the most important were:

- approaching ERP implementations from an excessively functional perspective;
- inappropriately cutting project scope;
- cutting end-user training;
- inadequate testing;
- not improving business processes initially;
- and underestimating data quality problems and reporting needs.

Approaching ERP implementations from an excessively functional perspective Cross-functional integration is still a new concept to many organizations. It is far more natural for them to approach implementing ERP on a module-by-module basis and to assume that ERP modules correspond to traditional functional departments in the organization (for example, accounting, manufacturing, sales). Configuration errors often follow when adopters set up project teams without appropriate cross-functional representation (for an example, see Koh, Soh, and Markus, 2000).

Inappropriately cutting project scope Savvy project managers know that exceeding project schedule is the major threat to project success (more so even than budget overruns). Therefore, cutting scope is a common tactic when the project shows signs of missing key milestones. Project managers are often tempted to cut scope according to what looks hardest to do; those who stay focused on 'what is the minimum functionality we can implement to get the desired business benefits?' are more successful. As mentioned earlier, several of our studied companies cut scope when schedule and budget ran short. These cuts often made it necessary for users to adopt inefficient manual processes in the Shakedown phase.

Cutting end-user training Schedule pressures affect training as well as scope, because end-user training is typically one of the last activities to occur in the project. Adopters frequently reported having underestimated the need for end-user training. In particular, they told us that users needed additional training and education in non-ERP areas, for example:

- Making the transition from 'green screen' (mainframe software) to 'client-server' (PC-based software). Surprisingly, this was a major hurdle in several adopter organizations.
- Understanding ERP and MRP concepts. Some adopters believed it necessary to conduct extensive education to accompany ERP training.
- Understanding cross-functional business processes. In many organizations, people understand what they do, but not how their work affects others. In the ERP setting, such a limited world view leads to errors and misunderstandings.
- Recovering from data entry mistakes. Because ERP systems are integrated, data entry errors have many more ramifications than do errors in traditional systems, and they are much harder to correct. Adopters reported suffering from lack of training that addressed recovery from data entry problems.

In some companies, training was not budgeted as part of the ERP project itself, but was left to the budgets and discretion of operating managers. This management policy increased the likelihood of inadequate end-user training.

Inadequate testing, particularly of interfaces, modifications, integrations, and exceptions Like scope and training, testing is often cut when the project schedule gets tight. Further, because many adopters lack extensive experience with integrated software and with cross-functional teaming, they are likely to overlook the need to conduct system (as opposed to module) tests. Areas where testing is most likely to be deficient include:

- ERP cross-module integrations;
- interfaces with legacy systems;
- modifications – especially those performed by external firms (adopters often assume that external provider work is properly tested);

- unusual business scenarios and scenarios involving the input of erroneous data.

Several adopters told us (they realized after the fact) that they had not adequately tested their ERP software.

Not first improving business processes where this needs doing Adopters naturally want faster implementations, and one of the best ways to shorten implementation schedules is to 'implement the software first, and reengineer the business processes later'. This is great advice when adopters have reasonably sound business processes to start with, but some adopters do not. Some companies have found that failure to change their business processes leads to:

- Inappropriate software modifications. One company we studied tried to implement ERP without changing *either* the software or its business practices. In the end, the company unnecessarily changed both. The software modifications could have been avoided through upfront business process improvements.
- Severe disappointment with ERP when managers realized that getting business benefits from ERP required change in business practices (Dolmetsch et al., 1998).

Underestimating data quality problems and reporting needs Our review of a few companies' detailed project plans revealed severe underestimation (even their implementation consultants missed this problem!) of project tasks associated with data. In the early days of a project, it is of course hard to know how many and which legacy systems will have to be retained. But, even when the ERP system replaces all legacy systems, data problems can be severe:

- Due to the nature of their businesses, adopters may need to retain legacy data for many years (for example, for regulatory compliance or because their products remain in service for many years).
- Adopters often underestimate the poor quality of their existing business records that will be input to ERP. Knowledgeable end-users often substitute for high-quality data in traditional systems: *they* know what the numbers really are. But because ERP systems are integrated, the data must be cleaner. Bad data may automatically trigger processes in distant areas where people lack the knowledge to override the system.
- Most large adopter organizations have extensive and complex data reporting needs. While these needs are best addressed with technologies other than ERP, adopters often believe that ERP will satisfy them. Therefore, ERP project plans often neglect reporting issues, and some adopters become very disappointed with ERP systems because their reporting needs are not well met.

In fact, our biggest surprise with the Shakedown phase was that, in the adopter's eyes, high-quality data and good reporting are *absolutely essential* for ERP success. End-users and line managers are unwilling to trust and use systems if they do not trust the data and reports. Lack of user acceptance of data and reporting can lead to de-installation of the system or unwillingness to invest in further upgrades. Note that achieving acceptance of a common source of data is often a highly political process, especially in large, complex organizations. But if these politics are not well managed during the Project phase, the success of the entire experience is at risk.

In short, the Shakedown phase reveals the unresolved or unrecognized problems of the Project phase. Many negative Shakedown experiences can be avoided by giving adequate attention *during the Project phase* to:

cross-functional configuration and testing of ERP software,
end-user training,
data conversion and management of legacy data,
reporting needs,
scenarios for recovering from data input errors.

At the same time, steps taken *during the Shakedown phase* to remedy these problems or their symptoms may fail to solve the problems and may actually make matters worse. For example, we found that because end-user training was inadequate and users did not understand how to back out of erroneous transactions, companies often began to rely heavily on 'key users' (project personnel) and IT staff to perform routine work that should have been done by users. As a result, the key users did not have time to conduct better end-user training, and IT staff did not have time to work out platform problems and upgrades. These companies later found themselves extremely vulnerable when key users and IT staff began leaving for better-paying jobs elsewhere. Similar observations about the persistence and negative consequences of 'workarounds' have been made by Tyre and Orlikowski (1994).

Onward and Upward phase problems

Different problems characterized the Onward and Upward phase. As with Shakedown phase problems, problems appearing during the Onward and Upward phase often had much earlier roots. The most important problems we observed in the Onward and Upward phase were:

Unknown business results Many adopters who had been using ERP long enough to have business results did not know whether they had realized improvements. In most cases, these companies had viewed ERP strictly as a technology replacement decision and had not prepared business cases justifying ERP in terms of business benefits.

Disappointing business results Some adopters in the Onward and Upward phase reported that their business results had not be achieved. In some cases, the absence of business results could be traced to inappropriate scope cutting decisions during the Project phase. In other cases, the organization did not have a culture of managing to results, did not collect and use metrics, did not demand business improvements, and so forth. The lesson is clear: ERP benefits are not automatic. They require human and organizational learning, both of which take time and require focused management attention.

Fragile human capital Many adopters were not in a strong position to go forward with ERP because of the fragile state of their ERP human capital. Many organizations lost and have difficulty replacing ERP-knowledgeable IT specialists and end-users. In some organizations, the only end-users who were ERP knowledgeable were those who participated on the project team. In addition, we saw IS specialists routinely doing work that belonged in end-user job responsibilities. This is a precarious situation for adopters. Not only may they fail to realize full business benefits from ERP, but also they may be unable to recover gracefully from future problems. Further, they may not be able to make future technology upgrades and business improvements without outside help.

Migration problems We spoke to several adopters who were on their second round of ERP implementation. Most reported having learned how poorly software modifications convert during implementation of later releases. In some cases, this was seen as a positive learning experience, because the organizations vowed never again to modify the ERP software but to make essential changes to their business processes. But we suspect that most companies that have difficulties in upgrading will simply stop enhancing their ERP systems. These organizations will in effect have implemented legacy ERP systems, obviating one of the major benefits of using packaged software – the ability to outsource to a vendor the ongoing maintenance and enhancement of software (Brehm and Markus, 2000).

In short, the Onward and Upward phase reveals the unresolved or unrecognized problems of earlier phases. In some cases, Onward and Upward phase problems could have been avoided by taking action during the Project phase:

- doing a much better job of end-user training during the Project phase;
- starting the Project phase with plans for long-term maintenance and migration;

- documenting the reasons for configuration decisions, not just the parameters, so that people not involved in the Project phase can get quickly up to speed;
- not disbanding the project team when the project goes live, but instead staffing a competence center to manage future evolution and learning.

In other cases, however, preventing and resolving Onward and Upward phase problems must occur well before the Project phase even begins. Elsewhere (Markus and Tanis, 2000), we have discussed the importance of what we call the Chartering phase, often unacknowledged in less-successful ERP adoptions, in which key business decisions related to the ERP system are made. In many cases, only senior executives (not project managers and team members) can address pre-existing organizational challenges that threaten ERP success. Among such challenges are the following, observed in several of our study companies:

- Lack of results orientation in the business is a key factor in failure to achieve business results. This is not something that an ERP project team can fix.
- A culture resistant to change is another big impediment to ERP success. Project teams can design and execute change management programs, but senior executives must work to make these efforts a success.
- When top managers do not buy in to the goals and plans of the ERP project team, the chances for success are weak. Good project managers can contribute to buy-in by good and frequent communication, but again success requires a concerted effort at the top, before and during the project.

Waiting to resolve these problems until the symptoms first appear – often as late as the Onward and Upward phase – can be a recipe for failure. Remedial actions taken late in the experience often fail to solve the problems. The more likely outcome when problem resolution is delayed is termination of the system.

1.5 Conclusions and Suggestions for Future Research

The implementation of ERP systems in organizations is an enormously complex undertaking. ERP systems can affect nearly every aspect of organizational performance and functioning, and measures of ERP systems success must reflect this fact. Our findings show that different measures of success are appropriate at different points in the ERP experience cycle, and that the outcomes measured at one point in time are only loosely related to outcomes measured later. This occurs because the experience cycle is a process (or really a set of processes), not a mechanical connection between starting conditions and final results. Over the course of this process several things can happen to influence the final outcomes observed: starting conditions can change, problems can arise (which may or may not be recognized), steps can be taken to address them (which may or may not be successful, possibly creating new problems in their wake).

In short, the connections between starting conditions, experienced problems, and outcomes in the ERP experience are not deterministic. While this can be construed as bad news for academic theory, it is good news for both ERP adopters and for information system researchers. For ERP adopters it means that it is possible to succeed with ERP despite bad luck, some mistakes, and even early failures. For researchers it means that there is much more work to be done to understand problem recognition and resolution behaviours and how they interact to result in successful and unsuccessful outcomes.

One particular area that deserves much future research is what we have called the Chartering phase – often unacknowledged and unfulfilled in the organizations we studied. In this phase, which should occur before a 'project' is 'chartered' (hence the name), senior executives in consultation with others make important business decisions about the objectives of the project, the decomposition of the project into manageable chunks, the level of budget to be allocated to the Project and Shakedown phases of each chunk, an appropriate project leader and/or implementation partner, and so forth. Further research is needed on how companies actually make or *avoid* making these decisions, what factors they consider and do not consider, whom they consult and follow, and the specific implications of these decisions for the problems and outcomes experienced later in the ERP experience cycle.

ACKNOWLEDGEMENTS

We gratefully acknowledged the support of Gordon Mosinho for this research. Andrew Martin and Chris Sauer provided helpful comments.

REFERENCES

Ballantine, J., Bonner, M., Levy, M., Martin, A., Munro, I., and Powell, P. L. (1996) The 3-D Model of Information Systems Success: The Search for the Dependent Variable Continues., *Information Resources Management Journal*, **9**(4), 5–14.

Bashein, B. J. and Markus, M. L. (2000) Data Warehouses: More than just Mining. Financial Executives Research Foundation, Morristown, NJ.

Bashein, B. J., Markus, M. L. and Finley, J. B. (1997) Safety Nets: Secrets of Effective Information Technology Controls. Financial Executives Research Foundation, Morristown, NJ.

Brehm, L. Heinzl, A., and Markus, M. L. (2000) Tailoring ERP Systems: A Spectrum of Choices and Their Implications. Under editorial review, available from Markus.

Brehm, L. and Markus, M. L. (2000) The Divided Software Life Cycle of ERP Packages. Proceedings of the 1st Global Information Technology Management (GITM) World Conference, Memphis, Tennessee, USA.

Bulkeley, W. M. (1996) A Cautionary Network Tale: Fox-Meyer's High-Tech Gamble. *Wall Street Journal Interactive Edition* (18 November).

Davenport, T. H. (1998) Putting the Enterprise into the Enterprise System. *Harvard Business Review*, **76**(4), 121–131.

DeLone, W. H. and McLean, E. R. (1992) Information Systems Success: The Quest for the Dependent Variable. *Information Systems Research*, **3**(1), 60–95.

Dolmetsch, R., Huber, T., Fleisch, E., and Osterle, H. (1998) Accelerated SAP: 4 Case Studies, Institute for Information Management, University of St. Gallen School for Administration, Economics, Law, and Social Sciences (HSG) (16 April).

Koh, C., Soh, C., and Markus, M. L. (2000) Process Theory Approach to ERP Implementation and Impacts: The Case of Revel Asia. *Journal of Information Technology Cases and Applications*, **2**(1), 4–23.

Larsen, M. A. and Myers, M. D. (1997) BPR Success or Failure? A Business Process Reengineering Model in the Financial Services Industry. Proceedings of the International Conference on Information Systems, Atlanta, GA, pp. 367–382.

Lyytinen, K., and Hirschheim, R. (1987) Information Systems Failures – A Survey and Classification of the Empirical Literature. In *Oxford Surveys in Information Technology*, vol. 4, Zorkoczy, P. I. (ed.), Oxford: Oxford University Press, pp. 257–309.

Markus, M. L. and Tanis, C. (2000) The Enterprise Systems Experience – From Adoption to Success. In *Framing the Domains of IT Research: Glimpsing the Future through the Past*, Zmud, R. W. (ed.), Cincinnati, OH: Pinnaflex Educational Resources.

Peters, T. J. and Waterman, R. H. (1982) *In Search of Excellence: Lessons from America's Best-Run Companies*. New York: Harper and Row.

Ross, J. W. and Vitale, M. (2000) *The ERP Revolution: Surviving Versus Thriving*. Information Systems Frontiers.

Sauer, C. (1993) *Why Information Systems Fail: A Case Study Approach*. London: McGraw-Hill.

Tyre, M. J. and Orlikowski, W. J. (1994) Windows of Opportunity: Temporal Patterns of Technological Adaptation in Organizations. *Organization Science*, **5**(1), 98–118.

Yin, R. K. (1994) *Case Study Research: Design and Methods*. Thousand Oaks, CA: Sage Publication.

2 Innovating with Packaged Business Software: Towards an Assessment

E. Burton Swanson

2.1 Introduction

In the 1990s many firms turned to software packages when they replaced the older, often home-built systems in their application portfolios. They chose to buy, rather than build, their new systems. Broadly, they followed the prevailing wisdom to downsize themselves so as to focus on their 'core competencies' and to outsource other tasks to the market. Many found no compelling logic for continued in-house development of their application software. Some, with their staffs already reorganized and reduced through reengineering, were simply no longer in a position to undertake major development tasks.

Replacement of systems had in the 1990s become a priority for many firms. These organizations felt increasingly burdened by their 'legacy systems'. Their support staff were often committed to a maintenance task for which the costs were all too apparent and the benefits appeared to be marginal and remedial. Their home-built software also often resided on expensive mainframes widely viewed as heading rapidly toward obsolescence. Further, users found this software increasingly cumbersome to work with in comparison to their newer PC-based tools with their graphical user interfaces.

It was in this context that the 1990s gave rise to ERP (enterprise resource planning), a significant innovation in the packaged software market. The vision for ERP was first articulated by the Gartner Group (Wylie, 1990). It took an 'enterprise-wide view' of traditional application software, allowing for internal integration of the technical and business core, as well as external integration with business customers. With its roots in manufacturing, ERP was first presented as a 'next-generation MRP II'. But as a concept it soon transcended industry boundaries. Viewed technologically, its promised benefits included: a single view of enterprise data; ease of software configuration and customization; a simplified client/server computing architecture;

and reduced initial and overall software costs (Keller, 1999). By the mid 1990s, major vendors included SAP AG, Oracle, Peoplesoft, Baan, and J.D. Edwards.

By 1995 the ERP movement was in full ascendance. Major consulting firms found it to be their largest new source of business. 'If SAP is the ten-ton messiah of enterprise-wide computing, the Big Six are its true disciples, with Andersen Consulting, ICS Deloitte, and Price Waterhouse leading the pilgrimage', *Fortune* magazine reported (Buchanan, 1995, pp. 122–123). ERP adoptions increased rapidly. Many companies were in the initial throes of their implementations.

Finally, as the new millennium approached, firms found one more compelling reason to jettison their old systems and replace them with packaged software, as the infamous Y2K problem at last achieved widespread recognition with the resulting spread of alarm. Now, many firms faced the painful prospect of spending large sums to bring their old systems into compliance, should they choose not to replace them with ready-made packaged solutions. And so, as time slipped away, still more firms made the leap to packages.

Summarizing, the 1990s were marked by what we term the *package transition* in the employment of business application software. While packaged business software had already long been in use, it was in the 1990s with the advent of ERP that it began to dominate enterprise decisions throughout the world. By he close of the decade, about half of the larger US and European companies had reportedly deployed an ERP system (Cap Gemini Ernst & Young, 1999).

In our view in the mid 1990s, this package transition represented a broad business change of significant importance, one in which firms appeared to face certain characteristic risks, which we discuss further below. In 1996, we accordingly initiated a research project to study the phenomenon. We sought to understand how firms confronted the associated risks and made their choices, and, where they adopted new packages, with what success they implemented them. The present article reports findings from a survey undertaken as one part of our research, which to date also includes one intensive case study (Hirt and Swanson, 2000). As we shall see, these findings suggest substantial success among firms in making the package transition and offer insights as to how this has been achieved.

Knowing Why, How and When to Innovate with Packaged Software

Broadly, in our view, the package transition in the 1990s entailed organizational and inter-organizational innovation. Firms undertook that which had not been before accomplished, learning directly and indirectly from each other, supported by vendors and consultants. The scope and scale of the collective effort was perhaps unprecedented. Davenport (2000) goes so far as to say, 'Successful implementation of ES does involve probably the greatest technological change most organizations have ever undergone . . . Even more difficult and important, however, are the major changes in business that come with an ES project. Business processes, the way work gets done in an organization, change dramatically' (pp. 5–6, Davenport prefers the term 'enterprise systems' – ES – to ERP).

Success with ERP and its complexities was therefore by no means assured. Firms confronted an innovation presenting substantial uncertainties and challenges. Making ERP successful would require most fundamentally the development and diffusion of new organizational and inter-organizational knowledge.

For the individual firm, the innovation challenge entailed a sequence of major steps. Whether a firm came to the innovation relatively early or late, in confronting the unfolding ERP phenomenon, each firm needed to *comprehend* ERP as an organizing vision (Swanson and Ramiller, 1997), that is, the firm needed to understand the broader sector's idea for the innovation, in terms of the innovation's rationale, organizational features, technology to be used, and the ways in which the innovation might be successfully accomplished. The firm might then consider whether and when to *adopt* the innovation. Where decided in favor, it would then need to *implement* it. Finally, it would need to *assimilate* ERP within the enterprise. The firm's likely overall success or failure in innovating with ERP would hinge on all of these steps.

In considering the adoption of ERP, each firm further faced three basic questions. First, *why* should it make or not make the move to ERP? Second, *when* if ever should it make the move? Third, *how* if at all should it undertake it? In answering these three questions, each firm drew not only upon its comprehension of the organizing vision for ERP, but upon its perception of its own readiness for it. The answers to these questions mattered greatly to the success of ERP implementation and assimilation. Miscomprehension and misperception would likely be costly.

For prospective early adopters, ERP would presumably seem to be a good fit to the enterprise, offering possible competitive advantage. However, innovating at the leading edge of practice would be risky, as not much knowledge about how to maximize on the ERP promise existed. Waiting for the ERP bandwagon posed different risks. While some knowledge of ERP would then be available in the market, it might be in extremely short supply, spread very thin among those now taking up the innovation. Finally, not to make the move to ERP entailed the risk of being left behind, potentially damaging the firm's competitive position. Thus, at each stage of ERP's diffusion as an innovation among firms, characteristic risks associated with the development and acquisition of new knowledge were faced.

In this theoretical context, three working hypotheses motivated our present study. First, we conjectured that at any particular time some reasons for adopting packaged business software would be better than would be others. Certain aspects of the rationale offered for ERP at a particular time might be well conceived, while other aspects might represent mostly wishful thinking. Regardless of when they make their moves, firms with the right *know-why* would be the more likely to be successful in their implementations. In the 1990s a variety of reasons motivated managers to undertake the package transition. We were curious to know which of these might have been the better ones, thus leading to successful outcomes.

Second, we conjectured that at any particular time some organizational capabilities for implementing packaged business software would be greater than would be others. Regardless of when they make their moves, organizations with the right *know-how* would be more likely to be successful in their implementations. In the 1990s, organizations with a wide variety of capabilities undertook the package transition. Many possessed substantial IS staff and long prior experience with in-house development, while others did not. Some already made sophisticated use of IT in supporting the core technology of the business, while many used it mainly for simpler administrative purposes. Some were centralized in their approach to IS, with highly placed managerial leadership, while many were decentralized across departments and locations. Different organizational capabilities would likely suggest different approaches to packaged software implementation. Firms lacking substantial IS staff might rely more heavily on the help of consultants, for instance. We were curious to know which organizational capabilities and implementation approaches might have been the more important ones in leading to successful outcomes.

Third, we conjectured that knowing both why and how to innovate with packaged business software would enable the firm to have the right *know-when* to do it. Because our study lacks the longitudinal time frame needed to adequately explore this notion further, we mention this conjecture here principally for completeness.

2.2 Undertaking a Survey

In the summer of 1998 we initiated a survey to assess how and why firms have been successful or not as they attempted to make the packaged software transition. Working with our own substantial CIO mailing list, we distributed our survey form in two waves, the first in the late summer 1998 and the second in the fall 1999. As the research is longitudinal and ongoing, additional waves are planned.

Because our CIO mailing list is peculiar to us (it is for instance biased toward our geographical location and toward larger firms with past interactions with us) we did not seek a randomly 'representative' respondent sample from it. Rather, we simply aimed for a respondent sample which was large and heterogeneous enough in its composition (across industries, for example) to suggest likely external validity to our findings.

We asked prospective respondents to report on their adoption and implementation of a business application package of significant importance to their firm or one of its divisions. We did not restrict respondents to ERP as it has been variously defined. However, our cover letter highlighted our interest in learning about the adoption of packages of major ERP vendors.

The survey form itself was structured into three parts. The first solicited basic background data on the enterprise. The second part sought certain factual data on the adoption and implementation of a selected package. The third part requested the CIO's assessment of the adoption and implementation experience, and included 55 judgmental statements, with each of which the respondent could simply agree or disagree on a five-point scale. The items were deliberately formulated so as to span the knowledge-related issues discussed in the previous section. They addressed both organizational know-why and organizational know-how, and they tapped the associated contributions of various participants, including IS staff, system users, top management, consultants and vendors, to both forms of knowledge. (A copy of the survey form is available from the author.)

The survey form was pre-tested with two firms for clarity and ease of completion before being sent out to the mailing list. From the two mailing waves, we received 90 responses from a wide variety of firms.

The Respondents

As seen in Figure 2.1, our 90 respondents represent a wide variety of industries, with the largest concentrations in manufacturing, financial services, health care, and utilities and power. These sectors together make up less than half the total. Thus, our sample is not dominated by only a few industries.

As with industry, the size of our respondent's firm (or division) and its IS unit also varies widely. So too does the custom-built portion of the firm's

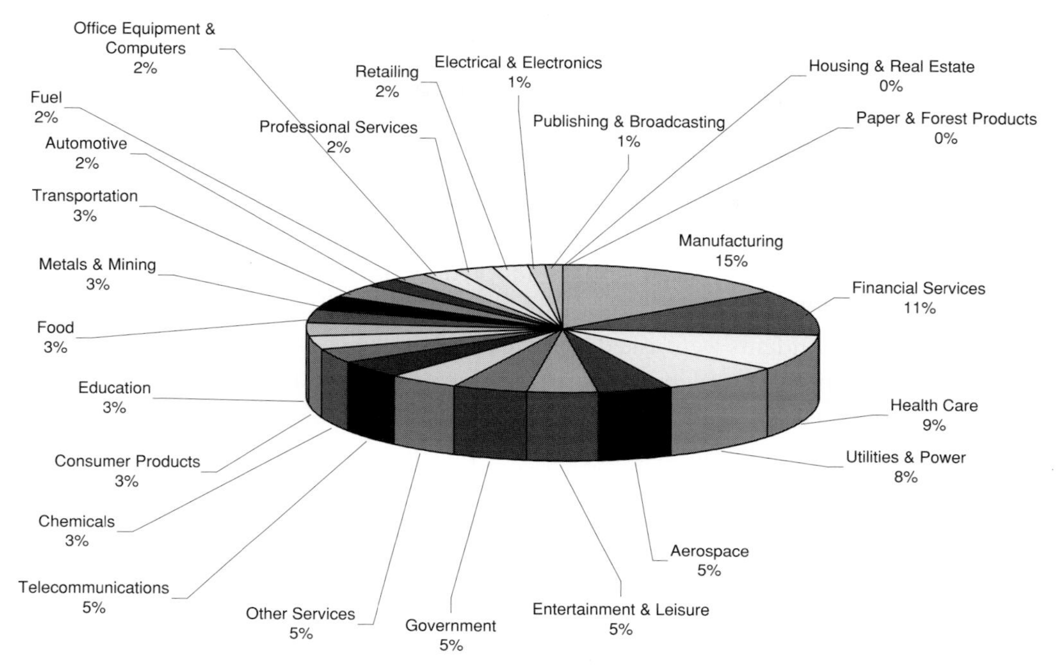

Figure 2.1 Research base by sector

(or division's) current application portfolio, which averages somewhat less than half the total (see Table 2.1).

Table 2.1 Firm characteristics

Firm characteristic	Median	Quartile range
Employees	5200	1500–15000
IS employees	150	30–400
Custom apps (percentage of total systems)	40	20–70

While the sample of respondents has substantial variety, it is probably not representative of all firms undertaking the package transition in the 1990s. As we shall see, it is likely biased toward those who have been relatively successful in their implementations. It tends to exclude, for example, those who abandoned their implementations altogether and would have little interest in responding to the survey.

Assessing the Package Transition

We report our findings in two parts. First, we summarize the factual data we requested on the packages reported as adopted and implemented by our respondents. Second, we report in detail the respondents' assessments of their adoption and implementation experience.

Package Adoption and Implementation

Respondents were asked to tell us about their adoption of a business application software package of significant importance to their firm or one of its divisions. The vendors of those packages adopted by our respondents include the five major ERP suppliers, comprising three quarters of the total (Figure 2.2).

Respondents were asked to indicate when the chosen package was adopted and whether implementation was phased by module and/or by site. They were further asked when implementation was completed in the first part (one or more modules put into production) and when it was (or would be) fully completed (all modules in production at all sites). They were also asked to report the number of support staff (FTE) currently allocated to implementing and/or maintaining the adopted package. Finally, they were asked to indicate the number of direct (hands-on) users of the package.

As shown in Table 2.2, the bulk of reported adoptions and implementations span the latter half of the 1990s. On average, implementation is seen to require about two to three years, from date of adoption to completion. The majority of implementations are multi-site. From data not shown, about half the implementations were phased by site (47%) and about half were phased by module (51%). The number of support staff averaged 20 FTE. We observe that support staff numbers can be relatively high, when compared with numbers of direct users. On average, one support staff member may support less than ten total users. (It would be interesting to compare these support data to equivalent data for the support of systems developed in-house. Pending such a comparison, the present data should perhaps serve as a warning to executives who

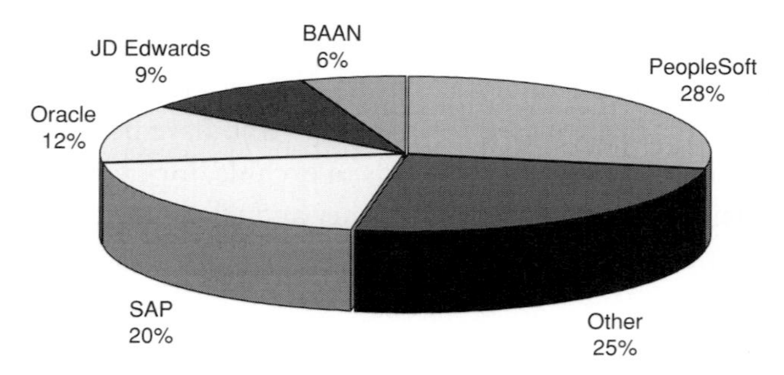

Figure 2.2 Packages chosen by researched firms

Table 2.2 Reported adoptions and implementations

	Median	Quartile range
Adoption date	Feb. '97	Oct. '95–Feb. '98
Implementation sites	2	1–13
First implementation	May '98	Oct. '96–Dec. '98
Completed implementation	June '99	Nov. '98–Nov. '99
Support staff	20	6–45
Daily users	75	25–250
Total users	150	35–805

might imagine that their support staff can be drastically reduced by moving from in-house systems to packaged software.)

Adoption and Implementation Assessment

The survey included 55 item statements relating to the respondent's experience with the adoption, implementation, and the use of the identified package. Respondents were asked in each case to indicate their agreement or disagreement with the statement. An illustrative statement and the distribution of our respondents agreement/disagreement with it pertains to their success in staying on track with their plans (Figure 2.3).

We observe that our respondents are on the whole on track with their plans. This markedly positive assessment is characteristic of our responses, as will be seen.

In the following tables we group and summarize the responses. We indicate the levels of agreement or disagreement as follows. Three asterisks (***) indicates that a majority of respondents agreed (or disagreed) and that they did so by a margin of four to one or better over those indicating the opposite. Two asterisks (**) indicates a margin of three to one or better. One asterisk (*) indicates two to one or better. Where there is no majority agreement or disagreement by a margin of two to one or better, we report the response as 'split'.

For presentation purposes, the assessment items are grouped as follows: success indicators, business benefits, other adoption motivation, package functionality and fit, adoption and implementation support, requisite knowledge and skills, implementation task complexity, and package reliability and vendor relationship. As a caution, these particular groups do not purport to distinguish sharply between know-why and know-how, and some individual items as well as groups may in fact reflect both, directly or indirectly.

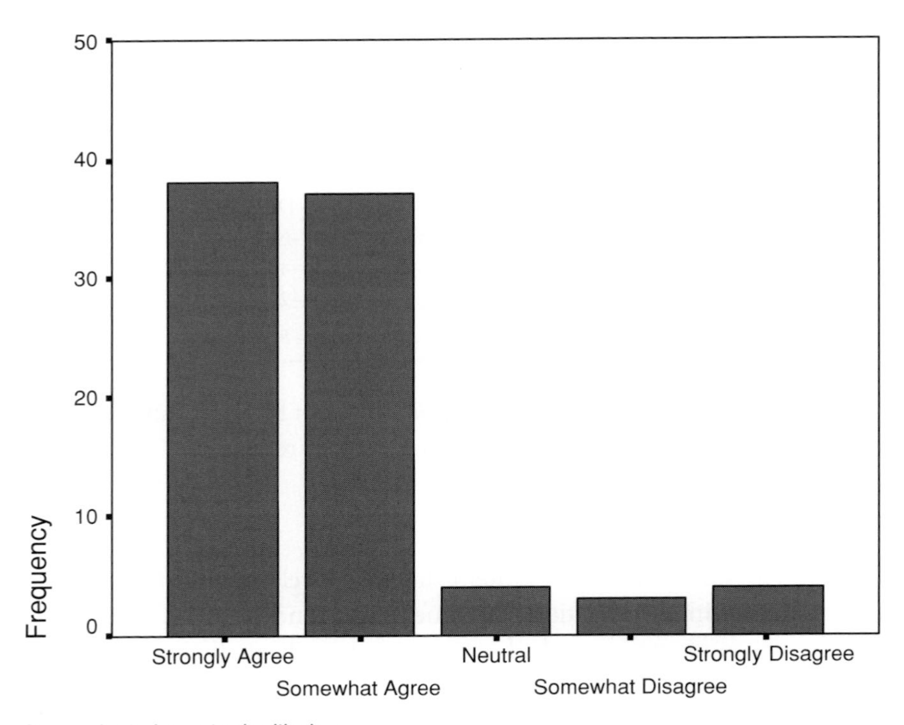

Figure 2.3 Success in staying on track with plans

Success Indicators

As shown in Table 2.3, our respondents tend to agree by substantial margins that they have been successful with their adopted packages. They do not regret their choice of package. They consider their implementation successful on the whole. Most users are happy with the package.

Among these success indicators, Item 55 deserves special note as it is designed to offer a single global measure of implementation success. Below, we

Table 2.3 Levels of success with adopted packages

Item no.	Agree or disagree	Level	Item statement
33	Agree	***	We are on track with our plans for the implementation and use of this package.
24	Agree	***	Most of our users of this package are on the whole happy with it.
55	Agree	***	On the whole, our implementation of this package has been successful.
27	Disagree	**	If we knew then what we know now, we would probably not have adopted this package.

report the correlation of our other assessment items to this metric. Significant positive success correlates are indicated by (+), negative success correlates by (−). Significance is at the 0.05 level or better for three separate one-tailed tests (Pearson, Kendall's τ b, Spearman's π) making different parametric assumptions. While we use only the single item as our overall success metric, we should mention that it correlates strongly at the 0.001 level with each of the other success indicators, suggesting its appropriateness for our purpose. We should also mention that Item 55 is *not* significantly correlated with basic firm characteristics reported above. Hence we do not control for these characteristics in our analysis. Nor is this success metric significantly correlated with the timing of package adoption or the duration of implementation. Curiously, it is positively associated with the total number of users.

Business Benefits

Our respondents also tend to agree that their package provides important business benefits, although the margins of their agreement vary across specific benefits, and some benefits are not agreed upon (see Table 2.4). Most

Table 2.4 Business benefits experienced

Item no.	Agree or disagree	Level	Item statement	Success correlate
2	Agree	***	This package provides features which enable us to work better with our suppliers and/or customers.	(+)
37	Agree	***	Our top management understands the business value of this package.	(+)
46	Agree	**	This package facilitates our user communication across departments.	(+)
8	Agree	*	Without this package we would be at a competitive disadvantage in our industry.	
43	Agree	*	This package enables us to engage in electronic commerce.	
5	Split		This package is becoming a standard for firms in our industry.	(+)
11	Split		An advantage of this package is that it incorporates many of the best practices in our industry.	
42	Split		An advantage of this package is that it enables us to integrate our international business operations.	(+)
50	Split		While we have gained operational efficiencies with this package, we have lost some of our organizational flexibility.	
53	Split		This package has enabled us to reduce our overall costs of business operations.	(+)

notably, there is strong agreement that top management understands this business value of the package, and further that the package enables the firm to work better with its suppliers and customers. Additionally, there is substantial agreement that the package facilitates communication across departments.

Agreement as to several other benefits is seen to be weaker. And, significantly, our respondents are split on some suggested benefits, including the reduction of overall business costs. There are thus important differences among our respondents in terms of reported business benefits.

Importantly, the majority of items in the business benefits category are found to be positive correlates of implementation success. Thus, where success is claimed, so too are the business benefits. In the absence of such benefits, implementation success is apparently less likely.

Other Adoption Motivation

Beyond business benefits, respondents were agreed that an older application was judged to have reached the end of its useful life and to be in need of replacement (see Table 2.5). They were further agreed that outsourcing their IS department was not an objective. They were largely split on other motivations for adopting the package.

Table 2.5 Motivations for adoption

Item no.	Agree or disagree	Level	Item statement	Success correlate
31	Agree	***	This package replaced one of our older applications that had obviously reached the end of its useful life.	
48	Disagree	***	Adoption of this package is a step toward outsourcing our IS department.	(−)
16	Split		Our IS function lacks the expertise needed to build an application equal to this package.	
28	Split		We adopted this package as part of a broader move toward a client/server computing environment.	
38	Split		We adopted this package as part of a broader initiative to resolve our Y2K problems.	
40	Split		We were one of the first in our industry to adopt this package.	
41	Split		We adopted this package in order to reengineer our business processes.	

Remarkably, only one item in this group is found to be a success correlate. And this item, outsourcing the IS department, correlates negatively with success. Thus, apart from business benefits, other reasons for adoption are not predictive of implementation success. Perhaps, we speculate, business benefits provide the 'only good reasons' for successfully staying the often problematic course of implementation.

Package Functionality and Fit

Our respondents further agreed that they had changed their business processes to take advantage of the package's functionality (see Table 2.6). At the same time, they felt that their chosen package required enhancements to better meet their needs. Interestingly, they did *not* view the package's functionality as dominating that of alternative, competing packages.

Interestingly, to the extent respondents felt that enhancement to their package was *not* required, this was predictive of implementation success. So too was willingness to change business processes as needed. Thus, adapting business processes to the package, rather than adapting the package to business processes, appears to be the more successful approach among our respondents. Of course, this is easiest where the package and its functionality provide a good fit to desired business processes.

Table 2.6 Functionality of the package

Item no.	Agree or disagree	Level	Item statement	Success correlate
3	Agree	**	We willingly changed our business processes to take advantage of the functionality of this package.	(+)
47	Agree	**	The package requires enhancements in order to better meet our needs.	(−)
10	Disagree	*	No other vendor offers a package comparable in functionality to this one.	
15	Agree	*	Custom-built software would offer us little advantage in functionality compared to this package.	
7	Split		The package enables us to bridge easily to other applications in our portfolio.	
17	Split		This package offers more functionality than we really need.	
23	Split		An advantage of this package is that it works well with packages from other vendors.	
34	Split		This package provides an easy-to-use graphical user interface.	

Adoption and Implementation Support

Respondents reported strong internal support for adoption and implementation. Notably, users were substantially involved in the decision to adopt the package (see Table 2.7). Top management also led and further provided the resources necessary for successful implementation. Respondents also agreed that the package was not pushed upon them by outside consultants.

However, user involvement in the adoption decision did not always translate into lack of key user resistance. Nor did top management leadership always translate into an understanding of the costs ultimately involved. On these points, our respondents were split.

Significantly, top management's understanding of implementation costs and its willingness to provide the necessary resources both prove to be success correlates. Other items are not success correlates, however. Apparently, implementation resources, more than initial support for adoption, predict implementation success. Interestingly, user involvement in the adoption decision is not a significant success correlate.

Table 2.7 Level of support for adoption/implementation

Item no.	Agree or disagree	Level	Item statement	Success correlate
21	Agree	***	The users of this package were substantially involved in our decision to adopt it.	
29	Disagree	***	This package was initially evaluated and recommended to us by a consultant.	
35	Agree	***	Top management has provided us with the necessary resources to successfully implement this package.	(+)
26	Agree	**	Top management led in our move to adopt this package.	
30	Split		Certain of our key users resisted adoption of this package.	
39	Split		Our top management understands the costs to implement and maintain this package.	(+)

Requisite Knowledge and Skills

Respondents agreed in particular that hiring, training, and retaining staff with requisite skills poses a significant challenge (see Table 2.8). Less consistently, they also reported that most of their users have found it easy to work with the package and that they have gained from the implementation experience of others. They are split on several other items.

Table 2.8 Requisite knowledge and skills

Item no.	Agree or disagree	Level	Item statement	Success correlate
20	Agree	***	It is hard for us to hire and retain people with good skills in this package.	
32	Agree	***	Maintaining this package has required our IS staff to learn a whole new set of skills.	
1	Agree	**	Most of our users have found it easy to work with this package.	(+)
45	Agree	*	We have gained from the experience of others in implementing this package.	(+)
49	Split		Our users make sophisticated use of this package.	(+)
51	Split		The consultants in our implementation of the package were not so knowledgeable as we had expected.	
54	Split		The task of maintaining this package in our organization will fall substantially to its users.	

Interestingly, among our respondents, the difficulty of hiring, training, and retaining staff is not found to be a success correlate. Rather it is apparently the common challenge to all firms. User sophistication and facility with the package, on the other hand, do correlate with successful implementation. (We note in this regard that user knowledgeability has also been associated with fewer problems in maintaining traditional home-built systems (Lientz and Swanson, 1981).)

With regard to consultants, our respondents' experiences have been mixed. However, consultant knowledgeability fails at the margin to be a success correlate.

2.3 Implementation Task Complexity

Respondents were largely split on a variety of items related to the complexity of their implementation task and to their abilities to plan and manage accordingly (see Table 2.9). Here our respondents report different experiences.

Significantly, the majority of items in this group are, however, correlated with implementation success. The significant correlates suggest on the whole that where task complexity is underestimated, the resources needed for successful implementation may fall short as a consequence.

This finding further relates to top management support for implementation, discussed above. From additional correlational analysis (not shown), we

Table 2.9 Complexity of implementation task

Item no.	Agree or disagree	Level	Item statement	Success correlate
25	Agree	*	We underestimated the time it would take us to implement the package.	
4	Split		We underestimated the amount of customization needed for this package.	(−)
6	Split		We underestimated the amount of training needed to implement this package.	
13	Split		We underestimated the amount of consulting we required to implement this package.	(−)
19	Split		Configuration of this package to our business processes has been straightforward.	(+)
36	Split		This package has substantial hidden costs for us to implement and maintain it.	(−)
52	Split		The scope of our implementation of this package has expanded beyond its original setting.	

find that to the extent task complexity is underestimated, top management is less likely to understand implementation costs and to provide the necessary implementation resources. Perhaps, in many instances, top management simply balks at providing additional resources beyond those associated with original task estimates.

Package Reliability and Vendor Relationship

Respondents are mixed in assessing the reliability of their package and their vendor relationship (see Table 2.10). They consistently find their packages to

Table 2.10 Reliability of package

Item no.	Agree or disagree	Level	Item statement	Success correlate
22	Agree	***	This package is highly reliable for us in terms of staying up and running.	(+)
9	Agree	*	This package is well maintained by the vendor.	(+)
12	Split		This package is well documented by the vendor.	(+)
14	Split		This package is relatively free of bugs.	(+)
18	Split		The vendor of this package is responsive to our requests for information and assistance.	(+)
44	Split		We are increasingly dependent on the vendor of this package.	

be operationally reliable, if not always relatively free of bugs. Less consistently, they report that the package is well maintained by the vendor. Elsewhere, there is a lack of consensus.

Significantly, if not surprisingly, most of the items in this group are, however, success correlates, underscoring the importance of both the package's operational reliability and the vendor relationship.

2.4 Conclusion

In summary, our survey findings indicate that many firms have been successful in innovating with packaged business software in the 1990s, although they have faced notable challenges, in particular in building and retaining staff with requisite skills. Importantly, our respondents report significant business benefits from their packages, most particularly in furthering external coordination with suppliers and/or customers and internal communication among different departments. They also report that top management understands these business benefits and has provided the necessary resources to successfully implement the package. On the whole, our findings provide a counterpoint to those ERP 'disaster stories' widely reported in the trade press.

While our respondents have thus been largely successful with their implementations, some have achieved more success than have others. Interestingly, business benefits prove in our analysis to be a key success correlate, while other candidate reasons for adoption do not. Apparently, business benefits are the important know-why for undertaking the package transition. Those who have warned that making the move to ERP should be a business decision, more than a technical decision, should find support here for their argument (see, in particular, Davenport, 2000).

Elsewhere among our findings, willingness to change business processes to take advantage of the package's functionality was predictive of implementation success. Kumar and Hillergersberg (2000) caution that, 'given the packaged nature of ERP, a disconnect can exist between specifying the organization's information requirements and the solution proposed by ERP.... [T]he solution finally implemented may be determined more by the capabilities and options inherent in the ERP package rather than the organization's underlying information requirements' (p. 25). While this may indeed be a likely and possibly troubling outcome, attempting to mold the package to idiosyncratic business process requirements may for many firms pose the greater risk. The failure to fully consider and engage business change, as distinct from software change, may lie at the heart of many ERP implementation failures (see again Davenport, 2000).

As might be expected, the package's operational reliability and the quality of its maintenance and documentation by the vendor were also predictive of implementation success in our study. User sophistication and facility with the package similarly correlated positively with success. Finally, where the complexity of the implementation task was underestimated, correlates suggested that the resources needed for successful implementation may fall short as a consequence.

The important know-how for successful implementation thus draws from a rich variety of contributors. IS staff, system users, top management, vendors and consultants all are involved in making the package transition successful. The importance of business know-how, in addition to technical know-how, is also underscored. In particular, the willingness to focus on changes to business processes, not just application software, is highlighted, as we have discussed. So too is the ability to achieve buy-in from management with reasonably accurate estimates of the organizational resources needed for implementation. Of course, the ability to make good estimates is probably itself reflective of having the broader necessary know-how.

In conclusion, while our respondents constitute a relatively small group, their reports suggest that firms can manage their package transitions with substantial success. Looking ahead, business executives can expect new challenges in extending this success into the next decade. Foremost among these challenges will be the integration of the Internet with the enterprise systems still being established and assimilated. Future research will need to assess this transition, much as we have attempted (and are still attempting) to assess the transition of the 1990s.

ACKNOWLEDGEMENTS

This research was sponsored by the Information Systems Research Program of the Anderson School at UCLA. It was originally presented at the SIM Academic Workshop in Brisbane, 9 December 2000. Several helpful comments and suggestions came from this workshop as well as from an invited talk at USC's Marshall School of Business. The author is especially grateful to Enrique Dans, David Firth, and Ping Wang for their research assistance.

REFERENCES

Buchanan, T. (1995) Here Comes SAP. *Fortune*, 2 October, 122 ff.

Cap Gemini Ernst & Young (1999) Enterprise Resource Planning: Survey 1999. Accessed 28 August. http://www.capgemini.de/news/erp.html

Davenport, T. H. (2000) *Mission Critical: Realizing the Promise of Enterprise Systems.* Boston, MA: Harvard Business School Press.

Hirt, S. G. and Swanson, E. B. (2000) Innovating with ERP: Siemens Power Corporation. Information Systems Research Program, Anderson School at UCLA, 22 March.

Keller, E. L. (1999) Lessons Learned. *Manufacturing Systems*, November, 44 ff.

Kumar, K., and Hillegersberg, J. (2000) ERP: Experiences and Evolution. *Communications of the ACM*, **43**(4), 23–26.

Lientz, B. P. and Swanson, E. B. (1981) Problems in Application Software Maintenance. *Communications of the ACM*, **24**(11), 763–769.

Swanson, E. B. and Ramiller, N. C. (1997) The Organizing Vision in Information Systems Innovation. *Organization Science*, **8**(5), 458–474.

Wylie, L. (1990) ERP: A Vision of the Next-Generation MRP II. Scenario S-300–339, Gartner Group, 12 April.

3 A Comprehensive Framework for Assessing and Managing the Benefits of Enterprise Systems: The Business Manager's Perspective

Shari Shang and Peter B. Seddon

3.1 Introduction

According to AMR Research, total revenue in the enterprise application and services market in 1999 was US$18.3bn. (Gilbert, 2000). Enterprise application software implementation costs are often reported to be five to ten times the cost of software licenses (Davenport, 2000). If so, organizations worldwide spent something like US$90–180bn. on enterprise systems in 1999. Most organizations that have implemented enterprise systems expect to continue using them for many years.

Enterprise systems are integrated, enterprise-wide, packaged software applications that impound deep knowledge of business practices accumulated from vendor implementations in many organizations. Today, enterprise systems are evolving to incorporate new technologies, such as e-commerce and the internet, data warehousing, and customer relationship management. Enterprise software is a semi-finished product with tables and parameters that user organizations and their implementation partners configure to their business needs. Implementation of enterprise systems therefore involves both business and IT managers who work together to define new operational and managerial processes.

The question we ask in this chapter is as follows: If organizations around the world spent US$100bn. or more on enterprise systems in 1999, *what sorts of benefits did they, or can they*, achieve? To answer this question, we present a comprehensive framework of benefits that organizations may be able to achieve from their use of enterprise systems. Our focus is on business benefits only, not on costs. Our framework establishes a base for managing the implementation and operation of enterprise systems. It can be used as a communication tool and checklist for consensus building in within-firm discussions on benefits realization, and as an assessment instrument for managing benefit realization issues. This chapter describes the steps we took to build the framework. It also shows how the framework was applied in four longitudinal case studies.

As defined in the introductory chapter to this book, enterprise systems (ES) are large-scale organizational systems, that is, systems composed of people, processes, and information technology, built around packaged enterprise application software. In this chapter, we focus on enterprise systems built around the most important class of enterprise application software, namely enterprise resource planning (ERP) software. ERP software is a set of packaged application software modules, with an integrated architecture, that can be used by organizations as their primary engine for integrating data, processes, and information technology, in real-time, across internal and external value chains. ERP software impounds deep knowledge of business practices accumulated from vendor implementations in many organizations. It is a semi-finished product with tables and parameters that user organizations and their implementation partners must configure to their business needs. Implementation of enterprise systems based around ERP software therefore involves both business and IT managers who work together to define new operational and managerial processes.

3.2 Defining Criteria for Enterprise System Evaluation

Seddon et al. (1999) recommend that anyone seeking to evaluate an IT investment should have very clear answers to each of Cameron and Whetten's (1983) seven questions on organizational effectiveness measurement. Cameron and Whetton's seven questions are shown in the left-hand column of Table 3.1. Our answers to their questions are shown on the right.

Answering the first question, 'From whose perspective is effectiveness being judged?', our goal in this paper is to develop an enterprise system benefits classification that considers benefits from the point of view of what Hammer and Champy (1993: 102) term 'process owners'. Process owners are senior middle-level managers responsible for what Anthony (1965) described as 'management control' and 'tactical planning'. Combining their detailed knowledge of operational issues with their thorough understanding of strategic goals, they manage the links between business strategy and business operations. Process owners are responsible for the processes that deliver value to their organization's customers.

Table 3.1 Seven questions to answer when measuring organizational performance

Seven questions for measuring organizational performance	Answers in this study for evaluating investment in enterprise systems
1 From whose perspective is effectiveness being judged?	Process owners
2 What is the domain of activity?	Enterprise systems
3 What is the level of analysis?	*Both* organization and function
4 What is the purpose of evaluation?	Planning, management, and improvement
5 What time frame is employed?	Years after the enterprise system went live
6 What types of data are to be used?	Objective and perceptual
7 Against which referent is effectiveness to be judged?	Markus' optimal success; stated goals of the organization, i.e., the business case; past performance of the organization

Source: Cameron and Whetten (1983: 270–274).

The reason we chose to focus on process owners and not other levels of managers in an organization are as follows. Anthony's (1965) planning and control systems framework is based on a three-level pyramid of management: strategic, tactical, and operational control. At the top of the pyramid, strategic planners (senior business executives and boardroom decision makers) tend to focus mainly on financial performance of their IT investments. The difficulty with evaluating investments in enterprise systems at this level, for example, Sircar, Turnbow, and Bordoloi (2000), is one of causality: one cannot be sure that investments in IT are the cause of observed changes in sales, corporate profitability, or market share.

At the bottom of Anthony's pyramid, operational managers (for example, foremen in a factory) evaluating an enterprise system are more likely to be interested in system attributes, such as information quality (are the data accurate, timely, etc.?) and ease of use. Instruments such as those from Doll and Torkzadeh (1988) and Davis (1989) could be distributed to a sample of employees to try to gauge their satisfaction with the parts of the enterprise system that they interact with. The problem with evaluating an enterprise system at this level is that perceptions of success are based on the needs of individuals. Such perceptions take no account of organizational goals, such as cost savings, improved productivity, improved customer service, etc. Yet achievement of these organizational goals is often the key to realizing benefits from investment in an enterprise system.

Because the views of strategic managers are too high-level for causal links to be identified between enterprise system investment and use decisions and benefits realization, and those of operational managers are too low-level to consider all relevant organizational goals, we argue that the most appropriate management level for evaluation of an enterprise system is at the level of the process owners (the upper-middle level of Anthony's pyramid). Process owners have a comprehensive understanding of both the capabilities of enterprise systems and their current plans for use of their enterprise system. It is therefore these process owners who we aim to serve in developing our framework of benefits from enterprise systems. Following Norris (1996), we imagine that the evaluation on behalf of the process owners might best be performed by a task force consisting of a number of business managers from functional areas served by the enterprise system, with the process owner coordinating their activities.

Returning to the other six questions in Table 3.1, Cameron and Whetten's (1983) questions 2 and 3 ask about the domain of activity and level of analysis evaluation. Our answers, shown on the right of Table 3.1, are that the domain of activity is the organization's enterprise system, and the proposed level of analysis is at both the organizational and functional level. Analysis at *both* the organizational and functional level is necessary because the way that, say, the financial module of an ERP system is implemented and used may be quite different from the way the firm's logistics module is used. The nature of the benefits for each function may therefore be quite different, so they need to be evaluated as distinct modules.

Our answer to question 4 is that the purpose of the evaluation is to help process owners plan, manage and improve benefits flowing from enterprise system use. Our answer to question 5 is that the time frame for evaluation is the years after the system goes live. In answer to question 6, we plan to use both objective (including financial) and perceptual data. The problem with restricting analysis to, say, financial measures such as Return on Investment (ROI), is that many of the benefits of enterprise system use are hard to quantify. For this reason we believe that perceptual data must also be included in the evaluation.

Finally, in answer to question 7, 'Against which referent is effectiveness to be judged?', we have three valid answers. If the task force were to compare the performance of the enterprise system, say, a year after implementation, with the business case for the investment, they would be using the 'stated goals of the organization' as the referent. The two other referents that may be useful in some cases are 'some other organization' (which in IT parlance is called benchmarking), and 'some ideal level of performance'. Markus and Tanis (2000) recommend use of the latter referent (some ideal level of performance) when they say:

To accommodate the multi-dimensionality and relativity of enterprise systems success from the adopting organization's perspective, we define a standard of 'optimal success'. Optimal success refers to the best outcomes the organization could possibly achieve with enterprise systems, given its business situation, measured against a portfolio of project, early operational, and longer term business results metrics.

All three possible referents could be valuable for evaluating investments in an enterprise system.

3.3 Developing a Framework for Classifying the Benefits from Enterprise Systems

We used the following four-stage process to develop our benefits framework for enterprise systems:

1 analysis of the features of enterprise systems;
2 a comprehensive review of the literature on information technology (IT) evaluation;
3 analysis of data from 233 enterprise system-vendor success stories published on the web;
4 interviewing managers in 34 enterprise system-using organizations, we have produced a consolidated enterprise system benefits framework with five dimensions.

The result of stage (1) was a clearer understanding of the capabilities of enterprise systems and reasons why organizations were interested in them. Apart from the brief section on the key attributes of enterprise systems above, details from stage (1) are not reported here.

Stage (2): Literature Review on Benefits from Investments in IT

The result of stage (2) is the proposed enterprise system benefits framework shown in Table 3.2. The benefits in Table 3.2 are grouped into five main benefits dimensions, each with a number of sub-dimensions. It was not our expectation that all enterprise systems would produce benefits in each dimension, but we did expect the list in Table 3.2 to provide a good starting point for comparison of benefits for different organizations. Each of these five types of benefit is now discussed in turn.

Operational benefits (dimension 1)

Information technology has a long history of use in cutting costs and raising outputs by automating basic, repetitive operations. There is evidence that

Table 3.2 Proposed enterprise system benefits framework

Dimensions	Sub-dimensions (21 in total at this stage)
Operational	1.1 Cost reduction
	1.2 Cycle time reduction
	1.3 Productivity improvement
	1.4 Quality improvement
	1.5 Customer services improvement
Managerial	2.1 Better resource management
	2.2 Improved decision making and planning
	2.3 Performance improvement
Strategic	3.1 Support business growth
	3.2 Support business alliance
	3.3 Build business innovations
	3.4 Build cost leadership
	3.5 Generate product differentiation (including customization)
	3.6 Build external linkages (customers and suppliers)
IT Infra-structure	4.1 build business flexibility for current and future changes
	4.2 IT costs reduction
	4.3 Increased IT infrastructure capability
Organizational	5.1 Support organizational changes
	5.2 Facilitate business learning
	5.3 Empowerment
	5.4 Built common visions

investment in information technology to streamline processes and automate transactions provides business benefits by speeding up processes, substituting labour, and increasing operation volumes (Weill and Broadbent, 1998; Weill, 1990; Lichtenberg, 1995; Blackburn, 1991; Smith, 1991; Morrison and Berndt, 1990; Brynjolfsson and Hitt, 1996; Brynjolfsson and Hitt, 1993). Since enterprise systems automate business processes and enable process changes, one would expect enterprise systems to offer benefits in terms of cost reduction, cycle time reduction, productivity improvement, quality improvement, and improved customer service. These five types of benefit are listed as points 1.1 through 1.5 of Table 3.2.

Managerial benefits (dimension 2)

Zani (1970), Gorry and Scott-Morton (1971), Ginzberg and Reitman (1982), Keen and Scott Morton (1982), and Rockart and DeLong (1988) have all focused on the managerial benefits to business managers of information systems. Enterprise systems with their centralized database and built-in data analysis capabilities could similarly provide decision and planning benefits to management. As shown in Table 3.2, points 2.1 through 2.3, informational benefits might help an organization achieve better resource management, improved decision making and planning, and performance improvement in different operating divisions of the organization.

Strategic benefits (dimension 3)

Strategic benefits are competitive advantages that arise from use of enterprise systems. Porter and Miller (1985) define three generic strategies where IT could be used to contribute to achieving competitive advantages in business: cost leadership, differentiation and focus. They also identify the strength that IT could add to the business value chain and market competition as Tallon, Kraemer, and Gurbaxani (2000) and Sethi and King (1994) suggested. McFarlan (1984) and Earl (1989) argue that IT has matured to become an integral part of the way enterprises conduct their business. Rackoff, Wiseman, and Ullrich (1985) expanded Porter's model to five strategic thrust areas where the company could make a major offensive or defensive move. These five thrusts are: differentiation, cost, innovation, growth, and alliance.

Integrated information systems present a new opportunity for achieving competitive differentiation by helping firms to customize products or services at a lower cost (Victor and Boynton, 1998; Pine II, 1993; Jaikumar, 1986; Ferdows and Skinner, 1987) and by directly supporting tight links with customers (Clemons and McFarlan, 1986; Vitale, 1986; Malone and Yates, 1987) and with all related business parties (Venkatraman, 1994; Mirani and Lederer, 1998). Enterprise systems, with their large-scale business involvement and internal/external integration capabilities, could assist in achieving the following strategic benefits: business growth, alliance, innovation, cost, differentiation, and external linkages (points 3.1 through 3.6 in Table 3.2).

IT infrastructure benefits (dimension 4)

IT infrastructure consists of sharable and re-usable IT resources that provide a foundation to enable present and future business applications (Keen, 1991; Duncan, 1995; Davenport and Linder, 1994; Earl, 1989; McKay and Brockway, 1989; Niedman, Brancheau, and Wetherbe, 1991; Truijens, 1990). Weill and Broadbent (1998) show that infrastructure building is one of the fundamental management objectives in IT investment, and that a large proportion of most organizations' IT budget is devoted to expenditure on IT infrastructure.

Although not as clearly identifiable as IT infrastructure as, say, investments in telecommunications networks and mainframe computers, enterprise systems are also significant investments in a firm's IT infrastructure. Systems with their integrated and standard application architecture provide an infrastructure that supports (1) business flexibility for future changes, (2) reduced IT costs and marginal cost of business units' IT, and (3) increased capability for quick and economic implementation of new applications. These benefits are summarized in points 4.1 through 4.3 in Table 3.2.

Organizational benefits (dimension 5)

In Peters and Waterman's (1982) observations of 43 successful US corporations, information technology was highly relied upon for its integrated processes and flexible system coordination in either supporting employee 'common vision' communications or facilitating a flattened organizational structure and empowering users. IT tools, accumulated information, and application knowledge are key factors that facilitate organizational learning behaviour (Garvin, 1993; Baets and Venugopal, 1998; Argyris, 1992; Andreu and Ciborra, 1996).

As summarized in points 5.1 through 5.4 of Table 3.2, the integrated information processing capabilities of enterprise systems could affect the growth of the organizational capabilities by: (1) supporting organization structure changes, (2) facilitating employee learning, (3) empowering workers, (4) building common visions.

Stage (3): Using Web Case Analysis to Refine and Extend Table 3.2

The literature reviewed in the previous section relates to IT investments generally; it lacks focus on the benefits that may flow from use of enterprise systems. Since few details of benefits of enterprise systems have been reported in the academic literature, we were forced to turn to the trade press and the web for examples of enterprise system-specific benefits. After reviewing literally hundreds of trade-press articles and vendor-published 'success stories' on the web, we decided to use web cases as the major source of data for enhancing Table 3.2. Vendor web cases were used because:

- They present a complete picture of the enterprise system investment scenario, including details of: business environment, background, objectives, competitive strategy, IS support, the enterprise system investment decisions, system implementation, and the realized benefits.

- They are traceable evidence with the organization's name and project sponsor's name and title, so follow-up verification is possible.
- They are reported from business users' points of view.

Although cases provided by vendors may exaggerate product strength and business benefits, and omit shortcomings of the products, the purpose of this study is *not to evaluate the degree of achievement of these benefits*. Our purpose is simply *to build a broad list of possible business benefits from a large range of enterprise system users*. Therefore, web-published business cases can be used as a starting point for understanding the benefits of enterprise systems. The seven-step process for case selection and review at this stage of our study is described in Table 3.3.

As summarized in Table 3.4, 233 cases were selected in step 2 of the case selection process, from 470 cases reviewed. The size of case organizations ranged from a $50 000 per annum consulting services firm in the USA to a $57bn. per annum consumer products company in Europe. Project sizes ranged from 20 users of a PeopleSoft system in a US financial services firm to 4000 users of an Oracle system in a UK petroleum company. Enterprise systems were implemented between 1995 and 1999. Implementation periods ranged from four months to three years. These cases were reported in their first or second year of enterprise system use. All cases provided details of business background, implementation of the enterprise system, and benefits from the enterprise system.

Stage 4: Confirming the Facts in a Sample of Cases

Thirty-four of the 233 firms were contacted through telephone, fax and e-mail. The stated benefits were confirmed and detailed by business system project managers in 32 of these firms. In one firm, the relevant person to discuss the enterprise system issues was unavailable due to frequent organizational restructuring. The other firm experienced constraints in developing new products.

3.4 Example Web Case and Analysis

To illustrate steps 2, 3, and 4 of the stage 3 process outlined in Table 3.3, we have selected a mid-sized case from the 233 cases. This case, presented in the Exhibit, is from the PeopleSoft web site. It contains an example of the least-mentioned benefit dimension: organizational impact (see benefit labeled 4 below).

Table 3.3 Case selection and review process

Step	Process
1	Visited enterprise system vendor websites for SAP, Oracle, PeopleSoft, Baan, Oracle, and JD Edwards for customer case studies (or customer success stories). Printed out case lists for: SAP, PeopleSoft, and Oracle. Lists from Baan and JD Edwards were not selected because they lacked complete information.
2	Reviewed these cases and selected qualified stories. Built three files with sets of qualified enterprise system product cases: SAP cases, PeopleSoft cases, and Oracle cases. The criteria for case selection were: They have applied enterprise systems to manage major enterprise resources. Cases with a single enterprise system module used by an organization to manage certain core processes and not linked with other core resource management processes were not selected. They have sufficient information about the case, with organization background, implementation descriptions, and benefit descriptions. They were focused on business benefits, not the product benefits. They have quantitative measures or precise business benefit descriptions.
3	Verified reliability of cases by contacting project managers in a convenient sample of 18 cases in Australia and randomly selected sample of 16 cases in the USA, Singapore, and Taiwan.
4	Built analysis table of cases with information about country, industry, user size (if available), modules installed, implementation stages and benefits achieved by the five categories. Three tables were constructed: SAP case analysis, PeopleSoft case analysis, Oracle case analysis.
5	Analysed benefit differences between industry, vendors, and firm sizes.
6	Assembled case benefits according to the dimensions in the framework in Table 3.2. Benefits were selected and highlighted in the file only if a similar benefit was achieved in more than two firms using products from different vendors. Three files organized by the enterprise system benefits framework were built: SAP benefits, PeopleSoft benefits, and Oracle benefits.
7	Consolidated benefit details from three benefits files into each benefit dimension. Modified the enterprise system benefits framework (Table 3.2) with some sub-categories removed or added. Prepared a list of benefits with (1) analysed results, (2) benefit descriptions, and (3) typical case examples. See the appendix for a three-page summary of enterprise system benefits at the end of this process.

Table 3.4 Summary of cases selected

Enterprise systems vendor	Cases published	Cases selected	Case scope
SAP	256	84	19 industries, 45 countries.
PeopleSoft	124	65	11 industries, 10 countries
Oracle	90	84	Approximately 12 industries, 6 countries
Total	**470**	**233**	

Start of Exhibit

Exhibit: PeopleSoft Case 34

Health First in Great Shape with PeopleSoft Financials and Human Resources[1]

'When we formed this integrated delivery system, the complexity of our organization increased tenfold overnight', says Rich Rogers, vice president and chief information officer of Health First, Inc. Located on Florida's 'Space Coast', Health First was formed in 1995, when Holmes Regional Medical Center and Palm Bay Community Hospital merged with Cape Canaveral Hospital. Today, the organization includes three hospitals, 29 primary care physician clinics, a commercial HMO, and a Medicare HMO.

From an information services standpoint, bringing together three large healthcare facilities was even more challenging because each had its own set of financial, payroll, and human resource systems. 'We had a number of different vendor systems and we were trying to piece them together to act as one organization', recalls Rogers. 'In this industry, we make very fast decisions. Our old systems just didn't have the flexibility to react to the market that quickly.'

Looking to consolidate its human resource and financial systems and gain functionality, Health First selected PeopleSoft in mid 1996. According to Rogers, PeopleSoft was chosen primarily because of its healthcare expertise. 'They understand the unique needs of our industry, and they build that knowledge into their products.'

Improved Resource Management

During the twelve-month implementation, Health First engaged CSC Pinnacle for project management and consulting support. After PeopleSoft was operational in:

1 October 1997, the most immediate productivity gain that the organization realized came in payroll processing, a task that had previously taken four days (processing 12 separate payrolls). With PeopleSoft Payroll that time was reduced to four hours.

PeopleSoft has also enabled Health First to track its 5000 employees across all its organizations. 'Salary is our biggest expense', states Rogers. 'Before PeopleSoft, we had no way of managing or tracking our people as they moved around our 70 locations. For example, we had nurses working a couple of days a week in our home care department as well as in one of our hospitals. They would be on two different payrolls.'

Faster Decision Making

With PeopleSoft, reports accountant Cindy Ward, the finance department can deliver more accurate and timely information in a fraction of the time that it took previously. 'We can now serve our customers ten times better. The quantity and quality of the information is much improved, and the detail level that we can go down to is phenomenal.'

2 For Health First's directors, that means having the financial information necessary to make critical decisions. 'It's been a big benefit on the productivity side', says Rogers. 'Our drill down capability has improved tremendously. We can zone in on a problem a lot more quickly than we did before. It's helped us evaluate the profitability of the business units.'

In addition, the department can now deliver accurate monthly cost allocations to each organization within Health First. Says Desmond Almarales, project manager, financial systems, 'We used to do it with spreadsheets, which was very cumbersome. With PeopleSoft, we can allocate our costs much faster. And we can change it on a dime.'

Creating a Corporate Culture

3 Beyond the improvements in productivity and information access, having PeopleSoft financial and human resources products throughout the enterprise has yielded another interesting advantage. Explains Rogers, 'Once you start using the same tools and the same reports, it goes a long way to establishing a corporate culture. It brings different parts of the organization together. They start to act as one, and work as a common unit.'

A Healthy Future

4 Now that they have PeopleSoft up and running throughout the organization, Health First will reengineer some of their processes, incorporating workflow to gain additional productivity. 'That's where we're going to see the biggest benefit from our investment in PeopleSoft',

says Rogers. 'It's going to have a direct effect on our bottom line, and will ultimately help us deliver better care because we'll have information at our fingertips a lot faster than we've ever had before.'

Health First also plans to upgrade to PeopleSoft 7.5 to deliver self-service functionality to its employees through the Internet. 'We're strong believers in the cost effectiveness of delivering information and functionality to occasional users through the Web,' says Rogers. 'It will alleviate having to continually upgrade all of our PCs, which is a big expense.'

Products: PeopleSoft General Ledger, Payables, Asset Management, Human Resources

[1] Source: http://search.peoplesoft.com/ and search for 'Health First in Great Shape'.

End of Exhibit

Step 2: Select cases. This case, from Health First in Florida, USA, was selected because the enterprise system was used to manage major enterprise resources – inventory and human – for this 5000 employee group of medical service providers. This case was then copied to the PeopleSoft case file as case number P34.

Step 3: Verify reliability. Vice president and CIO Mr. Rogers was contacted through phone and fax. An informative 30 minute telephone interview was conducted with the project manager who was previously the director of finance, but who chose to transfer to the project team after the system went live two years ago. The benefits reported in the case were further confirmed by two business users in finance and human resources through e-mail. Each of them sent two to three pages replying to questions regarding the enterprise system benefits. As users in this organization gained experience with the system, the flexibility of the system to support business growth in a changeable environment is perceived as the key achievement of the enterprise system in this 5000 employee hospital.

Step 4: Classify benefits. Four benefit dimensions are mentioned in this case:

1 **Operational benefit in payroll processing cycle time reduction:** Payroll processing time was reduced from four days to four hours (see block 1 in the Exhibit).
2 **Managerial benefit in resources management and decision making:** The human resources management capability of the PeopleSoft package was used to track the 5000 employee movements across 70 locations and produce

accurate salaries. Accurate, time-effective information delivered to managers improved the speed and quality of decision making and assisted with cost control (see block 2 in the Exhibit).

3 **IT infrastructure benefit in IT costs reduction and increased capability:** Perceived IT infrastructure benefits came from the confidence of being able to add new applications, conduct business changes, enable web services and save IT cost in PCs. These increased IT infrastructure capabilities could be described by the highlighted paragraphs (see block 3 in the Exhibit).

4 **Organizational benefit in building a consistent vision across organizations:** Organizational consistency across the 70 units and three newly merged organizations was enhanced through the utilization of the integrated system. The quote by the vice president in block 4 of the Exhibit suggests this benefit was obtained as a result of use of the PeopleSoft system.

After Step 4, the four blocks highlighted in the Exhibit were copied to benefit-dimension files for later consolidation in steps 5, 6, and 7 of the process described in Table 3.3.

3.5 Discussion

Business benefits from the other selected 232 cases were analysed in the same fashion as above. The resultant benefits framework at the end of the process (Step 7 of Table 3.3) is shown in the Appendix. Table 3.5 summarizes the types of benefits. The structure of the benefits listed in the Appendix is similar to that in Table 3.2, but differs in some important aspects. Eight comments on the overall analysis are presented below.

Comment 1: Validity of the framework verified

Examples of each benefit dimension were found in cases from each enterprise system vendor. Every business achieved benefits in at least two dimensions.

Table 3.5 Summary of results showing % of cases selected

	SAP	PeopleSoft	Oracle	Total
Total cases published	256	124	90	470
Cases *selected* in step 2	84	65	84	233
Operational benefits	75%	57%	83%	73%
Managerial benefits	57%	65%	45%	55%
Strategic benefits	62%	71%	38%	56%
IT infrastructure benefits	89%	80%	80%	83%
Organizational benefits	13%	23%	7%	14%

Operational and infrastructure benefits were the most quoted benefits: 170 cases (73% of the 233 cases) claim to have achieved operational benefits and 194 cases (83%) claim IT infrastructure benefits.

Comment 2: Enhancements to the framework

As a result of the web case analysis, the enterprise system benefits framework from Table 3.2 was expanded and enhanced with detailed descriptions. There was no need to change the five major benefit classes from Table 3.2, but the 21 sub-benefit dimensions from Table 3.2 had to be expanded to the 25 sub-dimensions in the Appendix. Under the heading 'Strategic benefits', two new sub-dimensions '3.7 Worldwide expansion' and '3.8 Enabling e-commerce' were added. Thirty four per cent of enterprise system users, across the three vendors, indicated 'enabling e-commerce in their business' as a major strategic benefit. Under the heading 'Organizational benefits', two more new sub-dimensions were added: '5.5 Changed employee behavior' and '5.6 Better employee morale and satisfaction'. Firms mentioned the benefit of using enterprise systems for shifting employee focus to core business functions in planning, managing and serving customers. Efficient support from the enterprise system created better morale in the work place. In addition to the new sub-dimensions, 78 dot-point examples were added to 'flesh out' (provide more details of) each benefit category. These examples are shown in the Appendix.

Comment 3: More benefits likely after additional experience with the system

Of the 34 firms contacted, 32 firms mentioned more benefits in addition to those mentioned in their case studies. Some of these benefits had become more apparent since the cases were written. Additional benefits were found especially in increased infrastructure capabilities for being able to extend their systems to new applications or support new strategies. One utility company in Australia was planning to establish a new business to provide enterprise system-enabled shared services to external customers.

Comment 4: Contingency factors

Although it was not the objective of this study to analyse the influences of contingency factors, some preliminary comments can be made:

- **Industry.** There were no apparent differences in types of benefits across industries.
- **Vendor**. Although products from the three enterprise system vendors provide similar functions, there were some differences (that might be due to the

style of case writing, but might also be more fundamental). First, SAP cases had an above average number of cases citing benefits from all five dimensions. Flexibility in supporting business changes was the most noted benefit for SAP users (89% of SAP users). Second, the PeopleSoft cases mentioned more strategic and organizational benefits than average (71% compared with 56%, and 23% compared with 14%, respectively). Third, Oracle cases mention more operational benefits than average (83% compared with 73%).

- **Firm Size.** Benefits gained by large and small organizations seem to be similar. All have gained benefits in the five dimensions, except that smaller organizations seem to have more quantified evidence of benefits than larger companies. This is probably because smaller companies gain tangible benefits more quickly than larger organizations. However, the degree of benefits realized could not be compared due to lack of standard measurements across cases.

Comment 5: Different organizations gain different benefits from the same applications

Consistent with Ragowsky and Stern's (1996) study of the value of packaged applications, different organizations using the *same* application packages achieved different benefits in different dimensions.

Comment 6: Criteria for selection of enterprise systems systems

Enterprise system product selection was based on the following factors (listed in order of frequency of citation): (1) Business fit, (2) Ease of implementation, (3) Vendor services and support, (4) Special industry or application capabilities, (5) Product affordability, (6) Compatibility with other systems.

Comment 7: Long expected system life for enterprise systems

Most organizations seem to expect a long-term return on the investment in their system. In the cases studied, the expected life of the enterprise system ranges from ten to 20 years. Other studies have reported much shorter life expectancies, for example, six years (Gartner Group, 1998). Expected longevity of enterprise system is probably because enterprise systems are implemented as a base for extension and expansion. In addition, regular vendor-supported system upgrades will keep the technology up to date.

Comment 8: Comparison to Davenport (2000)

Davenport (2000) was published after this study was completed. A two-page comparison table available from the authors was used to compare the benefits reported in his book (pages 7–235) to the framework in the Appendix. All of Davenport's benefits can be matched to benefits in our framework. No new benefit categories needed to be added.

3.6 An Application of the Enterprise System Benefit Framework

The framework presented in this paper (presented in detail in the Appendix) was developed to help process owners assess the value of their enterprise system. In this section, we show how the framework was used in case studies of four organizations. The four organizations are all recently privatized utility companies in two states in Australia. Prior to 1994, all were parts of state-government-owned electricity- and gas-producing organizations. After privatization, they found they needed more sophisticated information systems, and all elected to implement ERP systems. By the year 2000, all four organizations had been using their systems (three SAP systems and one PeopleSoft system) for at least three years.

Up to ten business managers were interviewed and tape recorded in each organization. In each interview, the framework in the Appendix was used to guide managers in their assessment of past, current, and future benefits from their enterprise system. The framework served as a checklist for retrospective and prospective thinking, with the detailed example benefits providing guidance about the meaning of each benefit class. Although some benefit items (for example, strategic benefits of building external linkages or e-commerce) were not relevant in some companies, they inspired business managers to ask 'If someone else has achieved this, why can't we?.' All interviewees agreed that these capabilities were available and achievable if they planned to use them. Executives in two companies requested copies of the framework for their own use. As one business executive commented: 'This framework gives us a chance to really think about the impacts of our ERP system.' After the interviews, differences between stakeholders' recollections were reconciled item-by-item to ensure the overall picture was consistent.

Based on the interviews, a summary table identifying business changes, benefits, and problems over the three years of enterprise system use was constructed for each case study organization. For example, Table 3.6 shows the summary table for one of the four case study organizations, Utility Company 2. The five benefit dimensions from our framework are shown in the five right-hand columns of Table 3.6.

Table 3.6 A summary of enterprise system benefits realized by utility company two (Utility 2)

Date	Business environment	ERP use	Operational benefits	Managerial benefits	Strategic benefits	IT infrastructure benefits	Organizational benefits
1994	Electricity Corporation Act passed						
1995	Utility2 established						
1996, July	Commenced full commercial trading						
1996, Dec.	Utility2 restructured into three main units: network, retail, and field service	• Board approved SAP investment					
1997, Dec.		SAP system went live					
1998, Jan.		• SAP user survey • Modified material management • Work planning reverted to old processes	• Payroll problems raised union issues • Deficient material delivery	No immediate managerial benefits	No strategic benefits noticed	Reduced cost in IT operation	• Low employee morale due to process errors. • Employees suffer from change pressure and lack of understanding of the system.
1998, June	Won 2nd 'Best Australian Customer Service' (from 39 utility companies)	• Role of Process Owner re-established • Increased on-site supports	Improved customer service				
1998, Oct.	Company disaggregated into three separate business entities	• Centralized user support activities • Additional training	• Reduced 20% inventory • Increased 10% productivity			Enabled business restructuring	

(cont.)

Table 3.6 (*cont.*)

Date	Business environment	ERP use	Operational benefits	Managerial benefits	Strategic benefits	IT infrastructure benefits	Organizational benefits
1999, June	Upgraded electricity infrastructure. Service expansion to rural areas	Conducted SAP Post-implementation review	85% of planned benefits achieved	• Improved resource management • Better decision making	• Established decentralized structure	Further reduction in IT expenses by replacing 5 more legacy systems	• Supporting organizational changes • Enhanced users' confidence in the system
1999, Dec.	Year 2000 changeover	• Enhanced user support • Continuous process improvement. • More relevant user training	100% of planned benefits achieved		Built a low cost structure	Support for business expansion	
2000, Jan.		• System Alignment Review • SAP User Survey	• Continuous productivity gains	• Enhanced performance control.	• Extended services • Standardization • Formalization of newly structured business practices	• Increased support capability • Smooth changeover for new GST	• Evolution of a different 'way of doing things'
2000, Mar.		• Resolved outstanding issues	• Expect to achieve more than planned in business case by Y2003–AUSIIm. per annum	• Quick response to user problems			
2000, July	New Australian Goods and Services Tax (GST) implemented	• Reinforced alignment between SAP processes and business objectives					

Table 3.7 Patterns of benefit development (based on findings from four case organizations)

Dimensions of ERP benefits	Operational benefits	Managerial benefits	Strategic benefits	IT infrastructure benefits	Organizational benefits
Path of ERP benefit development	years	years	years	years	years
Early benefits	Automation benefits from savings in labour and time.	Quicker decision making for real time information.	No immediate benefits.	Replacement of legacy systems.	Immediate downfalls in employee morale.
Problems	Extra time and labour in data entry to accommodate the business integration.	Rigidity in resource allocation because of tightly linked system integration.	Loss of competitive advantages when competitors are applying similar processes.	Inflexible system changes. Frequent system upgrades.	Low employee morale for extra work, mismatch processes, data errors and change pressures.
Explanations for benefits and problems	Business process change. ERP system modifications. Organizational learning.	Enhanced reporting functions. Accumulated data. Organizational learning.	Business strategic use of ERP systems. ERP technology upgrading.	Attain, expand, and extend ERP systems.	Business and system changes. Organizational learning.
Pace of benefit development	1–2 year plateau for business changes and organizational learning.	1–2 years plateau for system enhancement and organizational learning.	Depends on business strategies of ERP system use.	Gradually increased with system expansion. Significantly increased when system use reached economies of scale.	2–3 years for users to forget the initial process problems and to build system knowledge.

Across the four cases, some common patterns of benefit realization were noted. These are summarized in Table 3.7. The first row of Table 3.7 shows the overall pattern of benefit realization, with time on the horizontal axis and benefits on the vertical axis. Rows 2–5 describe the early benefits achieved, problems encountered, explanations of the early and further benefits, and comments about the pace of benefit realization in the four organizations. Use of the framework described in this chapter made it possible to track benefit development by asking managers to assess benefit development over time in each of the five dimensions.

3.7 Study Limitations

The major limitation of this study is that the primary result of the study, the framework reported in the Appendix, is derived from vendor web site data as classified by a single researcher (the first author). Second-hand data provided by enterprise system vendors in their web sites may not be reliable or may have been misinterpreted. However, since the main objective of this study was to understand comprehensively the possible benefits of enterprise systems, and all the benefits summarized in the Appendix were experienced by so many organizations (see Table 3.5), the schedule of benefits seems reasonable. In addition, the fact that 32 of the 34 organizations contacted directly confirmed the facts presented in the vendor success stories inclines us to think that the information in the cases is reliable enough to be useful. As an additional precaution against incorporation of unlikely benefits, this study selected the benefit items only if they were found in more than three cases from at least two different product vendors.

3.8 Conclusion

The objective of this study was to prepare a comprehensive list of business benefits of enterprise systems suitable for use by process owners seeking to evaluate the benefits of their enterprise systems. Instead of evaluating enterprise system success using some single subjective measure like, 'Overall, are you satisfied with your ERP system?', we have proposed a new, hopefully more objective, method for assessing the benefits of enterprise systems.

Our enterprise system benefits framework is presented in the Appendix. It has five main benefit dimensions, with a total of 25 sub-dimensions. It is not expected that all organizations will achieve business benefits in all five dimensions or 25 sub-dimensions. Rather, our framework provides a blueprint of benefits that have been achieved in other organizations that are using enterprise systems. In addition to the longitudinal case studies reported above, practical uses of this framework include:

- As a tool for enterprise system planning and management. In this role, the framework could be customized around an organization's goals and used to stimulate more effective communication within a business management team about their goals for the system.

- As a tool for benchmarking enterprise systems across different organizations.
- As a source of five distinctive dimensions in a balanced scorecard approach to evaluating the effectiveness of an enterprise system investment.
- As a technique for measuring the dependent variable in studies that seek to assess the impact of factors that influence enterprise system benefits, for example the influence of organizational characteristics on different benefit dimensions, or the impact of different implementation strategies on benefit achievement.
- As the basis for a mail survey of enterprise system user organizations, for example to explore possible linkages between costs, benefits (as measured by our framework), and bottom-line profitability.

ACKNOWLEDGEMENT

Earlier versions of this paper appeared in the Americas Conference on Information Systems (AMCIS), Long Beach, California, in August 2000, and as S. Shang and P. B. Seddon, Assessing and Managing the Benefits of Enterprise Systems: The Business Managers Perspective, *Information Systems Journal* (2002).

REFERENCES

Andreu, R. and Ciborra, C. (1996) Organizational Learning and Core Capabilities Development: The Role of IT. *Journal of Strategic Information System*, **5**, 111–127.

Anthony, R. N. (1965) Planning and Control Systems: A Framework for Analysis. Graduate School of Business Administration, Harvard University.

Argyris, C. (1992) *On Organizational Learning*. Oxford: Blackwell.

Baets, W. and Venugopal, V. (1998) An IT Architecture to Support Organizational Transformation. In *Information Technology and Organizational Transformation*, Galliers, R. D. and Baets, W. R. J. (eds), John Wiley & Sons, pp. 195–222.

Blackburn, J. D. (1991) The Quick Response Movement in the Apparel Industry: A Case Study in Time-Compressing Supply Chains. In *Time-Based Competition: The Next Battleground in American Manufacturing*, Blackburn, J. D. (ed.), Homewood, IL: Business One Irwin, pp. 167–172.

Brynjolfsson, E. and Hitt, L. (1993) Is Information Systems Spending Productive? New Evidence and New Results. In 14th International Conference on Information Systems, Orlando, Florida.

Brynjolfsson, E. and Hitt, L. (1996) Productivity, Business Profitability and Consumer Surplus: Three different Measures of Information Technology Value. *MIS Quarterly*, **20**, 121–142.

Cameron, K. S. and Whetten, D. A. (1983) *Organizational Effectiveness: A Comparison of Multiple Models.* New York: Academic Press.

Clemons, E. K. and McFarlan, W. (1986) Telecom: Hook Up or Lose Out. *Harvard Business Review*, July–August, 91–97.

Davenport, T. and Linder, J. (1994) Information Management Infrastructure: The New Competitive weapon? In 27th Annual Hawaii International Conference on Systems Science, Hawai.

Davenport, T. H. (2000) *Mission Critical – Realizing the Promise of Enterprise Systems.* Boston, MA: Harvard Business School Press.

Davis, F. F. (1989) Perceived Usefulness, Perceived Ease of Use, and User Acceptance of Information Technology. *MIS Quarterly*, **13**, 319–340.

Doll, W. J. and Torkzadeh, G. (1988) The Measurement of End User Computing Satisfaction. *MIS Quarterly*, June, 259–274.

Duncan, N. B. (1995) Capturing Flexibility for Information Technology Infrastructure: A Study of Resource Characteristics and Their Measure. *Journal of Management Information Systems*, **12**, 37–57.

Earl, M. J. (1989) *The Management Strategies for Information Technology.* London: Prentice-Hall.

Ferdows, K. and Skinner, W. (1987) The Sweeping Revolution in Manufacturing. *Journal of Business Strategy*, **8**, 64–69.

Gartner Group (1998) 1998 ERP and FMIS Study – Executive Summary. Gartner Group.

Garvin, D. A. (1993) Building a Learning Organization. *Harvard Business Review*, July–August, 78–91.

Gilbert, A. (2000) ERP Vendors Look for Rebound after Slowdown: Fourth-quarter Revenue Gains Indicate Possible Resurgence in 2000. *Information Week*, 14 February. http://www.informationweek.com/773/vaerp.htm.

Ginzberg, M. J. and Reitman, W. R. (1982) *Decisions Support Systems.* New York: North-Holland.

Gorry, G. A. and Scott-Morton, M. S. (1971) A Framework for Management Information Systems. *Sloan Management Review*, 13.

Hammer, M. and Champy, J. (1993) *Reengineering the Corporation: A Manifesto for Business Revolution.* New York: Harper-Business.

Jaikumar, R. (1986) Post-industrial Manufacturing. *Harvard Business Review*, pp. 69–76.

Keen, P. G. W. (1991) *Shaping the Future: Business Design through Information Technology.* Cambridge, MA: Harvard Business School Press.

Keen, P. G. W. and Scott Morton, M. S. (1982) *Decision Support Systems: An Organizational Perspective.* Reading, MA: Addison-Wesley.

Lichtenberg, F. (1995) The Output contributions of Computer Equipment and Personnel: A Firm Level Analysis. *Economics of Innovation and New Technology*, **3**, 4.

Malone, T. B. and Yates, J. (1987) Electronic Markets and Electronic Hierarchies: Effects of Information Technology on Market Structure and Corporate Strategies. *Communications of the ACM*, 30.

Markus, L. M. and Tanis, C. (2000) The Enterprise Systems Experience – from Adoption to Success. In *In Framing the Domains of IT Research: Glimpsing the Future through the Past*, Zmud, R. W. (eds), Cincinnati, OH: Pinnaflex Educational Resources.

McFarlan, F. W. (1984) Information Technology Changes the Way You Compete. *Harvard Business Review*, May–June.

McKay, D. T. and Brockway, D. W. (1989) Building I/T Infrastructure for the 1990s. *Stage by Stage*, **9**, 1–11.

Mirani, R. and Lederer, A. L. (1998) An Instrument for Assessing the Organizational Benefits of IS Project. *Decision Sciences*, **29**, 803–838.

Morrison, C. J. and Berndt, E. R. (1990) Assessing the Productivity of Information Technology Equipment in the US Manufacturing Industries. *National Bureau of Economic Research*, Cambridge, MA.

Niedman, F., Brancheau, J. C. and Wetherbe, J. C. (1991) Information Systems Management Issues for the 1990s. *MIS Quarterly*, December, 86–96.

Peters, T. and Waterman, R. (1982) *In Search of Excellence*. New York: Harper & Row.

Pine II, J. B. (1993) *Mass Customization: The New Frontier in Business Competition*. Boston, MA: Harvard Business School Press.

Porter, M. E. and Millar, V. E. (1985) How Information Gives You Competitive Advantage. *Harvard Business Review*, **63**, 149–160.

Rackoff, N., Wiseman, C. and Ullrich, W. A. (1985) Information Systems for Competitive Advantage: Implementation of a Planning Process. *MIS Quarterly*, **9**.

Ragowsky, A. and Stern, M. (1996) Identifying the Value and the Importance of an Information System Application. *Information and Management*, **31**.

Rockart, J. F. and DeLong, D. W. (1988) *Executive Support Systems: The Emergence of Top Management Computer Use*. Homewood, IL: Dow-Jones Irwin.

Seddon, P., Staples, S., Patnayakuni, R. and Bowtell, M. (1999) Dimensions of Information Systems Success. *Communications of AIS*, **2**.

Sethi, V. and King, W. R. (1994) Development of Measures to Assess the Extent to which an Information Technology Application Provides Competitive Advantage. *Management Science*, **40**, 1601–1627.

Sircar, S. L. Turnbow, J. and Bordoloi, B. (2000) A Framework for Assessing the Relationship Between Information Technology Investments and Firm Performance. *Journal of Management Information Systems*, **16**, 69–97.

Smith, F. W. (1991) The Distribution Revolution: Time Flies at Federal Express. In *Time-Based Competition: The Next Battleground in American Manufacturing*, Blackburn, J. D. (eds), Homewood, IL: Business One Irwin, pp. 237–238.

Tallon, P. P., Kraemer, K. L. and Gurbaxani, V. (2000) Executives' Perceptions of the Business Value of Information Technology: A Process-Oriented Approach. *Journal of Management Information Systems*, **16**, 145–173.

Truijens, J. (1990) *Infomation-infrastructure, een Instrument voor het Management*. Deventer, NL: Kluwer.

Venkatraman, N. (1994) IT-Enabled Business Transformation: From Automation to Business Scope Redefinition. *Sloan Management Review*, **35**, 73–87.

Victor, B. and Boynton, A. C. (1998) *Invented Here*. Boston, MA: Harvard Business School Press.

Vitale, M. R. (1986) American Hospital Supply Corp.: The ASAP System. *Harvard Business School Case Study*, March, 1–17.

Weill, P. (1990) *Do Computers Pay Off?*, International Center for Information Technologies, Washington, DC.

Weill, P. and Broadbent, M. (1998) *Leveraging the New Infrastructure: How Market Leaders Capitalize on Information Technology*. Boston, MA: Harvard Business School Press.

Zani, W. (1970) Blueprint in MIS. *Harvard Business Review*, November–December.

Appendix: Our framework of business benefits from enterprise systems

I Operational benefits

1.1 **Cost reduction**
- **Labour cost reduction.** The automation and removal of redundant processes or redesign of processes led to full-time staff reductions in tasks in each business area, including customer services, production, order fulfillment, administrative processes, purchasing, financial, training and human resources.
- Inventory cost reduction in management, relocation, warehousing, and improved turns.
- Administrative expenses reduction in printing paper and supplies.

1.2 **Cycle time reduction.** Measurable cycle time reductions were found in three kinds of activities that support customers, employees, and suppliers.
- **Customer support activities** in order fulfillment, billing, production, delivery, and customer services.
- **Employee support activities** in reporting, month-end closing, purchasing, or expense requisition, HR, and payroll.
- **Supplier support activities** in speed payments and combined multiple orders with discounts gained.

1.3 **Productivity improvement.** Products produced per employee or labour cost, customers served per employee or labour cost, or mission accomplished per employee in non-profit organizations.

1.4 **Quality improvement.** Error rate reduction, duplicates reduction, accuracy rate or reliability rate improvement.

1.5 **Customer services improvement.** Ease of customer data access and customer inquiries resolution.

II Managerial benefits

2.1 **Better resource management**
- Better asset management for improving costs, and of depreciation, relocation, custodian, physical inventory, and maintenance records control.
- Better inventory management for improving inventory turns, stock allocation, quick and accurate inventory information, just-in-time replacement, and having a variety of options dealing with various requests.
- Better production management for optimizing supply chain and production schedules.
- Better workforce management for improved manpower allocation, and better utilization of skills and experience.

2.2 **Better decision making**
- Improved strategic decisions for improved market responsiveness, *better profits*, cost control, and effective strategic planning.

- Improved operational decisions for flexible resource management, efficient processes, and quick response to work changes.
- Improved customer decisions with flexible customer services, rapid response to customer demands and quick service adjustments.

2.3 **Better performance control** in a variety of way in all levels of the organizations.
- Financial performance by lines of business, by product, by customers, by geographies, or by different combinations.
- Manufacturing performance monitoring, prediction, and quick adjustments
- Overall operation efficiency and effectiveness management

III Strategic benefits

3.1 **Support current and future business growth plan in:**
- Business growth in transaction volume, processing capacity and capability
- Business growth with new business products or services, new divisions, or new functions in different regions
- Business growth with increased employees, new policies, and procedures
- Business growth in new markets
- Business growth with industry's rapid changes in competition, regulation, and markets

3.2 **Support business alliances** by efficiently and effectively consolidating newly acquired companies into standard business practice

3.3 **Build business innovation** by:
- Enabling new market strategy
- Building new process chain
- Creating new business

3.4 Build cost leadership by reaching businesses' economies of scale.

3.5 Generate or enhance product differentiation by
- Providing customized product or services, for instance: early preparation for the new EMU currency policy, customized billing, individualized project services to different customer requirements, different levels of service appropriate for the varying sizes of customer companies.
- Providing lean production with make-to-order capabilities.

3.6 **Build external linkage** with suppliers, distributors, and related business parties.

3.7 **Enable worldwide expansion** with
- Centralized world operation
- Global resource management
- Multi-currency capability
- Global market penetration
- Solution quickly and cost effectively deplayed worldwide

3.8 **Enable e-business** by attracting new or by getting closer to customers through the web integration capability. The web-enabled ERP system provide benefits in business to business and business to individual in:
- Interactive customer service
- Improved product design through customer direct feedback
- Expanding to new markets

- Building virtual corporation with virtual supply and demand consortium
- Delivering customized service
- Providing real time and reliable data enquiries

IV IT infrastructure benefits

4.1 **Increased business flexibility** by responding to internal and external changes quickly at lower costs and providing a range of options in reaction to the change requirements.

4.2 **IT costs reduction** in:
- Legacy system integration and maintenance
- Mainframe or hardware replacing
- IT expense and staff for developing and maintaining the system
- Year 2000 compliance upgrade
- System architecture design and development
- System modification and maintenance
- Disparate information reconciliation and consolidation
- Technology R&D

4.3 **Increased IT infrastructure capability:** stable and flexible for the current and future business changes in process and structure.

Stability:
- Streamlined and standardized platform
- Global platform with global knowledge pipeline
- Database performance and integrity
- IS management transformation and increased IS resource capability
- Continuous improvement in system process and technology
- Global maintenance support

Flexibility:
- Modern technology adaptability
- Extendable to external parties
- Expandable to a range of applications
- Comparable with different systems
- Customizable and configurable

V Organizational benefits

5.1 **Support business organizational changes in structure, and processes**

5.2 **Facilitate business learning and broaden employees' skills**
- Learning by entire workforce
- Shorten learning time
- Broaden employees' skill

5.3 **Empowerment**
- Accountability, more value-added responsibility
- More pro-active users in problem solving
- Work autonomously

- Users have ownership of the system
- Middle management are no longer doers but planners
- Greater employee involvement in business management

5.4 Changed culture with common visions

- Efficient interpersonal communication
- Interdisciplinary thinking, differences coordinated and harmonized, and *common inter-departmental processes*
- Consistent vision across different levels of organization

5.5 Changed employee behaviour with shifted focus

- More critical managing and planning matters
- More concentration on core work
- Customer and market focus
- Move from back office to the front office

5.6 Better employee morale and satisfaction:

- Increased employee satisfaction with better decision-making tools
- Increased employee efficiency of field operations and services
- Satisfied users for solving problems efficiently
- Increased morale with better system performance
- Satisfied employees for better employee service

4 The Continuing ERP Revolution: Sustainable Lessons, New Modes of Delivery

Jeanne W. Ross, Michael R. Vitale, and Leslie P. Willcocks

4.1 Introduction

Recognizing the need to present a single face to global customers, to respond rapidly to customer demands, and to seek out economies of scale, business executives have been for some time regularly examining the capabilities offered by information technology (IT). As a result, from the mid 1990s on, many firms have been replacing the legacy systems that form the base of their information processing capabilities with enterprise resource planning systems (ERPs). ERPs replace a firm's disparate transaction processing systems with a single, integrated system that embodies the newly understood tight interdependencies among a firm's functional units. ERPs have offered much promise for revitalizing IT infrastructures and enabling global business process integration (O'Leary, 2000; Sauer and Willcocks, 2001), but as the research in this book demonstrates, implementation of these highly touted systems has regularly proven to be very expensive, and the rewards for implementation often appear to be elusive.

A Mckinsey report looking at the 1995–2000 period concluded that spending on IT in the USA had doubled, as had the labour productivity annual growth rate over the same period. However, despite claims to the contrary, little of that productivity growth could be attributed to IT, and in fact the majority of the US economy saw no or little productivity growth while accounting for 62% of the IT spending in that period. And what has been true generally for IT has been continuously and distinctively true of ERP: the technology has been no guaranteed passport to productivity (Goodwin, 2001; Norris et al., 2000; Willcocks and Graeser, 2001).

Will the advent of new modes of delivery – over the web and through third-party IT service providers – change this picture (Bennett and Timbrell, 2000; Hagel and Brown, 2001), or is this merely looking yet again for technology-based solutions, when business, organizational, cultural and managerial factors might prove more paramount (Weill and Vitale, 2001)? This chapter first analyses the experiences of 15 manufacturing companies that had completed significant ERP implementation and were subsequently attempting to generate a positive return on their ERP investments. This evidence will be used to establish the ground rules for successfully engaging with ERP systems. In fact this part of the chapter produces three major sets of findings – on motivations for ERP, ERP stages, and obstacles to look out for – from which are derived a number of sustainable lessons on how to implement ERP systems to achieve business leverage. These build on and extend the empirical findings and lessons found in other chapters in this book, particularly those of Chapters 1, 7, 8, and 10. The chapter then pinpoints developments towards ERPII systems and the extended enterprise, and the sourcing of ERP over the web, including through external applications and service providers. We discuss the old and new challenges these represent for implementing and using ERP effectively, and argue that the lessons learned from the earlier rounds of ERP implementations are still highly relevant, because ERP represents a 'backbone' set of applications, is critical to intra- and inter-organizational functioning, requires cultural and process change, frequently requires external vendor assistance, needs senior management and the business to be highly involved, and is highly dependent on continuous improvement if its business value is to be optimized.

4.2 First Wave ERP: The Research Study

The first set of research findings described here examined the impacts of packaged ERP systems on organizations. Specifically, this research was intended to identify how firms were leveraging their ERP environments to generate business value. Thus, we identified firms that had gone live with one of the leading ERP packages (SAP, Baan, PeopleSoft, or Oracle) firm-wide or within a major division. All implementations included manufacturing modules and some combination of financial, sales and marketing, and other modules.

Data were collected in August and September, 1998, and involved 40 hour-long telephone interviews at 15 different companies. Interviews at each firm sought three different perspectives on the ERP implementation: (a) an executive who sponsored the implementation, (b) a manager who headed up the implementation, and (c) a business executive whose function or division was impacted by the implementation. The sample intentionally sought diversity in the ERP packages. Eight of the firms had deployed SAP, three had implemented Baan, another three used Oracle, and one was a PeopleSoft client. Implementation costs, which included internal personnel costs in only half the firms, ranged from $2m. to $130m. Implementation time from the signing of the contract until the final 'go-live' ranged from one to five years. The firms ranged in size from $US125m. in sales to over $US 25bn.

4.3 Findings (A) Motivations for ERP

All respondents provided insight into why their firms had adopted an ERP system. In many cases, the reasons differed across respondents in the same firm, reflecting the multiple factors motivating decisions. The six most common reasons cited by respondents for ERP implementation were: (1) need for a common platform, (2) process improvement, (3) data visibility, (4) operating cost reductions, (5) increased customer responsiveness, and (6) improved strategic decision making. These reasons proved to be interrelated, as a new systems platform enabled new capabilities, which were in turn expected to generate important performance outcomes (see Figure 4.1).

The dominant motivation for ERPs was to provide a common systems platform. For one firm, Year 2000 compliance was the driving concern. But for the other firms that mentioned Year 2000 at all, this concern proved to be merely the catalyst to replace an aging IT infrastructure with one that was more manageable and a better enabler for new business processes. For example, at one firm, the cost of Year 2000 compliance was estimated at $30m., which was equal to the estimated cost of implementing SAP. This firm decided to implement SAP as the solution to both its Year 2000 problem and its need for a common systems platform to support the business.

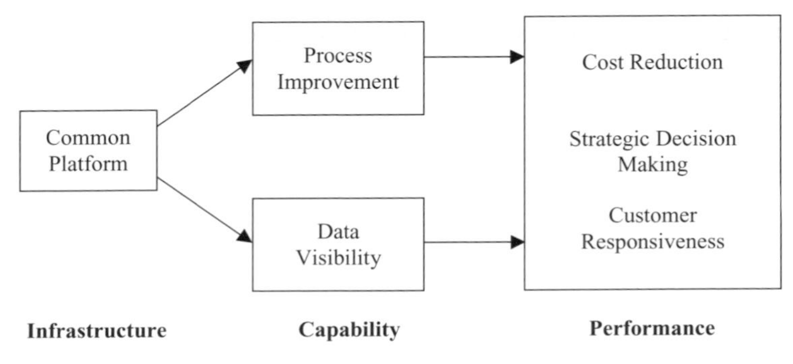

Figure 4.1 Motivations for ERP

Several firms deployed an ERP system to replace a melange of legacy systems that had been built on a myriad of outdated technologies, which had led to high support costs. The ERP system would be a way to standardize much of their IT environment, which they expected would lead to lower support cost. For some firms, the more critical issue was the difficulty of integrating the disparate systems that had grown up over time. At one firm, one newly built system had so many interfaces with existing systems that integration testing took nine months, and no one wanted the system by the time it was finally ready to go live. At several firms, the systems integration challenge resulted, at least in part, from mergers and acquisitions. Through business combinations, these firms had inherited systems that did not integrate effectively with similar systems in the parent company. One firm in the study had 23 different accounting systems; another had 14 bill of material systems. These multiple systems made firms' underlying information platforms highly inefficient and unreliable, and for some firms it had become difficult, if not impossible, to compete in a global environment.

A majority of the firms expected that their new ERP-based systems' environment would enable process improvements. In some cases, they wanted to improve specific processes, such as logistics, production scheduling, or customer services. These tended to be cost-driven reasons for ERP implementations. In other cases, management was more generally concerned with process standardization to ensure the quality and predictability of global business processes. Through process standardization, these firms anticipated reduced cycle times from order to delivery. These firms tended to be driven by concerns over customer responsiveness.

Another motivation for an ERP system was data visibility. Because ERPs are highly integrated, they have the potential to make much better decision-making information available to managers. This visibility, which gives an end-to-end view of supply chain processes, was expected to improve operating decisions. In addition, respondents viewed data visibility as key to their ability to present a single face to widely distributed customers and to recognize global customers as single entities. The impact of data visibility was expected to extend to strategic decision making. The online, real-time transaction processing characteristic of ERPs can provide current rather than historical information on a firm's performance, thereby facilitating increased responsiveness to market conditions and improved internal capabilities.

Although respondents were very clear about their motivations for adopting ERPs, they often had not established metrics for determining how well they were achieving their objectives. The most common metrics focused on cost reduction. Other metrics monitored customer responsiveness in terms of customer satisfaction, and measured process improvement in terms of inventory value, inventory turns, time to delivery, order accuracy, and related process outcomes. But most of the firms had been live with the system for less than a year, and felt that that they were not yet able to see measurable benefits from their new platforms (see Chapter 3 for a framework for assessing benefits). Nonetheless, the evidence that emerged from the interviews indicated that all firms were on the same general path and that success depended upon a firm's ability to traverse that path.

4.4 Findings (B) The Stages of ERP Implementation

The stages of ERP implementation resemble the journey of a prisoner escaping from an island prison. First, the prisoner plans an approach, carefully considering whether to follow through on his intentions and mapping out the path he will take. Second, the prisoner takes the dive off a cliff and heads toward the bottom of the sea. Third, he attempts to resurface, anxious to do so before he runs out of breath and hopeful that he will not be shot when he emerges. Fourth, the prisoner reaches the surface and starts to swim to freedom. Finally, if the diver is successful, he arrives at a distant shore, transformed from prisoner to free man.

The comparable stages in the ERP journey are (1) design, (2) implementation, (3) stabilization, (4) continuous improvement, and (5) transformation. These stages as they relate to organizational performance are shown in Figure 4.2. We will now briefly review each stage.

The approach – ERP design In the planning stage, firms made two important design decisions, one about process change and another about process standardization. First, the firm decided whether or not to accept the process

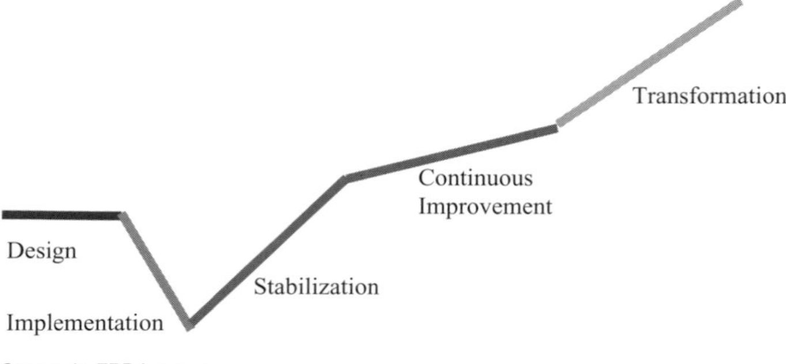

Figure 4.2 Stages in ERP journey

assumptions embedded in the software package. While all four ERP systems provided choices on how the software could be configured, each embodied some assumptions as to how data should flow through the system. This created a kind of 'technological imperative' relative to traditional systems development approaches, in which managers had decided on processes and then built systems to support them (for a detailed discussion of contrasting approaches in a further 27 projects, see Chapter 12).

One important assumption of any ERP system is process integration. Most firms had created linkages among their legacy transaction processing systems, but these linkages had been implemented as canals between the data pools of the individual systems. System users or operators controlled the access to the canals and determined when and how data flowed between systems. Within an ERP system package, data flows much like a river. Once data enter the system, users lose control over where it flows. Data are absorbed into the system and have a pervasive effect on organizational processes. In firms where managers wanted to change their technology platform, but not their organizational processes, management sometimes resisted process changes built into the system. As one respondent explained:

> We did not intend to change processes; we wanted a better system that allowed us to do things the way we'd always done them. (Cost accounting supervisor)

In contrast, other firms intended, when they implemented their ERPs, to adopt the 'best practices' embedded in the software. They portrayed a very different attitude toward process change:

> With Baan, we have to fit the organization around the software, so process change is inevitable. (Demand manager)

Firms that accepted the 'technological determinism' of their ERPs generally rejected requests to customize the software. A CEO whose firm had just completed its second ERP implementation noted that one difference between the firm's first and second implementations was its approach to customization:

> [During the second implementation] We told people that they could write down every change they felt they needed on a piece of paper and we would take it to the steering committee, who would reject it. (CEO)

Firms that wanted to limit process change, or to dictate the nature of such change, typically had to modify the software in order to achieve these goals. These modifications allowed the firms to customize their processes, but also led to concerns about their ability to support their systems later, and to manage new releases of the ERP software.

The second design decision firms made was regarding process standardization. Senior management had to determine the scope of process standardization, specifically whether processes would be standardized across the entire firm or only within certain sub-units. This was a key decision because it was difficult to change after the ERP system was in place. One respondent noted that an ERP system is like concrete – easy to mold while it is being poured, but nearly impossible to reshape after it has set. Generally, firms with related businesses found merit in standardization across business units, but not all adhered to an architecture that called for a single instance of the system. One implementation head whose firm had installed separate instances at each site explained the rationale:

> Culturally, it was not possible to do this any differently. It would have been cheaper to have a single instance throughout Europe, but it never would have gone in, due to our culture of autonomy. (Project Director)

The dive – implementation While most firms carefully planned for implementation, deploying implementation teams that trained users on the new system and, to some extent, on new processes, most found that 'going live' tended to be highly disruptive, because the new system tended to be linked to new processes. It was not possible to implement the new system and the new processes separately because they were highly interdependent. Consequently, 'going live' introduced major organizational change. One manager described the resulting confusion this way:

> It's like turning out the lights, people don't know where they are going – sometimes they end up in a different room. (Business Manager)

In most cases, managers involved with implementations found they had underestimated the extent to which individuals would be affected. Asked what they would do differently if they were to do the ERP project again, most responded that they would offer more training on how the system would change business processes. One respondent noted:

> People thought they'd implement the system and then go back to life as usual. They needed to understand that we aren't going back. With a big package like this, life is permanently changed. It's a commitment to a new way of life. They say 'go away and let me do my job,' but that's not going to happen. (Director, IS Operations)

Resurfacing – stabilization All fifteen firms in the study noted that immediately following implementation, there was a period of stabilization, during which the firm attempted to clean up its processes and data and adjust to the new environment. All firms acknowledged an initial performance dip, although some firms that had staged multiple implementations found they were able to minimize the impact in later implementations. While the typical stabilization period for a first implementation appeared to be four to 12 months, the intensity and length of firms' performance dips varied. Some firms had particularly difficult stabilization periods:

> When we first put it in, it was a disaster. The system was so slow that we had to limit the number of users in each area, so some transactions couldn't be entered. People were working crazy hours just to catch up on the data entry. Things weren't getting shipped. Some orders were wrong. Customers were furious. The really horrible pain lasted for about four months, until we replaced our database with Oracle. We are just now [18 months after going live] reaching stabilization, where we are operating at about the level that we were before we put the system in. (IT Manager)

Typical activities during stabilization included cleaning up data and parameters (sometimes referred to as business rules), providing additional training to new users, particularly on business processes, and working with vendors and consultants to resolve bugs in the software. At most firms, some early benefits of the ERP system started to emerge during stabilization. Many firms observed that end-of-period reporting cycles had been reduced. And the process of cleaning up the data led to significant improvements in the firms' understanding of their products and processes:

> Our bill of materials is in better shape. It is more accurate and provides greater detail than before. Thus, information is higher quality. We are getting better data on what we're shipping. For example, one assembly had 114 options, but 97% of customers ordered one of seven, so we have reduced the number of options available. (Director of Finance)

Swimming: continuous improvement Following stabilization, firms entered a stage in which they were adding functionality through new modules or bolt-ons from third-party vendors. In particular, they were implementing EDI, bar-coding, sales automation, warehousing and transportation capabilities, and sales forecasting. Approximately half of the firms in the study were in this stage (the other half were still in stabilization) and these firms were starting to generate significant operating benefits. One firm had cut inventory by 30% in the first year, another had increased inventory turns from eight to 26 times a year. A third firm had improved its order fill rate from 95% to 98%, and another had cut $35m. out of logistics expenses. Respondents also noted that they had closed plants, reduced headcount through consolidation, optimized transportation, and reduced working capital. Less tangible results include simplified system support, increased flexibility to adapt to external changes, such as the adoption of the Euro, increased system reliability, and improved sales force morale.

During this stage, firms were primarily focused on continuous improvement, but they were also starting to engage in process redesign to implement new structures and roles to leverage the system. In particular, most firms were adopting a process orientation. This typically meant that they had become more 'matrixed', so that new process teams or process executives were added to the firm's formal structures. A number of firms had added new roles such as 'data police', a process team with responsibility for ensuring the integrity of the firm's data and processes and for identifying opportunities for process change.

Transformation Although none of the firms in the study felt they had as yet transformed themselves, several were looking ahead to that stage. In particular, managers were anticipating that they would be able to leverage organizational visibility to gain increased agility. One senior manager noted:

> We will focus more on combinations of products and services to address customer needs. Over the last 50–60 years, we sold what we wanted to make. In the future, we will provide the products and services that customers need, relying, when necessary, on external sourcing. This means we will be increasingly connected with our suppliers, partners, and customers. (Business Vice President)

Other respondents noted that transformation would involve changing organizational boundaries, particularly with regard to systems. One CIO expected that his firm 'will see our customers' sales trends vs plans quicker than they will'. The extension of the ERP into customer and supplier systems was cited as a likely scenario by a number of firms.

In summary, ERPs appeared to hold the potential for significant improvement in business processes, customer responsiveness, and strategic decision making. However, realizing those benefits – from what this book call the second wave of ERP – required some persistence during wrenching organizational changes. It was not clear how many firms that implemented ERPs in our study would actually achieve the benefits, though we come to some conclusions below about how these benefits could be realized. What is clear is that there are a number of possible pitfalls that put the benefits at risk, and that careful planning can reduce the risk of failure.

4.5 Findings (C) Obstacles to Success

Despite the bountiful literature on the things that can go wrong in an ERP system implementation, we found that most of the study firms were not well prepared for the organizational changes that their implementations wrought. To add to the risk pictures that emerge from Chapters 5 and 6, this section will discuss four types of pitfalls reported by study participants:

1 failing to establish metrics;
2 resourcing the post-implementation stage inadequately;
3 ignoring management reporting requirements; and,
4 addressing resistance to change slowly or not at all.

Metrics The organizational changes that accompany ERP system implementation are enormous, and managers in most firms underestimated the scope of the impacts, even of those changes they could have anticipated. The business cases for the ERP system implementation tended to be vague, often for example referring to the need for a more solid infrastructure without quantifying anticipated business benefits. Consistently, firms without clear performance metrics that clarified expectations for their ERPs were unable to determine whether they were benefiting from the implementation (see also Chapter 3). For these firms the stabilization period was particularly discouraging, because everyone could see the problems and few people could see any benefits. Firms that had established metrics could identify some improvements even as they dealt with the confusion of major process change. In some cases, the actual performance benefits were different from those anticipated, but a complete lack of metrics proved to be a much bigger problem than metrics in the wrong areas:

> One thing we definitely should have done was to set some much more specific performance targets and establish a way of tracking them. We didn't identify tangible benefits like inventory turns, cycle time improvements, or cost reductions. If we were to do this again, we would have clear goals and an understanding of how we would achieve them. We need that clarity. (Director of IS Operations)

Ongoing resource requirements Because the cost of implementing an ERP system was high, management tended to be anxious to declare victory and move on to other projects. Respondents noted, however, that continued management attention during the post-implementation stage was critical to receiving value from the ERP system (see also Chapter 1). Implementation resulted not only in new processes, but – paradoxically – in some processes that had once been automated becoming manual. Consequently, resource requirements increased in some areas. In addition, the post-implementation stage involved some sorting out of the opportunities for process redesign. In this stage, firms were adding functionality and reengineering processes. Thus, the firms that were most excited about the benefits they were receiving were continuing to pour significant resources into their ERP programs. Although they knew that management must aggressively pursue ERP system benefits in order to obtain them, most firms found it difficult to sustain management attention:

There is some low-hanging fruit available immediately but some managers may not recognize it and thus it won't get picked. The problem is that most plant managers have no clue what they're getting into. They are very busy being prosperous right now. The changes required by the ERP could be viewed as hurting the business in the short run because the plants are already at capacity. (CIO)

Management reporting requirements A number of respondents noted that their ERP system was an effective on-line transaction processing system, but not a management support system. In these firms, the increased availability of data had not been translated into management information. In some cases, this meant that management could not determine how the business was performing after implementation of the ERP system. While many respondents recognized the potential for improved management reporting, they cautioned that querying an ERP system was generally not easy, and managers were inclined to rely upon formal printed reports.

Addressing resistance It is not surprising that the implementation of a system as large as an ERP system would encounter resistance. What was interesting about the resistance described by respondents in this study was the variety of forms it took. In most firms, there was the anticipated resistance to changes in individual jobs. Often these changes were significant:

Persons in mid-level positions have, in many cases, been the most susceptible to job changes and some are very uncomfortable with them. For example, the materials manager used to be 'action central'. Everybody gave him information and he made decisions. Now SAP accumulates the information so that everyone can make decisions locally. The materials manager will try to subvert the system rather than recognize that the nature of the job has changed. He should be identifying ways that people can make better decisions, and work at improving organizational processes. Instead he wants to keep sole access to information. (Functional Leader)

A more difficult form of resistance could be called intellectual resistance. Lasting success with an ERP system requires that employees understand general business processes well beyond their immediate responsibilities. Employees who had difficulty grasping how their behaviours could affect operations several processes removed from theirs could introduce contaminated data into the system. For example, in one firm goods that were in inventory could not be shipped because the system did not believe they existed. Thus, employees had to understand standardized process and comply with them:

> It is very hard for people to change from things they know well and are good at. We find that the people who were most effective in the old environment were those who knew how to 'beat the system'. With SAP beating the system is not good; what's good now is discipline. People have a lot of unlearning to do and it's very painful. (Business Vice President)

A third form of resistance was based on company culture and politics. In many organizations, it has become politically incorrect to speak of a technological imperative, so most managers involved in ERP system implementations talked instead of how the system would 'enable' change. However, the daily experience of persons actually using the system was that a computer was dictating how they would do things. At one firm an influential mill manager publicly insisted that 'no blankety-blank computer is going to tell me how to run my business'.

Firms addressed resistance in a variety of ways. New incentive programs that emphasized stock options were being used to focus attention on corporate goals and to help ease individual resistance. Training was particularly important for teaching new processes and gaining better understanding of the business. One firm, which allocated 20% of its ERP program budget to training, implemented a unique change management training program:

> We are putting everyone through a 2-hour change management session that is training people not only on systems and processes but also on how they will feel. In this session, we are explaining that people will have to think differently about the way they do their jobs, that there are new emotions they will have, and that it's ok to be 'ticked off', that this won't be easy, and that not only will the system not do everything they are accustomed to systems doing, it *will* do things for them that they were already comfortable with and wanted to continue to do. (CIO)

Firms that recruited new employees found that they were often attracted to the new systems and processes. One executive noted that SAP had had a positive influence on University recruiting, because it provided an image of being state-of-the-art.

4.6 Lessons from the Study

For the firms in this study, the major difficulty of implementing an ERP system was neither the introduction of a new system nor the simple fact of change. The real challenge was that, as part of their ERP system implementations, these firms were instilling discipline into relatively undisciplined organizations. This was a major cultural change – one that to many employees did not immediately seem an improvement. Thus, firms that were progressing to continuous improvement and possibly to an eventual transformation were doing more than addressing the obstacles to ERP system implementation. Typically, strong senior managers were demonstrating their commitment to their firms' ERP initiatives in a number of ways:

1 assigning their best people 100% of the time to the project,
2 developing a clear business case that clarified performance objectives,
3 demanding regular reports based on established metrics,
4 communicating goals and establishing program scope,
5 establishing a long-term vision.

Senior executives who were committed to generating benefits from ERP system implementations were dogged in their determination to instill a new culture of organizational discipline. These executives implemented mechanisms to sustain standardized processes and data. In addition, they communicated the paradox that standardization is key to flexibility. In doing so, they extolled the virtues of a disciplined organization as one in which the predictability around transaction processing would ensure quality processes and empower decision makers to respond to customer demands:

> In a way, we are slaves to the system, and we have accepted the technological imperative that that implies. We cannot improvise on processes because such innovations will ripple through the company and cause problems for someone else. It frees us in other ways to do the things we want to do. (CEO)

From this study, and the related case history of Dow Corning by Ross (1996a, b, 1999 – discussed in detail in Chapter 10), we can derive some critical success factors for getting ERP in and generating benefits:

- avoid customization;
- deploy cross-functional, full-time project teams with top talent (see also chapter 12);
- enlist senior management sponsorship;
- accelerate decision-making processes;
- manage scope and schedule;
- adopt a continuous improvement mind-set;
- evolve metrics;
- realize on-going training and selective recruiting are underestimated but we are key;
- establish clear IT governance structures;
- focus and persevere in complex processes of integration and use.

Our view is that these learnings are perennially applicable to ERP system implementation and usage. They reflect the centrality of cultural change in ERP system adoption. But they also reflect the complex, distinctive, critical characteristics of those ERP-based applications needed to deliver on visions of the future, integrated, extended enterprise. A cardinal mistake – made frequently elsewhere in IT management (Lacity, Willcocks, and Subramanian, 1998) – is to forget these learnings in the light of changes in technology – in this case moves to ERPII systems – and in the light of new modes of delivery – in this case over the web, either through in-house management or through using a mix of external service providers. This may occur due to supplier rhetoric about the degree to which the new technology is superior, and also more commoditised and under control than previous versions (Pozzebon, 2001). It may additionally occur because external sourcing using web-based delivery modes is presented as a way of taking away IT management issues and risks (Kern and Willcocks, 2002). As we shall see, in fact the lessons developed above still apply but extensive use of external 'netsourcing' of ERP also raises additional managerial challenges, and in itself creates some distinctive risks that need to be addressed (Kern, Willcocks, and Lacity, 2002; Kern, Lacity, and Willcocks).

4.7 Towards ERPII Systems

Through to at least 2005, organizations that have made large investments in ERP will be seeking to improve the usability and business returns of these current applications. Steenstrup (2001) has pointed out that before embarking on a migration to ERPII applications that support collaborative commerce and the extended, networked enterprise, this optimization phase is very necessary. Missed opportunities for ERP will become risks in ERPII. Common failings in ERP projects have been poor product training, process training, preparation of users for new roles, and inappropriate reward systems that did not motivate change. In ERPII such failures will heavily compromise success as the number of users and transactions expand beyond the enterprise, multi-process dependencies increase, and multiple enterprises are threatened by any failure to manage change properly.

Here ERPII is to be understood as an application and deployment strategy that builds on current deployment, but expands out from ERP functions to achieve integration of an enterprise's key domain-specific internal and external collaborative, operational and financial processes. This will allow internally and externally connected processes and information flows, value chain participation, and will mean a move away from web-aware, closed architectures to web-based open, componentized ones. However, it is unlikely that even the major vendors leading this field – SAP, Oracle, PeopleSoft and J.D. Edwards – will have all aspects of their ERPII applications in place before 2004/5. This means, on some estimates, that by 2005 less than 40% of large enterprises will use their main ERPII vendor's integration layers as their primary integration platforms (Steenstrup, 2001). Clearly ERPII will be slow coming, and organizations will need to approach integration proactively rather than over-depend on ERPII to solve all future integration problems. The path forward for an enterprise with large-scale ERP sunk investments would seem to be to:

(a) optimize and deepen use of current functionality;
(b) increase integratedness with other technical and application developments, e.g in supply chain management, different types of e-marketplaces and exchanges, portal technologies;
(c) monitor carefully vendor capability on ERPII, and the marketplace ability to deliver the netsourcing of rentable ERP applications, services and architectures;
(d) as migration to ERPII occurs, overemphasize in on-going project plans the lessons on optimization from previous rounds of ERP implementations.

4.8 New Modes of Delivery: Risky Business?

By 2001 many vendors were looking at providing ERPs, and many other applications, services and even infrastructure, over the web on a rentable basis – a development that has been called 'netsourcing' (Kern, Willcocks, and Lacity, 2002; Kern, Lacity and Willcocks, 2002). If this market had lived up to its advanced press, during 2000/1 many thousands of companies would have been accessing software applications, including ERP, over the internet from remotely managed and maintained servers. They would have torn up their software licences, redeployed the in-house IT team, stopped worrying about the costs of new hardware and software upgrading, and downgraded the importance of their capability to implement and manage software/business change projects. What was originally called the Application Service Provision (ASP) market rode on predicted revenues of up to $US15bn. by 2005. ASPs promised cost efficiency, accelerated application deployment, access to the latest technology and support, together with the transfer of ownership risk. ASPs also promised much to small and medium enterprises (SMEs) who were finding it hard to attract and retain IT staff, and had been previously excluded from applications such as ERP and customer relationship management (CRM) because they were too expensive (Bennett and Timbrell, 2000).

In practice 2000/2 were bad years for netsourcing, with commentators regularly predicting that the 1500 plus ASP companies extant in mid 2001 would shake down to a few hundred relatively quickly (Moran, 2000). At the very least, this put the onus on potential customers to vet carefully ASP business models and financial stability (*The Economist*, 2001). ASP dependency on a range of partners also underlined the importance of deep analysis of vendor positioning in its value chain. As one example, Cityreach International, one of Europe's largest web-hosting and co-location companies, went into receivership in September 2001 exposing several major clients including US service provider Usinternetworking. That said, as the studies by Kern et al. (2001), Kern, Willcocks, and Lacity (2002), Kern, Lacity, and Willcocks (2002) found, by early 2002 many companies and government institutions had adopted the netsourcing model relatively successfully, including for ERP software. In this section, based on our studies, we detail the risks emerging in such netsourcing arrangements, and how they can be mitigated. In the next section we will then derive further lessons from case studies of netsourced ERP system usage.

Risks from a Customer Perspective

Risks in the eyes of customers are those issues that could adversely affect their company and/or operations as a direct result of having entered into a netsourcing deal. A 2001 study of approximately 400 international customers either in a netsourcing deal or contemplating one, emphasized that there were real, recurring concerns over the potential and inherent risks of a netsourcing solution (Kern et al., 2001). Undoubtedly, this received particular emphasis because of the newness of the netsourcing market and many of its players, and because of the rapid downturn in the IT and the ASP market from the end of 2000, and indeed into 2002. As we shall see later, it is clear that the netsourcing market has some distinctive risks, but, for customers, most netsourcing risks that emerge are not so different from any use of the IT services market, and need pro-active management. Not surprisingly, therefore, the same rule applies: 'do not expect too much from the supplier or too little from yourselves' (Kern and Willcocks, 2002).

Of the risks that might come into play when looking at a netsourcing solution, potential customers had grave concerns for all issues rated 3 to 4 on a five-point scale (see Table 4.1). The primary risks for these customers, in order of impact, were:

1 a netsourcing service provider's service and business stability
2 security issues
3 reliability issues
4 netsourcing's longevity and existence
5 netsourcing's dependency on other parties.

This is quite a considerable worry list. How do we explain this? Remember that this is a list from *potential* customers. Customers with no experience of a netsourcing solution are likely to be more concerned, due to the surrounding uncertainty and perceived immaturity of the market. Developments in 2001 in particular – when netsourcing service providers were struggling to survive the sudden downfall of the technology stock market and the resulting drying-up of investment funds – did little to assuage these perceptions.

However, the rising number of netsourcing service providers with track records, growing customer bases, profitability and sufficient investment capital offered some reassurance on how these potential risks could be offset. Moreover, the big IT service, software and telecoms firms are much less likely to be confounded by issues of longevity, funding and stability. Furthermore, referring to Table 4.1, it is noticeable that those customers actually in a netsourcing deal rated the risks identified by potential customers as up to 30% lower (see for example ASP dependency on other parties). In other words, fear of these risks is a third higher than the actual experiences of customers indicate such fear needs to be. This was actually a common finding across the research, and indicates that netsourcing service providers needed to invest a lot of time in customer education, especially while the market was immature, and so much was volatile and not well understood.

Nevertheless, if potential customers tend to rate netsourcing risks as more serious than do existing customers, both worried about the same types of risk (see Table 4.1). And while experienced netsourcing customers suggest that many of the issues cited were less problematic than envisaged, they still on average, across the issues, regularly rated the risks as between 2 and 3 on a scale of 1–5. Moreover, none of the risks we investigated was actually rated by the existing customers as of no/little risk at all, that is 1 to 2. So it is apparent that even when in a netsourcing deal, customers still had many, considerable uncertainties and concerns – ranging from the technical factors such as security, reliability, and scalability to the managerial issues of lock-in, unproven business model, and dependency on others. All these issues remain, in fact, as attendant, characteristic netsourcing risks.

Table 4.1 Netsourcing risks cited by potential and existing customers

Drawback	Rating: 1 No risk %	2 %	3 %	4 %	5 major risk %	Average rating (1–5)
Higher initial capital outlay	4.2	19.0	53.6	17.9	5.4	3.01
	10.5	26.3	47.4	15.8	0.0	2.68
Unproven business model	6.0	16.1	38.1	28.6	11.3	3.23
	18.4	26.3	28.9	21.1	5.3	2.68
Company size too small/large for ASP use	16.7	21.4	41.1	11.3	9.5	2.76
	26.3	13.2	36.8	10.5	13.2	2.71
Loss of control	9.5	19.0	28.0	29.8	13.7	3.19
	10.5	31.6	39.5	10.5	7.9	2.74
Unclear on ASP concept and practice	16.1	16.1	38.7	19.6	9.5	2.90
	26.3	15.8	28.9	15.8	13.2	2.74
Not enough service and application choices	4.8	20.2	41.1	22.6	11.3	3.15
	5.3	31.6	39.5	21.1	2.6	2.84
Reduced flexibility	8.9	28.6	31.5	23.2	7.7	2.92
	13.2	28.9	26.3	18.4	13.2	2.89
Contract lock-in	2.4	14.3	40.5	29.8	13.1	3.37
	10.5	18.4	42.1	23.7	5.3	2.95
Higher IT costs over time	5.4	19.6	42.9	23.8	8.3	3.10
	15.8	15.8	31.6	31.6	5.3	2.95
ASP's dependency on other parties	4.8	7.1	26.2	37.5	24.4	3.70
	5.3	26.3	42.1	13.2	13.2	3.03
Scalability	7.7	22.6	35.1	25.6	8.9	3.05
	7.9	13.2	44.7	31.6	2.6	3.08
Integrating ASP with existing applications	3.6	11.3	30.4	33.3	21.4	3.58
	13.2	13.2	34.2	26.3	13.2	3.13
Security	1.8	9.5	22.0	26.2	40.5	3.94
	0.0	28.9	23.7	26.3	21.1	3.39
ASP's service and business stability	2.4	5.4	20.8	35.1	36.3	3.98
	2.6	21.1	23.7	34.2	18.4	3.45
Reliability	2.4	6.0	24.4	36.3	31.0	3.88
	0.0	21.1	26.3	28.9	23.7	3.55
ASP's longevity and existence	2.4	7.7	28.6	34.5	26.8	3.76
Sample 213	2.6	7.9	39.5	28.9	21.1	3.58

1 2 3 4 5
■ Expectation □ Experience

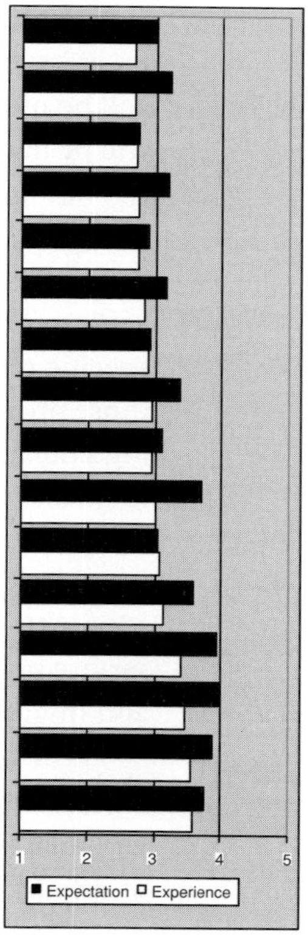

Note: Risks are sorted (ascending) on the average of the response of the respondents that come from companies currently sourcing from an ASP.

The gravest concerns for *existing* customers were:

1 netsourcing's longevity and existence,
2 reliability issues,
3 a netsourcing service provider's service and business stability,
4 security issues,
5 integrating the netsourcing solution with existing applications.

All of these will also inform and often shape customers' ongoing management issues in any netsourcing deal. For example, many of the technical issues will be monitored through service-level agreements, while contractual issues will be handled through carefully developed contracts.

The longevity and stability issue poses a significant risk to both potential and existing customers. There remained, into 2002, substantial uncertainty regarding future developments in the general marketplace (for example the speed and extent of supplier consolidation, levels of customer demand), but also for specific netsourcing service providers, even those with many customers. For those providers in 2002 yet to make a profit, much depended on the continuing availability of investment capital, and ability to manage cash flow and the financial position. In turn, it seems sensible for those customers already signed up with start-ups to ensure they keep themselves informed about their service provider's financial situation and be prepared on short notice to switch their services to a comparable provider. One suggestion for those unsure about their service provider's future, is that they need to prepare a preliminary transition and migration plan that will simplify the process if need be.

Those looking at signing netsourcing contracts will now be much more aware of the need to include clauses that cover exit terms, such as termination and guarantees and methods of providing continuity of service in the event of a failure. Neglect of these can prove very costly indeed. Thus, one large ten-year IT outsourcing deal with a poor termination clause cost the client nearly $50m. to get out of after 17 months.

Security and reliability of service are risk issues that run pretty much throughout any netsourcing venture. A sound policy of monitoring and ensuring that protocols are followed will assist customers in stopping these concerns turning into real threats. Much depends on the detailed attention given to constructing service performance measures before the netsourcing contract is signed. If these are sound, and are regularly reviewed and updated, much then depends on the quality of the client's and supplier's monitoring policies, and their ability to work together on uncovering and fixing any reliability and security problems. Close ongoing attention to a netsourcing deal's workings on these issues provides one approach to mitigating risks.

It will be up to individual managers and the specific application service provider firm to alleviate and mitigate such concerns, remembering here that all of the factors above have been actually perceived as genuine, sometimes serious, risks in the eyes of customers. According to both existing and potential customers the four core risks that need to be mitigated throughout any netsourcing venture are security, service and business stability, reliability, and netsourcing longevity. However, all the risks cited in Table 4.1 are felt, real risks associated with applying netsourcing models to client firms. As such, they all need to be addressed in any risk management strategy. From this review of customer and would-be customer perceptions, we can build towards a more comprehensive way of analysing and managing risks. We do this in the next section by developing a risk profiling framework.

Netsourcing: A Risk Profiling Framework

It is clear from the above discussion that the netsourcing of services and applications, including ERP and ERPII, carries distinctive risks that need to be addressed. In Figure 4.3 we provide a risk profiling framework that distills

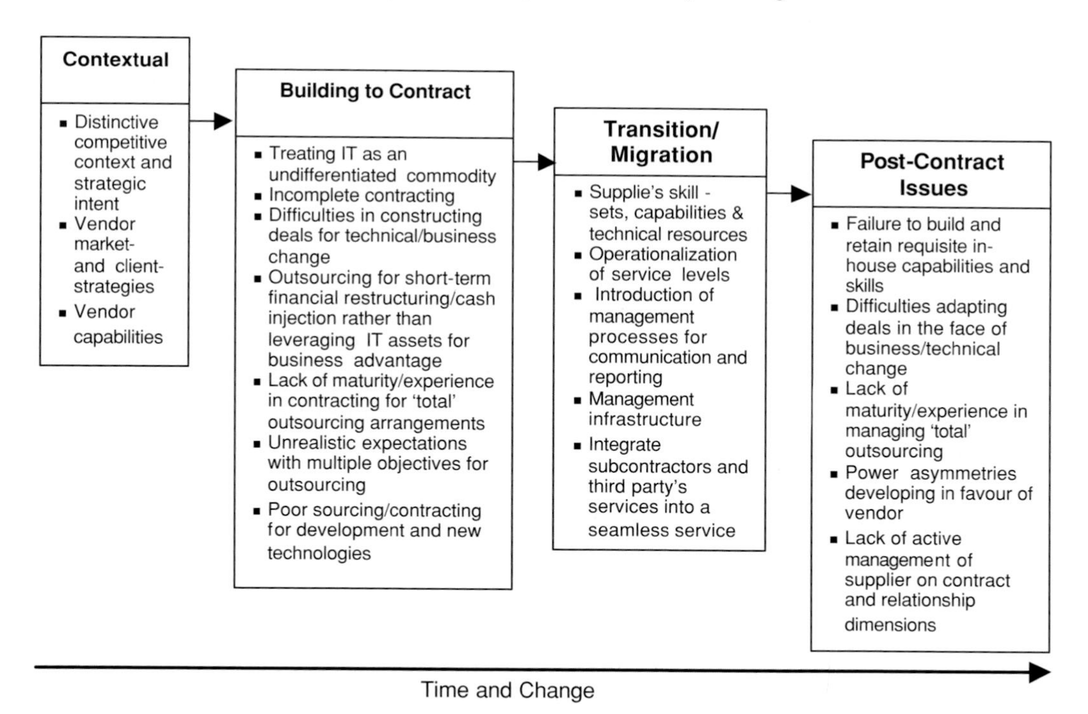

Figure 4.3 Netsourcing: a risk profiling framework

the lessons learned in detailed studies of traditional IT outsourcing and also netsourcing arrangements, including those covering ERP usage (Kern and Willcocks, 2002; Kern, Lacity, and Willcocks, 2002; Kern, Willcocks, and Lacity, 2002; for ERP netsourcing examples see also the case studies below).

As can be seen, these studies have consistently found risks materializing in four areas. The first is classified in Figure 4.3 as 'contextual'. Here the client's distinctive competitive context can determine which IT is likely to be core and differentiating and what speed of systems development and level of service are required. Strategic intent can determine the objectives for netsourcing – for example, should customers go for cost reduction, allow the in-house team to refocus on higher value development work, or hire a supplier to achieve global standardization in certain aspects of IT, for example in ERP? It is also important to be very clear as to the strategies the prospective suppliers are pursuing in the netsourcing market-place and the ramifications of those for the client company over the life time of the contract. The supplier's core capabilities also need very careful analysis, as we emphasized in the discussion on ERPII developments above.

The second area is 'building to contract'. Figure 4.3 delineates seven major practices that regularly materialized in negative outcomes. With ERP, the first and fourth can be particularly dangerous, given that such applications are increasingly mission critical. With the seventh practice, it is particularly dangerous to go down an outsourcing route with new ERP developments because they fit so closely with business operations. A 'buy-in', close relationship with the supplier is the preferred risk-mitigating practice, with internal business and IT people retaining management reponsibilities for ERP development and delivery.

The framework actually puts a great deal of weight on risk factors that occur before the contract is operationalized. However, the third grouping shows five distinctive major risks that regularly arise at transition and migration phases, while the fourth grouping shows the five most serious risks that regularly occur during post-contract operationalization of outsourcing/netsourcing arrangements. In earlier research, we actually found the most common and damaging external sourcing error was failure to retain enough in-house capability to maintain flexible control over the enterprise's IT destiny. In this respect, when netsourcing ERP systems, it is critical to retain technical architecting capability and the troubleshooting knowledge and ability to deal with technical problems that require non-routine solutions. Additionally, we have found that business managers must also be mature in their ability to take on distinctive roles and responsibilities relating to ERP development and its business optimization.

Some seven other IT capabilities can be regarded as core and also need to be retained (for a full description see Chapter 12).

4.9 ERP Through Netsourcing: Two Illustrative Cases

By 2002 many suppliers were offering netsourced ERP packages, invariably in partnership with a range of other suppliers in the netsourcing value chain. As one example, in 1999 SAP itself introduced one of the biggest – mySAP.com – a product line defining its Internet strategy. The mySAP.com offering consists in essence of collaborative horizontal and vertical e-business software solutions. Vertical solutions link different horizontal solutions tailored to specific industries. For this SAP has 22 industry templates, ranging from Aerospace & Defense, Automotive and Banking to Telecommunications and Transportation. These solutions allow organizations to not only fully integrate their operational business processes, but also to integrate customers and business partners along the supply and value chain.

The mySAP.com product line consists of four distinguishable internet-based building blocks:

- marketplace;
- business applications;
- workplace – an enterprise portal that provides employees in companies with a browser-based, personalized working environment from which they access all the functions they need daily in their particular role;
- application hosting – set of hosting services provided by SAP and partners jointly, includes application hosting, application service provision, application management and hosting of marketplaces.

The SAP ASP offering combines software, implementation, infrastructure, service, and support. It is backed by a rapid implementation philosophy that enables one-stop shopping at the lowest cost of operation and implementation. The ASP model generally is a one-to-many model with little customer configuration and therefore is easily adoptable by many customers. With the one-to-many SAP ASP model, customers can have a standardized full service package that is accessible via all types of networks, including the Internet. In 2000 SAP reached a major milestone when the number of registered licensed users of its mySAP platform reached one million. To early 2002, more than three million additional users had licensed the solution. Key customers as at that date included – Avaya Inc., Bristol-Myers Squibb Co., The Coca-Cola

Company, Colgate-Palmolive Co., Compaq Computer Corp., Dow Corning Corp., Eli Lilly and Company, General Mills Inc., Gillette Co., Hewlett-Packard Co., Honeywell Inc., Kimberly-Clark Corp., Compaq, Motorola Inc., Palm Inc., Philip Morris Companies Inc., Procter & Gamble Co. and Texaco Inc.

Clearly, despite the 2000–2002 market difficulties, the netsourcing model was developing viability amongst those customers requiring ERP on a different basis from more traditional modes of development and delivery. In this final section we illustrate many of the lessons and risks pinpointed in this chapter through two cases of ERP being delivered effectively through netsourcing. The cases are taken in shortened form from Kern et al. (2001) (Note 1).

Customer Case (A): Punch International

As at 2001, Punch International provided 'Electronics Manufacturing Solutions', integrating manufacturing services, from product and process engineering to the delivery of finished products. Punch International's range of products included audio and video equipment, multimedia products, office and industrial printing equipment, pay-terminals and controller boxes. Punch is a parts supplier to companies like Philips, Sanyo, and Sony. Geographically dispersed, Punch operated 11 facilities in seven countries: Belgium, United Kingdom, France, Slovakia, Hungary, USA, and Mexico. The Group was based in Ghent, Belgium. Listed on the Belgian stock market, in 2001 Punch employed approximately 1800 people with a turnover of approximately US$97m.

Punch decided early on that it needed a full-blown ERP system to integrate and manage the local efforts. Due to its rapid international growth, the resulting need for improved co-ordination and logistics, and the demand for company pervasiveness, ERP was to offer the necessary integrated solution:

> We had different systems in different plants. Our strategy was to be ready, in terms of IT systems, to have 20 to 25 plants. When you have five plants, and for example you need to consolidate your book-keeping you can use Excel, it will be sufficient. When you have 10 plants this is getting difficult, when you have 20 plants this is crazy. So, firstly, the idea was to be prepared for growth. We needed a global system, also to be able to check what is in stock in a certain plant. (Wouter Cortebeeck, Global ICT manager at Punch International)

Although the growth was also a reason to choose an ERP package by SAP, the most important reason was an external one: demand from customers. The trend toward sub-assembly and complete assembly was the result of a strong tendency toward outsourcing in the electronics industry. In its annual report, Punch referred to outsourcing as 'the logical solution for producers of electronics'. It indicated a number of developments that fostered this tendency, the most important being the shortening of the life cycle and the development time for electronic products. Additionally, many of its key customers increasingly demanded direct access to sub-assembly products, like Punch International's, to better coordinate production and sales. SAP, in turn, became a natural choice for interfacing with customers and business partners:

> Quite rapidly we chose SAP. We did have a close look at Navision, Scala and Remax. . . . More than half of all our customers work with SAP. In fact customers used it as selection criterion: they asked us 'do you have SAP, because we want to interface'. . . . We have everything except for project management and advanced production scheduling. So the SAP solution we chose is quite complete. (Wouter Cortebeeck, Global ICT manager at Punch International)

Moreover, since outsourcing was a core business practice for Punch International, it was only a small step to externally source a SAP solution. Since the company had no focus on IT and did not have the necessary capabilities to run a complex ERP solution internally – nor did it want to – an IT service supplier was sought:

> Because we have no SAP expertise in-house, the decision to source the solution externally was quickly made. For us this is very interesting because IT is not our core business. We do not have a big IT department, only about five persons centrally and a power user at every plant. For the same reason we source the maintenance of the desktop systems externally. (Wouter Cortebeeck, Global ICT manager at Punch International)

Being geographically widespread, Punch decided it needed greater control through a centrally managed system. Yet it wanted the flexibility to move the central location of its servers, as the company sold and acquired new plants internationally. In turn, it wanted the system to run somewhere remotely, independent from the location of the plants and at the same time accessible from all locations and at all times. The ASP model was thus a good match.

The selected IT partner would then take care of the complete process of implementing the solution, including rapid future implementations at new plants, and running the system internationally. Following a supplier evaluation period The Vision Web was identified as the preferred choice due to its dynamic character and its SAP expertise:

> We chose The Vision Web. A relatively small company, but very dynamic and flexible. That's what we like, we think like that too. We didn't opt for one of the BIG five, because

of experiences in the past. Firstly these big companies are very expensive and secondly, they are not flexible and not dynamic. It is more important to have the right people on the project than the name of the company. We chose for The Vision Web. They understood what we wanted very rapidly, and reacted on it very fast. After we decided for The Vision Web, we did a feasibility study, a mini blue print, on which the contracts were based. (Wouter Cortebeeck, Global ICT manager at Punch International)

Because all of Punch International's plants were different, it was decided that the solution was to be implemented in the plants sequentially rather than simultaneously. To ensure the solution would fit and be likely to complement future processes an extensive evaluation was undertaken. This so-called '*blueprinting*' resulted in a master solution entailing all the needed functionality to handle the different scenarios at the different plants within Punch International:

The SAP solution we chose is quite complete: it entails everything except for project management and advanced production scheduling. We had two project managers; one from Solvision, also part of The Vision Web, and one from within Punch International. Together they developed the blueprint on which the SAP solution is based. They visited the plants talked with the key users and went through all the key processes. Then they made the blueprint, it was tested and after some revisions it was approved. So the first contacts with The Vision Web were about one-and-half years ago, the creation of the blueprint took about nine months. However before the completion of the blueprint, we had already implemented the solution in our French plant, although it was a partial solution; only the financial modules. We needed the solution quickly in France. France did go live 1 November, blue printing was finished beginning 2001. (Wouter Cortebeeck, Global ICT manager at Punch International)

The nine months of blueprinting accounted for a large part of the overall costs. Under normal conditions, this phase would have entailed a substantial one-time investment. Yet The Vision Web largely took care of the evaluation and the resulting implementations. Punch partially pays for the individual implementations through its monthly fixed fee. The necessary training was also done by The Vision Web; it trained the power users who then trained the other users. The first SAP was rolled out in France:

At the moment (2001) there are 10 users in France. In the UK another 20 will be added. In France, implementing the solution was no problem. But there were no legacy systems over there, so it was relatively easy. On the other hand it meant we had to build it from scratch. In the UK it's different. It's a very mature plant, very well organized. Now The Vision Web will begin installing the system in the UK 1 April; it will be live this summer. It's also the first full implementation, but it will be no problem. For us the UK implementation is a very important benchmark. (Wouter Cortebeeck, Global ICT manager at Punch International)

When asked about the first experiences so far with the 'ASP' SAP model, it was described as generally very positive by the responsible IT manager. It allowed the company to focus on its own business instead of worrying about the ERP application solution. So far, the experiences matched the expectations, as Wouter Cortebeeck, the Global ICT Manager for Punch International explained:

> The first experiences are very good. No fuss, no hassle. We do our job and they do theirs, that's it. We don't want and we don't need to worry about routers, firewalls, and servers. That's their business. We have an SLA. Every month we review the performance. If they do not meet the SLA standards, we will get a payback. What is important to me, is the response time and the uptime, how they do it, that's not my problem. This is all settled in the SLA. When there is a crash, it should be resolved within eight hours. If it's with a back up server or something else, that's not my problem. The agreed upon up time is like 99.9 and something percent. The down time is something like half a day per year. We have one point of contact: The Vision Web. For example, we do not have a contract with PSINet; the lodge deals with PSINet. For us that is what the ASP concept is about: one point of contact.... Another application we are considering to source via this type of arrangement is MS Exchange, among others. MS Exchange is actually in the pipeline already, also via The Vision Web.

Customer Case (B): RealScale Technologies

RealScale Technologies is a start-up founded in 1997 by Serge Dujardin and Jean-Christophe Pari. The company has developed an innovative scalable platform that enables Service Providers to offer Internet applications to a high number of users. The RealScale™ architecture builds on experiences from the telecom industry, where different hardware and management systems are used. The company decided it needed an ERP solution that integrated finance, cost control, enterprise consolidation and purchasing functionality:

> I started with package selection. I had a look at three established broad scoped ERP packages. A broad scope, because for us the financial and cost controlling functionality is just the beginning. Sales and distribution, and production planning are to follow. So most of the packages did not qualify at all. We quickly decided for mySAP.com for its functionality; it has the broadest scope, but also because of pricing reasons. SAP has positioned mySAP.com at a very competitive price in comparison to a lot of other packages. We evaluated some other ERP packages from other ISVs, like Navision and IFS, both Scandinavian products. IFS was very sexy and very promising. That was also the biggest downside of these offerings: no expertise around, no references. Too much risk. Navision is a good product, but was perceived to be too slim. Everything is tailor-made.

We wanted standard.... This project was a consultant's dream; it's a Greenfield implementation. We have no legacy systems, no data to convert, no bad habits. We can adapt our processes to the system. (Olav Claassen, Vice President Information Services Europe, Real Scale Technologies)

For a company this size, it was quit unusual to implement a high end ERP system such as SAP offers. SAP software is traditionally targeted at large enterprises, that is, most Fortune 500 companies use SAP. However, RealScale Technologies in early 2001 was expecting to grow rapidly in the next 12 months as it products gained exposure:

Real Scale Technologies has about 40 employees (in early 2001). Hardly the size for a full-blown SAP solution. We are penetrating the market and we are in contact with potential clients, but it is in an early stage. We expect radical growth. If an ERP solution has to be implemented during such a growth period, it will cost considerably more time and money, because everybody will be tied up in operational issues. That's why we wanted an ERP system implemented that would support us in our growth. Not the current state of the company, but the future state of the company made us decide for SAP. (Olav Claassen, Vice President Information Services Europe, Real Scale Technologies)

Traditionally, SAP packages are financially beyond the reach of smaller companies. First, the traditional pricing model requires paying for the licences upfront – an investment that poses a barrier for smaller companies, especially start-ups and high-growth firms. Secondly, the implementation of such solutions requires special expertise and, more importantly, involves high costs. Lastly, to run these applications in-house requires specialized skills and again involves high costs. Smaller companies often lack both the needed money and skills. Nonetheless, the ASP model, and in this case The Vision Web's hosting model opened up the opportunity for smaller companies to seriously consider high-end ERP packages. The Vision Web's model allows customers to spread the cost of the software licences and hardware over three years, making the high up-front investments superfluous:

The Vision Web's proposal included a guaranteed live on the 2 January (2001) for a fixed price... If I would have had to hire the needed experts to run the SAP systems in-house, that would have cost me over € 200000, because these experts would have had to be extremely knowledgeable. That's a lot of money. Especially since we are a start-up, we need our cash; we don't want to invest this purely in software. In the current scenario, I don't have to spend that money, because I am paying the ASP to run the systems for me, to help me with technical problems and to help me with functional problems. For me that was an easy decision. (Olav Claassen, Vice President Information Services Europe, Real Scale Technologies)

The Vision Web was the only company that could guarantee RealScale Technologies to be up and running with SAP within the restricted timeframe. In addition, The Vision Web enabled RealScale Technologies to source the SAP via a hosting arrangement, combined with a leasing agreement, including licenses, infrastructure, hardware, connectivity, and implementation services. Moreover, the ASP allowed the company to focus on its core business and its foreseen growth, instead of having to worry about their IT systems:

So for cost reasons and for quality reasons I chose The Vision Web. For me that was an easy decision at that time. Of course we had some discussions about security, and reliability issues. For example, it uses PSINet as a connectivity partner. I wanted to be guaranteed that in case PSINet would go bankrupt, they would have a backup partner that could immediately take over and ensure that we can still run our business. The Vision Web has arrangements with backup partners, and that was good enough for me. In the end it's their problem not ours. I buy the service; they have to deliver. How they do it and with whom, I don't care. (Olav Claassen, Vice President Information Services Europe, Real Scale Technologies)

The demanded deadline implied a challenging timeframe of two to three months. This is an unusual timeframe for a complete SAP implementation, illustrated by the fact that The Vision Web was the only company to commit to the proposed timeframe beforehand. The on-time implementation was made possible in part through the low degree of customization RealScale Technologies demanded. The process of customization in the RealScale scenario took less than a month compared to the nine months of blueprinting in the Punch International scenario. The difference can be explained through the complexity of the organization structures to which the solution had to be adapted and the lack of a legacy system:

We started customizing the system in November and early December we started testing and training. During the implementation there were two full-time consultants assigned and a project manager from The Vision Web. I had a meeting with the project manager twice a week about the progress. The ERP solution went live on the 2 January, right on schedule. It all went unbelievably fast. Especially since it was a multi-country implementation: France and Belgium. (Olav Claassen, Vice President Information Services Europe, Real Scale Technologies)

Up to late 2001, RealScale Technologies' experiences with the ASP SAP solution had been very positive. Both the offered functionality and the technical performance of the Internet delivery method were good. The only down-side was attributed to RealScale Technologies internally. The adoption of the solution by its users was not yet optimal. The level of commitment would determine the success of the solution in the long run. This shows that netsourcing arrangements never take away completely the need for internal commitment and management.

4.10 Conclusions

The two final case studies illustrate the conditions under which relatively small companies can now gain access to ERP packages relatively cheaply over the net. The cases show the distinctive risks that emerge in netsourcing ERP in relatively small, but growing enterprises, and how these risks can be mitigated. However, the cases also illustrate the importance of working very closely with a vendor, and of how the key lessons about ERP system implementation learned from earlier, even large company ERP system implementations, apply directly into netsourced ERP usage. In this respect we come full circle and can talk of the continuing ERP revolution, and how lessons from the past translate equally well into future implementations. These lessons also include those learned about using ERP vendors. Invariably, from the late 1990s, implementing ERPs has used the external market extensively in some form or another. Since then, invariably, those enterprises that failed to build a strong informed buying capability, and that failed to take internal responsibility for distinctive, limited, but critical business and technical aspects of ERP system implementation, have continued to bolster the, as at 2002, still relatively high ERP system disappointment rate.

An ERP system implementation can lead to a totally new organizational environment, characterized by an increased emphasis on process, strategic vendor alliances, and constant change. In this environment, whatever the mode of delivery, the focus on organizational processes will require some internal technical and process expertise that will allow constant reassessment of those processes and the systems they depend upon. While processes will be standardized, they will not be static. Firms will introduce the process changes mandated by new software releases. Simultaneously, they will identify other process changes to respond to new customer and market demands, and look for software modules or bolt-ons that support requirements for new processes. Unlike the legacy systems of the past, which constrained organizational process change, ERPs, their migration to ERP II, together with new modes of delivery based on the net, will force continuous change, and pro-active management, on already dynamic organizations.

ACKNOWLEDGEMENTS

We gratefully acknowledge the work of Thomas Kern and Ramses Zuiderwijk of Erasmus Universiteit, Rotterdam in researching and preparing this case for publication in Kern et al. (2001, 2002). The cases are presented here in shortened, abstracted form for the limited purpose of illustrating successful ERP implementation over the net, using an external service provider.

REFERENCES

Bennett, C. and Timbrell, G. (2000) Application Service Providers – Will They Succeed? *Information Systems Frontiers*, **2**(2), 195–212.

Davenport, T. H. (2000) *Mission Critical – Realizing the Promise of Enterprise Systems*. Boston, MA: Harvard Business School Press.

Economist (2001) A Survey of Software. Special Report, 14 April.

Goodwin, B. (2001) IT: No Passport to Productivity. *Computer Weekly*, **18** (25 October).

Hagel 111, J. and Brown, J. (2001) Your Next IT Strategy. *Harvard Business Review*, October, 105–113.

Kern, T. and Willcocks, L. (2002) *The Relationship Advantage: Technologies, Sourcing and Management*. Oxford: Oxford University Press.

Kern, T, Lacity, M., and Willcocks, L., (2002) *Netsourcing: Renting Business Applications and Services over Networks*. New York: Prentice Hall.

Kern, T., Lacity, M., Willcocks, L., Zuiderwijk, R., and Teunissen, W. (2001) *ASP Market-Space Report 2001*. Netherlands: CMG.

Kern, T., Willcocks, L., and Lacity, M. (2002) Service Provision and the Net: Risky Application Sourcing? *MISQ Executive*, **1**(2), 113–126.

Lacity, M., Willcocks, L., and Subramanian, A. (1998) Client Server Implementation: New Technology, Lessons from History. *Journal of Strategic Information Systems*, March.

Manchester, P. (2001) Enterprise Application Integration. *Financial Times Special Report: ERP and Beyond*. 19 July, 7.

Markus, M. L. and Tanis, C. (2000) The Enterprise Systems Experience – from Adoption to Success. In *Framing the Domains of IT Research: Glimpsing the Future through the Past*, Zmud, R. W. (ed.), Cincinnati, OH: Pinnaflex Educational Resources.

Moran, N. (2000) ASP Outsourcing. *Financial Times* Special Report: ERP and Beyond, *Financial Times*, London, p. 6.

Meta Group (2001) *Worldwide IT Trends and Benchmark*. New York.

Norris, G., Hurley, J., Hartley, K., Dunleavy, J., and Balls, J. (2000) *E-Business and ERP: Transforming the Enterprise*. Chichester: Wiley.

O'Leary, D. (2000) *Enterprise Resource Planning Systems: Systems, Life Cycle, Electronic Commerce and Risk*. Cambridge: Cambridge University Press.

Pozzebon, M. (2001) Demystifying the Rhetorical Closure of ERP Packages. Proceedings of the Twenty Second International Conference on Information Systems, 17–19 December. New Orleans, USA.

Ross, J. (1996a) Dow Corning Corporation (A): Business Processes and Information Technology. Center for Information Systems Research, Sloan School of Management, MIT.

Ross, J. (1996b) Dow Corning Corporation (B): Reengineering Global Processes. Center for Information Systems Research, Sloan School of Management, MIT.

Ross, J. (1999) Dow Corning (C): Transforming the Organization. Center for Information Systems Research, Sloan School of Management, MIT.

Sauer, C. and Willcocks, L. (2001) *Building the E-Business Infrastructure.* London: Business Intelligence.

Steensrup, K. (2001) Living with Packaged Application Support. Session 21f, Gartner Group Australian Symposium, Brisbane, Australia (30 October–2 November).

Weill. P. and Vitale, M. (2001) *Place to Space: Migrating to New Business Models.* Boston, MA: Harvard Business Press.

Willcocks, L. and Graeser, V. (2001) *Delivering IT and E-Business Value.* London: Heinemann.

Part II

From Risks to Critical Success Factors

5 Enterprise System Implementation Risks and Controls

Severin V. Grabski, Stewart A. Leech, and Bai Lu

5.1 Introduction

Some organizations report success and significant process gains resulting from enterprise system[1] implementation, while many others encounter significant losses. It is readily apparent that the implementation of packaged enterprise system software, and associated requisite changes in business processes, has proved not to be an easy task. As many organizations have discovered, the implementation of enterprise systems can become a recipe for disaster unless the process is carefully handled. The following headlines provide a glimpse of the troubles encountered by organizations, 'SAP: Whirlpool's Rush to Go Live Led to Shipping Snafus' (Collett, 1999), 'Delays, Bugs, and Cost Overruns Plague PeopleSoft's Services' (Olsen, 1999) 'ERP Project Leads to Court Fight' (Stedman, 1999) and 'PeopleSoft Problems Persist, Cleveland State Looks for a New Project Manager' (Olsen, 2000). In 1995, US firms incurred an estimated $59 million in cost overruns on information system projects and another $81 million on cancelled software projects (Johnson, 1995). Furthermore, it is estimated that approximately 90% of enterprise system implementations are late or over budget (Martin, 1998).

Enterprise system (ES) implementations are different from 'traditional' system analysis and design projects (Davenport, 2000). Among the significant differences are the scale, complexity, organizational impact, and the costs of ES projects and subsequent business impact if the project does not succeed. An ES implementation impacts the entire organization, whereas a traditional project impacts often only a limited area of the organization. Furthermore, ES projects are often associated with the reengineering of business practices. This results from the desire to adopt the 'best practices' inherent

[1] We use the term enterprise systems rather than enterprise resource planning (ERP) systems. We agree with Davenport (2000, p. 2) that the term ERP reflects the manufacturing origins of these systems that has since been transcended and is no longer appropriate.

in the chosen software solution rather than changing the software to match current business practices (although a number of organizations simply implemented the basic ES package to solve year 2000 issues and did not undertake significant process reengineering). Historically, traditional analysis and design projects had no (or minimal) reengineering and the software was written to match current practices. Additionally, an ES implementation nearly always requires personnel to learn new programming languages and may also result in a shift in the organization's computing paradigm, from mainframe-based to network centric. Finally, the cost of ES projects is significantly higher than traditional projects, and failure can result in the demise of the organization (for example, the FoxMeyer Drugs bankruptcy – Scott, 1999).

The aim of this study is to identify the risks and controls used in ES implementations, with the objective of identifying the ways organizations can minimize the business risks involved. This is applicable in the current environment as many firms now seek to either reengineer their business processes to obtain the gains that should have resulted from the implementation of their ES. Additionally, many organizations will be faced with the need to upgrade their ES as the software vendors release new versions (and possibly eliminate support for the version used by the organization). Other organizations will need to implement a new ES because of acquisition or merger, or because their previous ES vendor has ceased to exist.

We argue that by controlling and minimizing the major business risks, organizations can increase the likelihood of successful ES implementation. This chapter was motivated by the significance, for both the research and practice communities, of understanding the risks and controls critical for the successful implementation of ES. The chapter proceeds as follows. In the next section, we identify the business risks associated with system implementations from both the ES literature and the traditional system analysis and design literature. Section 5.3 examines each of the risks in more detail and identifies the controls that can be utilized to minimize each risk. A model of risks and associated controls is developed in this section. In Section 5.4, we present a case study of the controls put in place by a specific organization for the successful implementation of an ES. Section 5.5 presents the results from the case study. A number of controls not previously identified or emphasized in the literature were discovered in the case study. Section 5.6 reports the results of a survey of the Chief Information Officers of major organizations in Australia. Finally, concluding thoughts and ideas for future research are presented.

5.2 ES Implementation Business Risks

Much research has been conducted on the identification of risks in information system projects (Jiang, Klein, and Balloun, 1996; Zmud, 1980). McFarlan (1981) presented a portfolio approach for managing software development risk. Other research has looked at risk from a technological perspective (Anderson and Narasumhan, 1979) or from a software development perspective (Barki, Rivard, and Talbot, 1993). Jiang and Klein (1999) examined risk as it related to a multidimensional concept of information success that included satisfaction with the development process, satisfaction with system use, satisfaction with system quality, and the impact of the information system on the organization. They found that the risks associated with the overall success of a system project were application complexity, lack of user experience, and lack of role clarity of role definitions of individuals on the project. Lack of user support was significant for the organizational impact, while technological newness affected system quality satisfaction. System development was affected by the team's general expertise, application complexity, and user experience, and system use was affected by role clarity and user experience. Reel (1999) discussed the importance of understanding user needs and mitigating user resistance, proper project management including project scope definition, top management sponsorship, and having a project team that possesses the appropriate skills. While these studies are important, none explicitly examined the ES environment. The risks and controls in these studies were more technological, focusing on software development rather than the implementation of packaged solutions with minimal modifications and reengineering of business processes.

In order to maximize the likelihood of success, task risks must be minimized (Barki, Rivard, and Talbot, 1993; Jiang and Klein, 1999). Audit firms minimize the risk of audit failure through the identification of inherent, control and detection risks followed by the establishment of an acceptable, specified level of overall audit risk that is a function of those other risks (Arens and Loebbecke, 1997). The same rationale holds for the implementation of an ES. In order to maximize the probability of success, the risks must be identified and appropriate controls put in place to minimize those risks.

The lack of alignment between the organization strategy, structure, and processes and the chosen ES application is one risk that is repeatedly identified in the literature (see, for example, Davenport, 1998, 2000). Both the business process reengineering literature (Hammer, 1990; Hammer and Champy, 1993) and the ES literature suggests that an ES alone cannot improve the company performance unless an organization restructures its operational processes (Bingi, Sharma, and Godla, 1999; Davenport, 1998; Davenport, 2000). Further, the ES implementation project must be a business initiative. This requires the organization to gain strategic clarity (that is, know the business, how it delivers value, etc.) and a constancy of purpose. Finally, an outcomes orientation is required to achieve these goals.

Within an ES project, the loss of control over the project is another major risk. Loss of control can arise in at least two ways: the lack of control over the project team, and the lack of control over employees once the system is operational. Risks associated with controlling major projects existed prior to the development of ES software, and much has been written on project escalation (see, for example, Brockner, 1992; Kanodia, Bushman, and Dickhaut, 1989; Keil, 1995; Sharp and Salter, 1997; Staw, 1976, 1981; Staw and Ross, 1987). The first risk is the lack of control over the project team. This lack of control results from the decentralization of decision making and subsequent ineffective ratification of decisions. Within the setting of an ES project, in order to ensure the collocation of knowledge with decision rights, it is common for an organization to form an ES implementation project team that involves individuals who have some relevant specific knowledge associated with the implementation of an ES (such as information technology knowledge or change management skills). Decision rights are then assigned to the team. However, where the project team has complete control over the ratification of its own decisions creates a potential business risk that the project team would act in their own interests rather than act in the best interests of the organization. The second risk is that an operational ES most always results in the devolution of responsibility and empowerment of lower-level employees. A lack of adequate controls over this increased responsibility, either within the ES itself or in the processes followed by the organization, is a potential business risk.

Another major risk is project complexity (see, for example, Barki, Rivard, and Talbot, 1993). An ES implementation involves relatively large expenditures for the acquisition of the hardware, software, implementation costs, consulting fees, and training costs (Davenport, 2000; McKie, 1998), and can last for an extended period of time. Also, an ES implementation project has a wider scope compared to most other information system implementations, and may cause a significant number of changes within an organization (Davenport, 2000). The scope and the complexity of the project are a source of significant business risk.

Lack of in-house skills is another source of risk in the implementation of ES. Lack of project team expertise has often been associated with software development risk (Anderson and Narasumhan, 1979; Barki, Rivard, and Talbot, 1993; Holland and Light, 1999; Jiang and Klein, 1999). An ES implementation project requires a wide range of skills (that is, change management, risk management, business process reengineering (BPR)) in addition to technical implementation knowledge (Davenport, 2000; Glover, Prawitt, and Romney, 1999). Organizations often lack change management skills and BPR skills required for an ES implementation. Further, an ES is often based on programming languages and concepts that are most likely new to existing IT staff (Kay, 1999). Thus, a lack of in-house skills required for the ES implementation is a potential business risk.

When an organization moves to a complex ES environment, changes in staff relationships will most likely occur. Employees may need to create new working relationships, share information among departments, acquire new skills and assume additional responsibilities (Appleton, 1999). These changes can lead to resistance, confusion, and fear among users of the new system (Glover, Prawitt, and Romney, 1999). Unwilling users increase implementation risk (Anderson and Narasumhan, 1979). Staff turnover and other types of user resistance create additional business risks associated with an ES implementation.

In summary, based on a review of the ES literature, there are five major business risks associated with the implementation of ES: the lack of alignment of the new information system and business processes; the possible loss of control due to decentralization of decision making; risks associated with project complexity; the potential lack of in-house skills; and user resistance.

We next examine each of the above risks in more detail and specify controls that can be utilized by organizations to minimize that risk.

5.3 ES Implementation Controls

Lack of Alignment of the ES and Business Processes

In order to minimize the risk associated with a lack of alignment of the ES and business processes, organizations should engage in business process reengineering, develop detailed requirements specifications, and conduct system testing prior to the ES implementation. Subsequent to implementation, they need to closely monitor system performance to identify any alignment problems that may have occurred and were not apparent until post-implementation.

First, the rethinking and radical redesign of business processes (Hammer and Champy, 1993) enables an organization's operational processes to be aligned with the ES and allows an organization to better obtain the full benefits offered by the ES. It also results in the ES implementation originating as a business initiative. Further, strategic clarity as well as constancy of purpose is attained (Davenport, 2000). Second, a detailed requirements specification for ES software selection increases the probability that the ES will meet the organization's system requirements and support the required operational processes. While the detailed planning is occurring, baseline metrics on current processes can be obtained that are requisite for the evaluation of the project's outcomes (Davenport, 2000). Third, system testing prior to system implementation and monitoring of the system after implementation are seen as critical to ensure that the ES operates smoothly and is able to provide adequate support for the organization's operational processes (Callaway, 1997; Davenport, 2000).

Loss of Control Due to Decentralization of Decision Making

An organization can minimize the loss of control associated with decentralization of decision making through the formulation of a steering committee, appointment of a project sponsor, and involvement of internal audit.

A steering committee enables senior management to directly monitor the project team. The steering committee will monitor the decisions made by the project team and will retain ratification and approval rights on all significant decisions. This ensures that adequate controls over the project team's decision-making processes exist (Davenport, 2000; Whitten and Bentley, 1998). In addition to the formulation of a steering committee, a project sponsor is assigned direct responsibility for the ES's progress and often is responsible to secure funding (especially when more funds are needed than originally budgeted). The project sponsor is directly accountable for the project (Davenport, 2000; Whitten and Bentley, 1998). Internal audit's involvement in the ES implementation also helps ensure the adequacy of controls and that all parties are performing the appropriate tasks in a timely manner. While often overlooked in the ES implementation literature, internal audit has extensive knowledge about an organization's control environment, business operational process and weaknesses existing in the current internal control system. This knowledge may not be available to the project team, managers, and external auditors (Glover, Prawitt, and Romney, 1999). Glover, Prawitt, and Romney argued that internal audit should be involved in an ES implementation early rather than later. They suggested that, at a minimum, auditors should stay informed throughout the system implementation process. This would enable internal audit to be aware of the changes due to the new system and to adjust the audit program accordingly.

Project Complexity

The minimization of the risks associated with project complexity largely depends upon the formulation of a steering committee, senior managers' support, appointment of a project sponsor, the development of a detailed implementation plan, project management, a project team with adequate skills, and involvement by both consultants and internal audit.

First, in ES implementation projects, senior managers are often involved via appointment to a steering committee (Cameron and Meyer, 1998; Clemons, 1998; Davenport, 2000). Senior management's direct involvement in the system implementation project often increases the projects perceived importance within the organization (Raghunathan and Raghunathan, 1998) which encourages employees, system users, and the IT department to be actively involved in, and provide support for, the ES implementation. Senior management commitment is needed because of the organizational changes that result from the implementation of ES (Bingi, Sharma, and Godla, 1999; Davenport, 2000; Holland and Light, 1999). Second, by appointing an executive-level individual with extensive knowledge of the organization's operational processes to be the project sponsor, senior management is better able to monitor the ES implementation. The project sponsor has direct responsibility and is held accountable for the project outcome (see, for example, Cameron and Meyer, 1998; Clemons, 1998; Davenport, 2000). The appointment of the project sponsor ensures that adequate accountability exists, thereby reducing project risk. Third, in order to retain control over the project, many organizations develop a detailed system implementation plan that provides direction for the project team by setting out the project goals and targets (Davenport, 2000; Deutsch, 1998). The detailed system implementation plan (which includes performance metrics for subsequent evaluation) assists in the identification of potential risks resulting from identified delays in a timely manner (Bingi, Sharma, and Godla, 1999; Deutsch, 1998; Holland and Light, 1999). Fourth, the detailed requirements specification forces the organization to identify, up front, the project specifics and understand the level of complexity associated with the project. Fifth, strong project management is crucial to the success of any large endeavor, and this is especially so in ES implementation projects that can span several years and cost millions of dollars (Davenport, 2000).

Sixth, identification of the skills and knowledge of the project team is important, as is the employment of consultants to provide expertise in areas where team members lack knowledge (Barki, Rivard, and Talbot, 1993; Cameron and Meyer, 1998; Clemons, 1998). Seventh, it is critical for the project team and consultant to be assigned to the project on a full-time basis (and have a separate office for the ES project), thereby ensuring they can focus completely on the project (Deutsch, 1998). Finally, internal audit's involvement is also vital for identification of the potential project risks, managing the risks, and ensuring the effectiveness of the internal controls. All of these types of controls are important in minimizing the risks associated with project complexity.

Lack of In-house Skills

In order to mitigate the business risks associated with the lack of in-house skills in an ES implementation, external consultants are often required, a close working relationship between the consultants and project team is also needed, and adequate training is critical.

Consultants are able to use their previous ES implementation experiences; consequently, they can act as knowledge providers who lower the knowledge deficiencies existing within organizations (Arens and Loebbecke, 1997). An organization, however, cannot completely rely on consultants to implement an ES system, as consultants have limited specific knowledge of the organization's operations. Thus, a close working relationship between consultants and the organization's project team can lead to a valuable knowledge transfers in both directions (Bowen, 1998). Additionally, training that is available through the consultants, the vendor, or through some third party provides a valuable resource to develop skills that are lacking in house. These controls are seen as important in minimizing the risks associated with a potential lack of in-house skills. Often a new group, 'super users', is created during the ES implementation. These individuals acquire detailed knowledge of the new business process and also technical system knowledge through their implementation activities and the training they receive (Davenport, 2000).

User Resistance

Business risks associated with possible user resistance to ES implementation can be reduced through managerial skills, user involvement, training, top management support, and communication.

User resistance has been associated with almost any type of system change, and even more so for ES projects that are combined with BPR (since the users are worried that their job may at worst be eliminated, or at best be changed from their 'usual' way of doing things). Workers who are reengineered out of a position and are subsequently redeployed within the company may enter a grieving process resulting in low productivity (Arnold, Hunton, and Sutton, 2000). Consequently, organizations often implement some risk management strategies to minimize user resistance. Appleton (1999) argued that managers' soft skills (such as communication and team-building skills) are among the most important skills required for a successful ES implementation. User involvement in the ES project was also identified as important to gain the users 'buy in' for the project (see, for example, Cameron and Meyer, 1998; Clemons, 1998). Involving users in the project enables the project team to be aware of user requirements and address user concerns (Best, 1997). In addition to involvement, user training enables users to acquire the requisite skills to utilize the ES. To ensure that users are aware of the impact the ES project will have on their responsibilities, many organizations develop formal communication plans and issue regular reports (Cameron and Meyer, 1998). Finally, when users perceive that top management really supports a particular project (and is willing to provide adequate resources), they will have a higher level of acceptance for that project.

In summary, the ERP implementation risks and their associated controls are shown in Table 5.1. The marked columns indicate when the control is applicable for minimizing the implementation risk. In the next section we present a case study of how a specific organization implemented various controls to minimize the business risks associated with the implementation of an ERP package.

5.4 The Case Study

Data Collection

The case study used in this research was the implementation project of a university's New Financial Management System (NFMS). The project provided a rich setting for the investigation of an enterprise system implementation and its interaction with various organizational factors. In order to gain an in-depth understanding of the NFMS project, multiple data collection methods, including interviews, survey, and archival data were used in this study.

Table 5.1 ERP implementation risks and associated controls

Risks controls	Lack of alignment between IS and business processes	Loss of control due to decentralized decision-making	Project complexity and mismanagement of complex projects	Lack of in-house skills	Users' resistance
Business process reengineering	×				
Consultants' involvement			×	×	
The close working relationship between the project team and consultants				×	
Senior managements' support		×	×		×
Project sponsor		×	×		
Steering committee		×	×		
The project team			×		
Detailed requirements specification	×		×		
Detailed implementation plan			×		
Frequent communication with the system users					×
Managerial 'People' skills					×
User involvement					×
Training				×	×
Internal audit involvement		×	×		
System testing prior to the system implementation	×				
Close monitoring the system after the system implementation	×				
Project management		×	×		

Note: The controls that are italicized are not associated with traditional systems implementations.

Interview sessions were conducted with the director of financial services and the business analyst; the system analyst and the information technology services (ITS) manager; the consultant; and the internal auditor. The director of financial services, the business analyst, the system analyst, and ITS manager were senior employees of the university. The consultant was from a Big-Five consulting firm. The university outsourced its internal audit services to a different Big-Five accounting firm, and the senior internal auditor (a director in that Big-Five firm who was involved in auditing the NFMS) participated in the study.

Prior to each interview a script was developed, which allowed a semi-structured data gathering technique with sufficient flexibility to pursue interesting information when disclosed by the participant. Additionally, an information sheet and an interview agenda (the interview script in bullet point) were sent to each participant prior to each interview. This allowed each participant to focus in advance on the issues and activities performed during the NFMS project. In order to attempt to control for possible collaboration among interviewees internal to the university, the first two interview sessions were scheduled within one day of each other and the interview agenda sent no more than twenty-four hours prior to each interview.

The interview scripts for sessions with the director of financial services and the business analyst, the system analyst and the ITS manager, and the consultant were based on archival information provided by the university and the risks and controls identified in the literature. The interview script for the session with the internal auditor was based on both the survey response from the internal auditor (see below) and archival sources. Some questions were asked to multiple interview groups to determine whether differences in opinions and perceptions existed among those groups.

The interview with the director of financial services and the business analyst focused on project initiation, sponsorship, management, what they would do differently (the same), project issues, and perceived success. The interview with the ITS manager and the system analyst focused on technical IT issues associated with the NFMS installation, communication, what they would do differently (the same), project issues, and perceived success. The interview with the consultant focused on the consultant's role, the critical success factors and controls within the project, and the perceived project risks. The interview with the internal auditor focused on the involvement of internal audit in the project, the perceived project risks, what the internal auditor would have done differently (the same), and perception of the project's success. All participants agreed at the start of the interview to allow the session to be (audio) taped.

The participants were also informed that they could stop the tape at any point during the interview (none did).

In addition to a scripted interview form, a questionnaire was developed and sent to the internal auditor 14 days before the interview and returned within seven days. The objective of the questionnaire was to obtain more detailed information than would normally be possible during the interview. As compared to the interviews with the other groups, it was decided that a questionnaire was required before the interview of the internal auditor due to the technical nature of the questions on risk assessment and audit. The questionnaire was broadly based on a questionnaire developed by the Institute of Internal Auditors, United States, which appeared on their web site (http://www.theiia.org/survevr.htm, 16 February 1999). The questionnaire was adapted to the university environment and pilot tested with two experts in auditing – an academic and a partner in a major accounting firm. The questionnaire was revised as a result of the pilot test before being sent to the internal auditor.

The university web site was accessed for archival data dealing with the NFMS project. The director of financial services also provided copies of all internal reports that were given to the steering committee and copies of other documents associated with the project. This information served to provide additional background and richness, clarified concepts to be introduced during the interviews, and provided additional quantitative data that helped improve the understanding of the processes employed in this setting.

Case Setting

In the late 1980s, the university had installed a packaged financial software – Old Financial Management System (OFMS). The software was heavily customized to meet the specific needs of the university. During the 1990s, it became apparent that the OFMS was no longer able to meet the university's requirements. Feedback from users indicated that the OFMS did not provide timely and accurate information; in addition, it was user unfriendly (character rather than windows-based) and provided low-quality reports. Moreover, the OFMS was unable to be integrated with any other information systems operated within the university, and it was unable to handle the increasingly complex and evolving university organizational structure.

Under funding pressure, the university's administration searched for alternatives to improve operational efficiency. Additionally, the information technology services (ITS) division undertook a preliminary year 2000 (Y2K) evaluation of the university's administrative systems. As a result of this review process, the OFMS was identified as one of the most critical Y2K problem areas. Due to the potential technical difficulties and the significant costs associated with upgrading the OFMS software, the university decided to replace it.

The senior management of the university decided that the NFMS project presented an opportunity to review and significantly reengineer the financial operational processes to improve efficiency. The directive associated with the financial operational process review and redesign was to:

- further decentralize and devolve the financial management responsibility and financial transaction processing functions to each individual school and division level;
- restructure of all central financial management functions under one head and redefine the role of the financial services division towards client services and decision support;
- review workflows, redesign and eliminate non-value added activities associated with the processing of financial information, and simplify the financial operational processes;
- implement the NFMS, with the resultant elimination of paper flows, provision of client/server and graphical user interface technology, and improvement of responsiveness to the end users.

In the early stages of the project, the requirements documentation was approved by the NFMS steering committee, which consisted of the university's senior management. The university tendered for the NFMS and six responses were received. The tenders were evaluated against detailed system selection criteria that had been developed by the project team and approved by the steering committee.

After the evaluation of the tenders, a short list of two products remained. The subsequent evaluation of both products emphasized product suitability, implementation cost and strategy, the product's ability to change the university's financial operational processes; its capability for paperless systems and electronic transaction generation; PC and Macintosh support; and the potential consultancy arrangement.

The contract was awarded to a 'Big Five' consulting firm who proposed the installation of the current version of an ES software package. In addition, the consulting firm also agreed to provide implementation support, on-going consultancy support, and to organize the training sessions for the project team and system users.

After the system selection decision was made, the director of financial services selected the NFMS project team members, consisting of:

Director of financial services	Information technology services
(Project team leader)[2]	Manager
Business analyst	Accountant
Senior analyst programmer	Systems analyst
	Consultant (project manager)

The vice-president for finance and operations was appointed as the project sponsor. Along with primary oversight responsibility, the project sponsor was also responsible for the project funding.

The first phase of the project was an analysis of current business practices (an 'as is' analysis). This was followed by a 'could be' analysis in which all the operational constraints were removed and business process reengineering was conducted. The last analysis conducted was a 'to be' analysis in which the 'could be' analysis was combined with the capability of the unmodified ES software to arrive at what the operational processes would be when modified.

It was planned to take two to three years to complete the project. Overall, the project team spent 12 months to complete the business process review and redesign, and a further six months to complete the implementation of the ES.

At the NFMS project planning stage, the internal auditor was identified as a NFMS project stakeholder and was invited to the initial project meetings. However, the internal auditor was not involved in an oversight role in the NFMS project until some seven months later when the project started to encounter delays in the planned schedule. From that time on until the completion of the project, the internal auditor provided several types of assurances about the NFMS project. First, monthly quality assurance reports relating to the project were made to the steering committee. Second, the internal auditor evaluated the adequacy of the internal control procedures in the NFMS. Prior to the system implementation, the project team members 'walked through' the NFMS with the internal auditor. Finally, the internal auditor continually performed risk assessment throughout the life of the NFMS and after the system was implemented. After the internal auditor was involved, actionable items that were the responsibility of the consulting firm dropped from items over three months old to only current month actionable items.

Training sessions were held so that the NFMS operatives and users could acquire the skills and knowledge required for operating the new NFMS. Given the different needs of the user groups, the training sessions were separated into general accounting and financial management training, NFMS 'hands

[2] For expositional purposes the director of financial services will be referred to as the project team leader. He was also the co-project manager on the NFMS with a representative from the consulting firm. The consulting firm representative will be referred to as the project manager.

on' training, NFMS reporting training, technical training, and system administration training.

The NFMS was implemented three months later than initially scheduled. The project manager and senior managers of the university were generally satisfied with the NFMS project. As a consequence of the NFMS implementation, the financial transaction processing responsibilities were devolved to each individual school and division level. The university's financial policies are now designed to eliminate the paper-intensive and labor-intensive operational processes of the past, and are sensitive to the needs of the end users.

5.5 Results

The controls used by the university are shown in Table 5.2, and are compared to the controls identified in the literature. These controls (discussed below) were identified by the interviewees as critical to ensure that the project risks were minimized. Consistent with the ES implementation literature, the interviewees suggested that BPR; the project team members' skills and knowledge; the consultant's involvement; post-implementation review; internal auditor's involvement; formulation of the steering committee; managerial 'people' skills; and training sessions were vital to minimize risks perceived to be associated with the NFMS project. All the critical controls identified in the literature and the associated minimization of business risks were relevant for the NFMS project success. The interviewees, however, also identified, change management skills, in-depth project planning and ownership by users as vital for minimizing the risk of lack of success with the NFMS project.

Also consistent with the ES literature, the need for project management skills became salient in this case setting and was vividly recalled by all participants. Midway through the project, the consulting organization reduced the project manager's hours on the job to approximately 8–12 hours per week. At that point in time, the project team leader tried to compensate and take over the project management role. Unfortunately, the project team leader also had day-to-day university activities that precluded full-time project management activities. Additionally, the project team leader did not have the same depth of project management skills as the consultant. It was at this time that the project experienced slippage and the internal auditor was called in. Once the consulting firm assigned a new full-time project manager, no additional slippage occurred and the project was completed within the revised timetable.

The consultant believed that project ownership by the users was vital in order for the project to gain support and acceptance. Users were pro-actively

Table 5.2 Comparison of the critical enterprise system implementation controls

Critical ERP system implementation controls identified in the literature	Critical NFMS project controls identified by interviewees
Business process reengineering	*Business processes reengineering*
Consultants' involvement	*Consultants' involvement*
Close working relationship between the project team and consultant	*Close working relationship between the project team and consultant*
Senior management support	Senior management support
Project sponsor	Project sponsor
Steering committee	Steering committee
Project team:	Project team:
members' skills and knowledge	members' skills and knowledge
dedication of team members	dedication of team members
Detailed requirements specification	Detailed requirements evaluation
Detailed implementation plan	Detailed implementation plan
Frequent communication	Frequent communication
Managerial 'people' skills	Managerial 'people' skills
User involvement	User involvement
Training	Training
Internal auditor's involvement	Internal auditor's involvement
System testing prior to the system implementation	System testing prior to the system implementation
Post-implementation review	Post-implementation review
Project management	Project management skills
	User project ownership
	Change management and transition management
	In-depth up front project planning

Note: The italicized items are not associated with traditional systems implementations, and the bold items were identified in the case study.

involved in the project by joining various working parties and participating in training sessions and surveys. Prior to system implementation, training sessions were held for all the NFMS operatives to ensure they acquired the needed skills. Detailed user procedure manuals were distributed and a help desk was established to provide direct assistance for the NFMS operatives. Two-way communication channels were also established between the system users and project team, thereby ensuring the project team was aware of user requirements. By having a sense of project ownership, it was easier to gain user acceptance for the changes associated with NFMS project. Although the literature recognizes the importance of managerial 'people' skills, user involvement, training, frequent communication, and user acceptance for the success of the ES implementation, the literature failed to identify the importance of *ownership* by the users, especially as it was expressed by all the parties involved in this implementation.

Similarly, the literature acknowledges the importance of managerial skills; however, the various types of project management skills that are critical to minimize risks in an ES implementation were not identified. During interview sessions, the interviewees identified a range of project management skills, in particular, change management skills, as vital to reducing the business risks associated with the NFMS project.

Due to the scope of the NFMS project and risks involved, in-depth project planning was identified by the consultant as critical for the success of NFMS. If more detailed in-depth project planning had been conducted by the NFMS project team, the consultant argued that many NFMS project risks would have been identified at an earlier stage of the project. The reviewed literature failed to identify the importance of in-depth up-front project planning.

The consultants and project team worked closely and held regularly weekly meetings where the comprehensive meeting minutes and action items were recorded in a Lotus Notes database. Those meetings were identified as 'crucial to success' of the project by all the parties involved in the project; it enabled the team member to identify problems in a timely manner. In addition to the Lotus Notes databases, a NFMS newsletter was issued on a monthly basis to keep the university staff reasonably informed about the project progress. This latter item was one way that user resistance was mitigated along with building a sense of ownership. Due to a bad experience the university previously had with system customization, the NFMS project sponsor decided to implement the ES with a minimum amount of customization of the software. Consequently, the university's operational processes were redesigned to fit with the ES. The university realized that BPR was essential to obtain the full benefits from the ES. Similarly, the consultant emphasized the critical role of business process redesign and change management.

Several limitations apply to the case study. First, a single site was utilized. Consequently, the limitations associated with the case study approach are applicable (Yin, 1994). Second, the investigation of the NFMS project was carried out shortly after the NFMS implementation. The research findings therefore reflect the interviewees' retrospective perceptions of the NFMS implementation. Interviewees may have had different views as to what were the critical controls and risks during the NFMS implementation project. Third, the study was conducted within three months of the implementation of the NFMS and the results are limited to the perceived short-run success. Fourth, the NFMS was limited to the implementation of one module of, albeit, a major ES software package.

5.6 Survey

In addition to the case study, the Chief Information Officers of major organizations in Australia were contacted as to the risks, controls, and success of the implementation of ES in their organizations. A total of 211 organizations were contacted and 32 usable responses from those who had implemented ES systems were obtained. The CIOs generally believed that the business processes were aligned with their ES (average response was 4.56 on a seven-point scale anchored at 1 = very poor fit and 7 = very good fit). Further, they believed the that ES implementation had a positive impact on their organization (average of 4.7, on a scale anchored at 1 = very negative and 7 = very positive). They expected that users were unsure that the system was a success (average of 4.0, on a scale anchored at 1 = very unsuccessful and 7 = very successful). They, however, were much more enthusiastic about the ES project themselves and considered it a success (average of 5.1, on a scale anchored on the same scale as user belief of success).

When asked whether the control factors identified in the case study were present, they responded that all were present (a minimum average score of 4.5 for business process reengineering, on a scale of 1 = no presence and 7 = significant presence; and a maximum average score of 5.8 for consultant's involvement). It seems reasonable to conclude that the factors associated with mitigating the risk of an enterprise system implementation were present and that this lead to the reported perceived success of the ES by the CIOs. The survey has the usual limitations associated with survey research, and is limited by the relatively low response rate and to the perceptions of the respondents.

5.7 Concluding Thoughts

The purpose of this study was to identify the risks and controls related to the implementation of ES in organizations. The extant literature on enterprise systems and the literature on systems development risks were used to investigate the risks and controls related to the successful implementation of an enterprise system in a case study of an organization. The case study provided support for risks and controls identified in the literature. It also allowed for the identification of several new controls. An expanded model of risks and controls appears as Table 5.3. The newly identified controls have an effect on

Table 5.3 Expanded enterprise system implementation risks and associated controls

Risks controls	Lack of alignment between IS and business processes	Loss of control due to decentralized decision-making	Project complexity and management of complex projects	Lack of in-house skills	User resistance
Business process reengineering	×				
Consultants' involvement			×	×	
The close working relationship between the project team and consultants				×	
Senior managements' support		×	×		×
Project sponsor		×	×		
Steering committee		×	×		
The project team: members' skills and knowledge dedication of team members			×		
Detailed requirements specification	×		×		
Detailed implementation plan			×		
Frequent communication with the system users					×
Managerial 'People' skills					×
User involvement					×
Training				×	×
Internal audit's involvement		×	×		
System testing prior to the system implementation	×				
Close monitoring the system after the system implementation	×				
Project management skills		×	×		
Change management and transition management		×	×		×
User project ownership					×
In-depth up-front project planning	×	×	×	×	×

Note: The controls that are italicised are not associated with traditional systems implementations. The embolden items were identified in the case study.

all risk areas. CIOs whose organizations had recently implemented an ES also reported a presence of the control factors and they also perceived that the ES implementation was a success. It is interesting to note that the two lowest-rated control factors were business process reengineering (4.5) and development of user project ownership (4.6). It makes sense that the users are perceived to be neutral about the ES implementation success; they never took ownership and they see that the organization is still doing things the 'same old way.'

An ES implementation project is different from other system development projects. The prior literature had identified significant risk factors that included technological change, organizational change, and project complexity. These factors are the hallmarks of most (if not all) ES implementations. Consequently, it is important to understand how these risk factors can be mitigated. In this research, controls required to minimize five types of risks in an ES implementation were identified. The results of this research provide support for the proposition that the success of an ES implementation is dependent, in the first instance, on identifying the major business risks and the controls that need to be put in place to minimize those risks.

Several avenues for future research result from this study. While the research findings provided additional evidence on risks and associated controls in an ES implementation, further case studies would be valuable to gain an understanding as to whether these risks and controls exist across organizational settings. Second, future research can examine the correlation among the risks and controls identified in this study through a survey of organizations that implemented ERP systems. Third, longitudinal studies of ES implementations, and subsequent use in organizations, would provide evidence of the persistence of risks and controls. Fourth, further work is needed on the contribution of business risks and associate controls to the success of implementing ES. Finally, further research can examine the timing and extent of internal audit involvement in enterprise system implementations.

ACKNOWLEDGEMENT

An earlier version of this chapter appeared previously in the *International Journal of Digital Accounting Research.*

REFERENCES

Arnold, V., Hunton, J. E., and Sutton, S. G. (2000) On the Death and Dying of Originality in the Workplace: A Critical View of Enterprise Resource Planning Systems' Impact on Workers and the Work Environment. Working Paper, University of South Florida.

Anderson J. and Narasumhan, R. (1979) Assessing Implementation Risk: A Technological Approach. *Management Science*, **25**(6), 512–521.

Appleton, E. (1999) How to Survive ERP. *Datamation*, March.

Arens, A. A. and Loebbecke, J. K. (1997) *Auditing: An Integrated Approach.* Upper Saddle River, NJ: Prentice-Hall.

Attewell, P. (1992) Technology Diffusion and Organisational Learning: The Case of Business Computing. *Organisation Science*, **1**(2), 1–19.

Barki, H., Rivard, S., and Talbot, J. (1993) Toward an Assessment of Software Development Risk. *Journal of Management Information Systems*, **10**(2), 203–225.

Best, C. (1997) Integrated System Builds on Human Foundation, *Computing Canada*, **23**, December.

Bingi, P., Sharma, M., and Godla, J. (1999) Critical Issues Affecting an ERP Implementation. *Information Systems Management*, **16**(3), 7–14.

Bowen, T. (1998) Committing to Consultants: Outside Help Requires Internal Commitment and Management Skills. *Info World.*

Brockner, J. (1992) The Escalation of Commitment towards a Failing Course of Action: Towards Theoretical Progress. *Academy of Management Review*, **17**(1), 39–61.

Callaway, E. (1997) ERP: Test for Success. *PC Week* online, 22 December.

Cameron, D. and Meyer, L. (1998) Rapid ERP Implementation – A Contradiction. *Management Accounting (USA)*, **80**(6), 56–60.

Clemons, C. (1998) Successful Implementation of an Enterprise System: A Case Study. *Proceedings of the AIS Conference Americans.* Baltmore Maryland: Association for Information Systems, pp. 109–110.

Collett, S. (1999) SAP: Whirlpool's Rush to Go Live Led to Shipping Snafus. *Computerworld Online News.* http://www.computerworld.com/cwi/story/0,1199,NAV47_STO29365,00.html

Davenport, T. (1998) Putting the Enterprise into the Enterprise System. *Harvard Business Review*, **76**(4), 121–133.

Davenport, T. (2000) *Mission Critical: Realizing the Promise of Enterprise Systems.* Boston, MA: Harvard Business School Press.

Deutsch, C. H. (1998) Some Tips on Avoiding the Pain. *The New York Times*, 8 November.

Glover, S. M., Prawitt, D., and Romney, M. (1999) Implementing ERP. *Internal Auditor*, 40–47.

Grabski, S., Leech, S. and Lu, B. (2001) *Risks and Controls in the Implementation of ERP Systems*, **1**(1), 1–29.

Hammer, M. (1990) Reengineering Work: Don't Automate. Obliterate, *Harvard Business Review*, **68**(4), 104–112.

Hammer, M. and Champy, J. (1993) *Reengineering the Corporation: A Manifesto for Business Revolution.* New York: Harper Business.

Holland, C. and Light, B. (1999) Critical Success Factors Model for ERP Implementation. *IEEE Software*, **16**(3), 30–36.

Jiang, J. J. and Klein, G. (1999) Risks to Different Aspects of System Success. *Information and Management*, **36**, 263–272.

Jiang, J., Klein, G. and Balloun, J. (1996) Ranking of System Implementation Success Factors. *Project Management Journal*, **27**(4), 50–55.

Johnson, J. (1995) The Dollar Drain of IT Project Failures. *Application Development Trends*, **2**(1), 41–47.

Kanodia, C., Bushman, R., and Dickhaut, J. (1989) Escalation Errors and the Sunk Cost Effect: An Explanation Based on Reputation and Information Asymmetries. *Journal of Accounting Research*, **27**(1), 59–77.

Kay, E. (1999) Desperately Seeking SAP Support, *Datamation*, March.

Keil, M. (1995) Pulling the Plug: Software Project Management and the Problem of Project Escalation. *MIS Quarterly*, **19**(4), 421–447.

Martin, M. (1998) An ERP Strategy. *Fortune*, 2 February, 95–97.

McFarlan, F. (1981) Portfolio Approach to Information Systems. *Harvard Business Review*, **59**(5), 142–150.

McKie, S. (1998) Packaged Solution or Pandora's Box? *Intelligent Enterprise*, November, 39–43.

Olsen, F. (1999) Delays, Bugs, and Cost Overruns Plague PeopleSoft's Services. *The Chronicle of Higher Education*, 24 September, A32, A33–34, A36.

Olsen, F. (2000) PeopleSoft Problems Persist, Cleveland State Looks for a New Project Manager. *The Chronicle of Higher Education*, 4 February, A49.

Raghunathan, B. and Raghunathan, T. (1998) Impact of Top Management Support on IS Planning. *Journal of Information Systems*, **12**(1), 15–23.

Reel, J. (1999) Critical Success Factors in Software Projects. *IEEE Software*, **16**(3), 18–33.

Scott, J. (1999) The FoxMeyer Drugs' Bankruptcy: Was It a Failure of ERP? *Proceedings of AMCIS 1999 Americas Conference on Information Systems*, 223–225.

Sharp, D. and Salter, S. (1997) Project Escalation and Sunk Costs: A Test of the International Generalizability of Agency and Prospect Theories. *Journal of International Business*, **28**(1), 101–122.

Staw, B. (1976) Knee-Deep in the Big Muddy: A Study of Escalating Commitment to a Chosen Course of Action. *Organizational Behavior and Human Performance*, **16**(1), 27–44.

Staw, B. (1981) The Escalation of Commitment to a Course of Action. *Academy of Management Review*, **6**(4), 577–587.

Staw, B. and Ross, J. (1987) Behavior in Escalations Decisions: Antecedents, Prototypes, and Solutions. In *Research In Organizational Behavior*, Cummings, L. L. and Staw, B. (Eds), Greenwich, CT: JAI Press.

Stedman, C. (1999) ERP Project Leads to Court Fight. *Computerworld Online News*. http://www.computerworld.com/cwi/story/0,1199,NAV47_STO36333,00.html

Whitten, J. and Bentley, L. (1998) *Systems Analysis and Design Methods*, 4th edn. Boston, MA: Irwin/McGraw-Hill.

Yin, R. K. (1994) *Case Study Research, Design and Methods*, 2nd edn. Sage Publications.

Zmud, R. W. (1980) Management of Large Software Development Efforts. *MIS Quarterly*, **4**(2), 45–55.

6 Risk Factors in Enterprise-wide/ERP Projects

Mary Sumner

6.1 Risk Factors in Enterprise-wide/ERP Projects

In the past several years many organizations have initiated enterprise-wide/ ERP projects, using such packages as SAP, PeopleSoft, and Oracle. These projects often represent the single largest investment in an information systems project in the history of these companies, and in many cases the largest single investment in any corporate-wide project.

These enterprise-wide/ERP projects bring about a host of new questions, because they represent a new type of management challenge. The management approaches for these projects may be altogether different from the managerial approaches for traditional MIS projects. Some of these questions and issues are:

- What are the major risk factors associated with implementing traditional MIS projects?
- What are the major risk factors associated with enterprise-wide information management projects?

What are the differences? What new risk factors need to be addressed in ERP projects? What are some of the risks in ERP projects that are not factors in non-ERP projects?

Most organizations have extensive experience managing traditional MIS projects, but these new ERP projects may represent new challenges and present new risk factors that must be handled differently. This paper will provide case studies of seven organizations implementing enterprise-wide/ERP projects and will provide insight into each of these questions based upon their experiences.

6.2 Risks in Implementing Information System Projects

A simple definition of 'risk' is a problem that has not happened yet but could cause some loss or threaten the success of your project if it did (Wiegers, 1998). A number of research studies have investigated the issue of the relative importance of various risks in software development projects and have attempted to classify them in various ways. Much has been written about the causes of information system project failures. Poor technical methods is only one of the causes, and this cause is relatively minor in comparison to larger issues, such as failures in communications and ineffective leadership.

Studies dealing with risk factors in information system projects describe issues of organizational fit, skill mix, management structure and strategy, software system design, user involvement and training, technology planning, project management, and social commitment (see also Table 6.1).

Organizational fit. In their paper, Barki, Rivard, and Talbot propose a variety of risk factors associated with organizational environment, including task complexity, extent of changes, resource insufficiency, and magnitude of potential loss (Barki, Rivard, and Talbot, 1993). In a framework developed by Keil et al., the risks in the environment quadrant deal with issues over which the project manager may have no control, such as changing scope/objectives and conflicts between user departments (Keil et al., 1998). Robert Block, in his text on factors contributing to project failure, points to resource failures (conflicts of people, time, and project scope) and requirement failures (poor specification of requirements) (Block, 1983).

Skill mix. Lack of expertise, including lack of development expertise, lack of application-specific knowledge, and lack of user experience, contributes to project risk (Barki, Rivard, and Talbot, 1993; Ewusi-Mensah, 1997). Risk factors in the execution quadrant of Keil's framework include inappropriate staffing and personnel shortfalls (Keil et al., 1998).

Management structure and strategy. In their study of the factors that software project managers perceive as risks, Keil et al., addressed the risks associated with customer mandate, which deals with the lack of senior management commitment (Keil et al., 1998). Ewusi-Mensah also points to lack of agreement on a set of project goals/objectives and lack of senior management involvement

Table 6.1 Summary of risk factors in information system projects

Organizational fit	Organizational environment (resource insufficiency, extent of changes), Barki, Rivard, and Talbot, 1993; Block, 1983.
	Changing scope and objectives, Keil et al., 1998.
Skill mix	Lack of technical expertise, Ewusi-Mensah, 1997.
	Lack of application knowledge, Ewusi-Mensah, 1997; Barki, Rivard, and Talbot, 1993.
	Inappropriate staffing, personnel shortfalls, Keil et al., 1998; Boehm, 1991; Block, 1983.
Management structure and strategy	Lack of agreement on project goals, Ewusi-Mensah, 1997; Block, 1983.
	Lack of senior management involvement, Ewusi-Mensah, 1997; Keil et al., 1998.
Software system design	Misunderstanding requirements, changes in requirements, Keil et al., 1998; Boehm, 1991; Block, 1983; Cash et al., 1992.
	Lack of an effective methodology, poor estimation, failure to perform the activities needed, Keil et al., 1998; Block, 1983.
User involvement and training	Lack of user commitment, ineffective communications with users, Keil et al., 1998; Block, 1983.
	Conflicts between user departments, Keil et al., 1998.
Technology planning	Lack of adequate technology infrastructure, Ewusi-Mensah, 1997.
	Technological newness, strained technical capabilities, failure of technology to meet specifications, Barki, Rivard, and Talbot, 1993; Boehm, 1991; Block, 1983; Cash et al., 1992.
	Application complexity (technical complexity), Barki, Rivard, and Talbot, 1993.
Project management	Unrealistic schedules and budgets, Boehm, 1991.
	People and personality failures, Lack of effort, antagonistic attitudes, people clashes, Block, 1983.
	Lack of measurement system for controlling risk, inadequate project management and tracking, Ewusi-Mensah, 1997; Block, 1983.
Social commitment	Inability to recognize problems; tendency to keep pouring resources into a failed project; unrealistic expectations (Keil and Montealegre, 2000; Ginzberg, 1981; Willcocks and Margetts, 1994).

(Ewusi-Mensah, 1997). Block describes goal failures (inadequate statement of system goals) and organizational failures (lack of leadership) (Block, 1983).

Software system design. Risks associated with scope and requirements include misunderstanding requirements and failing to manage change properly. Lack of an effective methodology and poor estimation can lead to cost and time overruns (Keil et al., 1998). In his paper, 'Software Risk Management: Principles and Practices', Boehm identifies ten software risk factors, including developing the wrong functions, developing the wrong user interface, 'gold-plating', a continuing stream of changes in requirements, shortfalls in externally furnished components, shortfalls in externally performed tasks, and performance shortfalls (Boehm, 1991).

User involvement and training. Lack of user commitment, ineffective communications with users, and conflicts among user departments are all sources of risk (Keil et al., 1998; Block, 1983).

Technology planning. In a study of issues that contribute to the cancellation of information system development projects, Ewusi-Mensah points out that lack of adequate technical expertise and lack of an adequate technology infrastructure to support project requirements contribute to escalating time and cost overruns and are associated with project abandonment (Ewusi-Mensah, 1997). Risk factors include technological newness (need for new hardware, software), application size (project scope, number of users, team diversity), application complexity (technical complexity, links to existing legacy systems) and failure of technology to meet specifications (Barki, Rivard, and Talbot, 1993; Block, 1983).

Project management. Project cost and time overruns can occur because of a lack of a measurement system for assessing and controlling project risk (Ewusi-Mensah, 1997). McFarlan developed dimensions of project risk assessment based upon project size, experience with the technology, and project structure (McFarlan, 1981). Project management and control failures, caused by inadequate planning and tracking, can contribute to unrealistic schedules and budgets and project failure (Block, 1983; Boehm, 1991).

Social commitment. Risk factors and risk outcomes need to take into account distinctive human and organizational practices and patterns of belief and action, as well as traditional project-related factors (Willcocks and Margetts, 1994). In information technology projects, there is a tendency to discount problems and their severity may remain unknown for a long period of time. When projects run into difficulty, there is a tendency to escalate projects because of societal norms (for example, needing to save face) and to keep pouring resources into a failing project. This may augment risk. To minimize problems, it is essential look for opportunities to use external feedback to recognize the problem and then to redefine the problem. This may entail considering alternatives to accomplishing the project's goals and preparing key stakeholders for the decision – especially if the decision is an exit strategy (Keil and Montealegre, 2000).

Ginzberg conducted a longitudinal study of user expectations as predictors of project success or failure, and his findings suggest that system implementation failure is more likely when there are unrealistic expectations about a system. Users who have more realistic expectations are more likely to be satisfied with the outcomes (Ginzberg, 1981).

6.3 Managing Large-Scale Commercial Off-the-shelf Software Projects

The existing research on managing commercial off-the-shelf software projects provides insight into factors associated with project success and failure. In their analysis of implementing packaged software, Lucas, Walton, and Ginzberg (1988) suggest that package implementation is different from custom implementation because the user may have to change procedures to work with the package, the user is likely to want to change some programs in the package to fit his/her unique needs, and the user becomes dependent upon the vendor for assistance and updates. Some of the variables associated with the successful implementation of packages are: (1) greater vendor participation in implementation and support; (2) a higher rating of user/customer capabilities by the vendor; and (3) a higher rating of user skills by MIS management. A highly skilled workforce is important for successful package implementation.

The experience implementing large-scale integrated packages, including manufacturing resource planning (MRP) systems provides a better understanding of the challenges associated with commercial off-the-shelf software implementation. In their research on success factors in MRP projects, Duchessi, Schaninger, and Hobbs (1989) concluded that commitment from top management and adequate training are 'critical' success factors for implementation. In another study of the problems encountered during MRP implementation, Ang, Sum, and Yang (1994) found that lack of training leads to difficulties in MRP system implementation.

6.4 Research in Managing ERP Projects

A number of recent research studies have focused on the difficulties encountered in managing ERP projects and strategies that can be used to minimize these difficulties. In their study dealing with the difficulties involved in managing ERP projects, Willcocks and Sykes noted that even though the information technology function is expected to deliver these systems on time and within budget, ERP projects pose great difficulty because of the lack of internal IT skills, the complexity of linking legacy systems, and the challenge of implementing new technologies. Perhaps the greatest challenge is finding and retaining individuals with ERP skills in the context of an IT workforce crisis (Willcocks and Sykes, 2000).

In their research, Willcocks and Sykes identify effective strategies for managing ERP projects, including senior-level sponsorship, reengineering business processes, creating cross-functional teams, in-house development of technical expertise, effective partnering with suppliers, and the ability of the chief information officer to perform a role as a strategic business partner. Skilled project management is critical, with a clearly defined management structure and project staffing and an effective champion of the project. The success of ERP projects depends upon IT leadership which is willing to build relationships with functional area executives, to demonstrate business system thinking, to view the ERP as an investment in business innovation, and to create the architectures needed to support future technology requirements for ERP.

A number of studies of ERP projects focus on defining the critical success factors for managing these projects. In their research on critical success factors for ERP implementation, Holland and Light note that CSF's are top management support, clear business vision, aligning business processes, and factors related to ERP implementation, including integration with legacy systems, business process change, and effective software configuration (Holland and Light, 1999). Technical factors associated with successful ERP implementation include client acceptance, effective communications with users, and effective project monitoring and control. In their analysis of ERP implementations within two companies, Holland and Light emphasize the need to align business processes with the software during implementation.

In their study of success factors and pitfalls in ERP implementations, Bancroft, Seip, and Sprengel (1998) cite critical success factors for ERP projects, including top management support, presence of a champion, good communications with stakeholders, and effective project management. Other factors specific to ERP implementations are reengineering business processes, understanding the importance of corporate cultural change, and using business analysts on the project team (Bancroft, Seip, and Sprengel, 1998).

In another study of critical success factors in ERP implementations, Bingi, Sharma, and Godla (1999) note that top management commitment, the ability to recruit and retain qualified ERP consultants, and the ability to train and retain in-house ERP technologies are all critical to project success. The reason why top management commitment is so important is because ERP implementations involve significant alterations in business processes as well as a considerable financial investment. The authors add the importance of dealing with a financially stable ERP vendor, the ability to successfully manage consultants, and the willingness to train employees to successfully use the new system. Finding the right people and retaining them is a major challenge of ERP implementation because these projects require multiple skill sets, including functional, technical, and interpersonal skills. Consultants with industry-specific knowledge are also critical (Bingi, Sharma, and Godla, 1999).

Several new studies of experiences with ERP implementations break ERP projects into phases and analyse problems encountered and success factors associated with each of the phases that have been defined. In their study of problems encountered with ERP projects, Markus et al. (2000) break ERP projects into three phases, including the project phase, during which the ERP system is introduced; the shakedown phase, during which the company integrates the ERP system with normal operations; and the onward and upward phase, during which the company realizes the benefits of the ERP system and plans the next steps for business improvement. Success in each of these phases is measured by different metrics. In the project phase, for example, success is measured by project time and cost completion within budget and schedule. In the shakedown phase, success is measured by changes in key business indicators (for example, operating costs), and in the onward and upward phase, success is measured by actual business results (for example, reduced inventory costs) (Markus et al., 2000).

According to Markus et al., each of these phases poses specific problems. In the project phase, attempts to customize software can present problems, because it is difficult to integrate vendor modifications and updates. Another set of problems deals with system integration, including integrating the ERP system with existing hardware platforms and retaining some functions from existing legacy systems. Another important set of issues deals with turnover among key ERP systems personnel and inconsistent staffing of consultants.

In the shakedown phase, when the system is integrated with normal business operations, system performance can be affected by errors in data entry, lack of functionality, and lack of management reporting capabilities. Problems occur when managers try to maintain old procedures, rather than learning the new business processes which are supported by the new system. Turnover among key users, and among key system support personnel, also creates problems. It is important to retain people who understand the implementation and use of the ERP system and who take responsibility for improving their own understanding and skill levels.

Finally, in the onward and upward phase, lack of top management commitment and a culture which is resistant to change may make it difficult to realize significant returns in improved business performance.

In their earlier research Parr, Shanks, and Darke (1999) observe that top management support and role of a champion are critical to success because users must support the reengineering of business processes. Based upon interviews with senior members of ERP implementation teams, Parr, Shanks, and Darke identify a number of critical success factors for ERP projects, including the ability to partition large projects into smaller, 'Vanilla' implementations, organizing a balanced team consisting of technical experts and business experts, and making a commitment to change (Parr and Shanks, 1999).

In their subsequent study, Parr and Shanks use two case studies to define critical success factors in the context of each of four project phases, including the planning phase, project phase, installation phase, and enhancement phase. In the planning phase, in which the ERP system is selected, critical success factors are the role of the champion, top management support, commitment to change, and commitment to a 'Vanilla' ERP implementation, which means reengineering business processes to fit the software.

In the project phase, in which the current business processes are re-designed to support the business model of the ERP, critical success factors are top management support, definition of scope and goals, and a balanced team. In the detailed design, and configuration and testing phases, CSF's are top management support and a 'Vanilla' ERP implementation. During the installation phase, in which desktops and networks are put into place and users are trained to use the new system, management support, a balanced team, and commitment to change are critical to success (Parr and Shanks, 2000). Throughout their case studies, top management support and implementation of a 'Vanilla' ERP system stand out as important strategies for success across multiple phases.

Since one of the major challenges associated with managing ERP projects successfully is their size and scope, the research of Frederic Adam and Peter O'Doherty regarding smaller-scale ERP implementations in Ireland is particularly interesting. Adam and O'Doherty conclude that the duration and the complexity of large ERP projects within large organizations may be due as much to the complexity of the organizations in which these systems are being implemented as to the complexity of the ERP systems themselves. Smaller-scale implementations are more straightforward and often more successful (Adam and O'Doherty, 2000). They re-iterate some of the experiences of other researchers and note that the trade-off between the software implementer wanting a 'Vanilla' implementation and the client wanting functionality and customization requires extensive coordination, collaboration, and communication among the software implementer and the managers.

The issue of managing 'misfits', or incompatibilities between ERP software and the functional requirements of end-users has been addressed in detail in the research of Soh, Kien, and Tay-Yap (2000). In their analysis, they categorize misfits as data misfits (for example, incompatibilities between requirements and the data formats the ERP supports); functional misfits (for example, incompatibilities between requirements and the processing procedures the ERP supports); control misfits (for example, the lack of validation procedures in the ERP system); and operational misfits (for example, incompatibilities between the ERP's embedded business practices and operating procedures of the firm). Resolving these issues means analysing the trade-off's between the amount of organizational change and the amount of customization required (Soh, Kien, and Tay-Yap, 2000).

The key issue in ERP implementations is how to manage the trade-off between implementing a Vanilla ERP system and providing the functionality which users want. Strategies for success, in Soh, Kien, and Tay-Yap's view, are involving users in evaluating the appropriateness of the ERP software and selecting vendors with significant industry knowledge.

This review of current research which deals with critical success factors in ERP implementations confirms the importance of top management support, reengineering business processes, finding a balanced mix of technical and business specialists, and the presence of a champion throughout ERP projects. A summary of these findings is included in Table 6.2.

6.5 Managing Client–Server Information Systems

The implementation of ERP systems often entail the use of client–server technology, and this may cause further complications. It is often critical to acquire external expertise, including vendor support, to facilitate successful implementation. Also, the costs of training and support are often underestimated, and these costs may be many times greater than originally anticipated. Client–server implementations often bring 'surprises' with respect to cost, because of the costs of decentralized servers, system integration software, technical support, and software updates and version control. In actuality, the total cost

Table 6.2 Research on success factors in enterprise resource planning (ERP) projects

Critical success factors	Research sources
Top management support	Parr, Shanks, and Darke (1999)
	Parr and Shanks (2000)
	Willcocks and Sykes (2000)
	Bingi, Sharma, and Godla (1999)
	Holland, Light, and Gibson (1999)
	Markus et al. (2000)
	Bancroft, Seip, and Sprengel (1998)
Use of business system analysts	Willcocks and Sykes (2000)
	Parr, Shanks, and Darke (1999)
	Parr and Shanks (2000)
Implementation of a 'Vanilla' ERP (reengineering business processes)	Parr, Shanks, and Darke (1999)
	Parr and Shanks (2000)
	Willcocks and Sykes (2000)
	Holland and Light (1999)
	Davenport (1998)
	Markus et al. (2000)
Obtaining/retaining the right ERP skill set	Willcocks and Sykes (2000)
	Bingi, Sharma, and Godla (1999)
	Markus et al. (2000)
	Adam and O'Doherty (2000)
Creating projects with smaller scope	Parr and Shanks (1999)
	Adam and O'Doherty (2000)
Obtaining a balanced team (e.g., a mix of technical and business skills)	Parr, Shanks, and Darke (1999)
	Parr and Shanks (2000)
Effective use of consultants	Bingi and Sharma (1999)
Effective management of project scope and objectives	Bingi and Sharma (1999)
	Holland and Light (1999)
	Parr, Shanks, and Darke (1999)
	Parr and Shanks (2000)
Effective communications and coordination skills	Adams and O'Doherty (2000)
	Bancroft, Seip, and Sprengel (1998)
Effective management reporting capabilities (in shakedown phase)	Markus et al. (2000)
Effective management of supplier relationships	Willcocks and Sykes (2000)
	Bingi and Sharma (1999)
	Soh, Kien, and Tay-Yap (2000)
Strategic partnering between software implementers and business partners	Willcocks and Sykes (2000)
	Adam and O'Doherty (2000)
Effectively managing 'misfits', including incompatibilities in data, incompatibilities in processes, and incompatibilities with operating procedures	Soh, Kien, and Tay-Yap (2000)

Table 6.2 (*cont.*)

Critical success factors	Research sources
Effective integration with legacy systems	Holland and Light (1999)
A champion	Willcocks and Sykes (2000)
	Parr, Shanks, and Darke (1999)
	Parr and Shanks (2000)
	Bancroft, Seip, and Sprengel (1998)
Effective training of end-users	Bingi and Sharma (1999)
User involvement	Soh, Kien, and Tay-Yap (2000)
Commitment to change by all stakeholders	Parr, Shanks, and Darke (1999)
	Parr and Shanks (2000)

of a client server implementation can be three to six times greater than for a comparable mainframe-based system. Even though there are great cost reductions possible through moving off the mainframe, the costs of learning the new technology and of acquiring technical support are substantial (Caldwell, 1996).

6.6 Research Objectives

The purpose of this study is to develop a better understanding of the major risk factors associated with enterprise-wide/ERP projects. These case studies will examine these risk factors. The case studies describe the experiences of seven companies implementing enterprise-wide information management systems using SAP, PeopleSoft, and Oracle. The case studies were developed using in-depth structured interviews with the senior project managers responsible for planning and implementing enterprise-wide/ERP systems within their respective organizations. In each of the interviews, a structured interview format was followed. The questions dealt with project characteristics (purpose and scope, project duration, project justification), project management issues (project sponsorship, project team makeup, mix of internal/external team members), technical challenges, critical success factors (organizational factors, people factors, technology factors), and lessons learned. In addition to identifying the critical success factors and the risk factors associated with technology, organizational fit, and people factors, the project managers provided insight into the unique factors associated with successful project management and control of ERP projects.

Table 6.3 Company profiles

	Nature of business	1998 sales	No. of employees worldwide	No. IT (info technology) employees	No. of project employees
Beverage manufacturer (BV)	Manufactures food and beverage products	$12 832m.	25 123	1 100	50 internal, 25 external*
Military aircraft manufacturer (ML)	Manufactures military aircraft	$15 000m.	60 000	850	80–100 internal, 20 external*
Electrical manufacturer (EL)	Manufacturer of electrical and electronic products and systems	$12 298 600 000	100 700	90 (one division)	25 internal, 50–60 external* (one division)
Investment brokerage firm (IN)	National investment brokerage firm	$1 135 000 000	13 690	725	25 internal
Pharmaceutical manufacturer (PH)	Manufactures and markets high-value agricultural products, products, pharmaceuticals, and food ingredients	$7 514 000 000	24 700	600	25 internal, 10 external*
Consumer product manufacturer (CP)	Manufactures dog/cat foods and dry cell battery products	$4 653 000 000	23 000	750	100 internal, 20 external*
Chemical manufacturer (CH)	Manufactures and distributes biochemicals, organic chromatography products, and diagnostic reagents	$1 127 000 000	6 000	200	20 internal, 10 external*

*Note** External consultants.

6.7 Company Profiles

The findings describe the experiences implementing ERP systems within seven large organizations with sales ranging from $1 billion to $15 billion annually. The firms represent a variety of industries, as you can see from Table 6.3.

6.8 The Case Studies: Findings

The findings describe the project justification and the risk factors identified by the project managers responsible for the SAP, PeopleSoft, and Oracle projects within these seven organizations. The first area of discussion was project justification. The risk factors identified in the interviews were organized into the categories of organizational fit, skill mix, management structure, software system design, user involvement, user training, technology planning, and project management.

Project Justification

In these case studies, the ERP projects were justified in terms of cost effectiveness and business benefits. Beginning in 1996, the pharmaceutical manufacturer started a corporate-wide SAP project. The business justification for the project was operational excellence, for example, cutting the costs of core transactions-processing systems, such as order processing and inventory management. In addition, an integrated package could support worldwide business operations and replace division-level systems. Before SAP, the pharmaceutical firm had four purchasing packages – one for each business unit. SAP provided economies of scale in development, maintenance, and operations. Its overall costs were divided by a much larger number of users. For example, buying a $100 000 package to support 5000 users is less expensive than buying a $25 000 package to support 100 users. In addition, the SAP project enabled the pharmaceutical company to reduce its information systems development staff from 500 to 50 people.

Some of the 'business drivers' for the SAP implementation at the pharmaceutical manufacturer included: data integration, standardization, access to timely and complete information, leverage gained in purchasing, and globalization. SAP cut the costs of operational systems, improved the reliability of customer service, and assured timely delivery and follow-up.

The original project justification for the SAP project at the beverage manufacturer was similar. There were extensive economies of scale associated with consolidating four MIS projects into one, and SAP offered an integrated, corporate-wide solution. The business justification entailed major cost savings from reducing the costs of operational level information systems. SAP provided hard-dollar savings, based upon integration of data and processes, a common database, and increased leverage in purchasing and buying.

The major sources of justification for the SAP project at the chemical manufacturer were the need to integrate a number of different order-processing systems, the need to improve and integrate financial systems, and the ability to reduce the workforce through systems integration. The major motivation behind the project was to gain a 'competitive advantage' by providing 'seamless' order processing to customers in a global marketplace. This meant that any customer in the world could place orders using one integrated order-processing system, as opposed to using many different systems for different product lines.

The PeopleSoft project at the military aircraft manufacturer was justified in terms of better information, cost reduction, and data integration. Between 70 and 80 systems were replaced by a single, integrated system. While the original intent was to implement an integrated human resources/payroll system using PeopleSoft, the first phase of the project involved completing the human resources (HR) component and creating an interface to the existing payroll system. After the completion of the firm's merger with a commercial aircraft manufacturer, the plan was to integrate both HR and payroll, using the PeopleSoft software. As you will learn later, this 'phased-in' approach created significant problems in system implementation.

The major justification for the PeopleSoft project at the investment brokerage firm was data integration, a common system approach, and hard dollar savings through integration. The Oracle project at the consumer products' manufacturer was also justified in terms of data integration and cost reduction through the reengineering of business processes.

The major purpose of the Oracle project at the electrical products' manufacturer was to implement Oracle financial, distribution, and manufacturing systems. The business justification included: inventory reduction, headcount savings, and reduced lead times through on-time delivery. Table 6.4 summarizes the basis for project justification for the various SAP, PeopleSoft, and Oracle projects.

Risk Factors

The findings provide the risk factors associated with ERP systems implementation which were mentioned by senior project leaders and which identify risks which actually materialized. These factors are represented in order of how frequently each one was mentioned.

- **Failure to re-design business processes to fit the software.** Based upon their experiences, all of the project managers learned to avoid customization. Many companies 'go to war' with the package and try to make it meet

Table 6.4 Project type and justification

	System	Justification	Project Initiation
Beverage manufacturer	SAP	Cost reduction of operational systems	1996
Military aircraft manufacturer	PeopleSoft	Cost reduction; data integration	1994
Electrical products manufacturer	Oracle (financials, inventory, etc.)	Cost reduction; inventory reduction; headcount savings	1996
Investment brokerage firm	PeopleSoft	Data integration; common systems	1996
Pharmaceutical manufacturer	SAP	Cost reduction of core operational systems	1996
Consumer products' manufacturer	Oracle (financials, inventory)	Cost reduction; data integration	1996
Chemical manufacturer	SAP	Cost reduction; system integration	1996

their business process requirements, only to lead the way to cost overruns and project failure in some cases. Rather than attempting to modify the software, the chemical manufacturer reengineered its business processes to be consistent with the software, and this proved to be critical to the project's success. In contrast, the military aircraft manufacturer customized human resources, payroll, and benefits modules in a PeopleSoft ERP package and experienced significant cost and time impacts. The creation of a 'bridge' between the human resources module of the ERP system and a legacy payroll application resulted in extensive time and cost delays (7).

- **Lack of senior management support.** Without question, top management support is critical. It is important to achieve the support of senior management for accomplishing project objectives and for aligning these goals with strategic business goals (6).
- **Insufficient training and re-skilling.** A number of firms learned that the investment in training and re-skilling the IT workforce was higher than expected. 'Growing' internal IT staff members with needed technical skills, especially in application-specific modules, was a strategy followed by four of the organizations (4).
- **Lack of ability to recruit and retain qualified ERP system developers.** Many of the organizations found it difficult to recruit and retain good ERP specialists because market rates for these people are high. Management must understand and appreciate the criticality of high-tech worker

turnover, recruitment, and retention issues. Four organizations developed recruitment and retention programs specifically designed to address the need for ERP system professionals. In their experience, the loss of trained ERP analysts to consulting firms was particularly frustrating (4).

- **Insufficient training of end-users.** Most firms emphasized making a major commitment to training end-users in system uses. This meant re-skilling the end-users in new technologies and applications and supplementing 'generalized' user training with training in the use of specific application modules. Several firms emphasized user training in reporting applications, including the use of report generators to design and generate custom reports (4).
- **Inability to obtain a full-time commitment of 'customers' to project management and project activities.** It may be difficult to get managers to commit to project management roles, because they may be uncertain about what responsibilities will still be open to them once they are transferred back to their functional areas. Getting the 'business' areas to dedicate people to the management of the project is a key priority, and some of the project managers found this to be difficult (3).
- **Lack of integration.** In terms of factors conducive to project failure, one of the main factors associated with failure is lack of integration. The project needs to be based on an enterprise-wide design. One project manager argued, 'you can't start with "pieces", and then try to integrate the software component's later on.' Another said, 'It is important to use a "federal" approach; define what is needed at the enterprise level, and then apply it to the business unit level' (3).
- **Lack of a proper management structure.** Without central project leadership, there is excessive duplication of effort. The pharmaceutical manufacturer put someone 'in charge' and centralized the management structure of the project in order to avoid duplication of effort. In implementing a 'centralized' system, a centralized management structure should exist. At the military aircraft manufacturing company, several senior executives had equal authority over the project, and this contributed to conflicts and lack of problem resolution (2).
- **Insufficient internal expertise.** When they did not have needed expertise internally, most firms brought in the consultants they needed to overcome technical and procedural challenges in design and implementation. It is important to obtain consultants who are specialists in specific application modules. This was emphasized by the managers representing the electrical manufacturer and the consumer products manufacturer – both of whom

were implementing Oracle Financials within various operating units of their respective companies (2).

- **Lack of a champion.** The project leader for an ERP project is clearly a 'champion' for the project, and this role is critical to marketing the project throughout the organization (2).
- **Lack of 'business' analysts.** One of the critical workforce requirements for an ERP project is the ability to obtain analysts with both business and technology knowledge. Instead of 200 'programmers' with average skills, the manager of the ERP project within the chemical manufacturer argued that ERP systems can best be accomplished with 20 'business' analysts who have specialized expertise, the ability to learn quickly, and effective communication skills (2).
- **Failure to mix internal and external personnel.** Using a mix of consultants and internal staff to work on a project team enables internal staff members to 'grow' the necessary technical skills for ERP system design and implementation. The project manager for the electrical manufacturer argued that extra external consultants were needed because of insufficient time lines to 'grow' internal staff, and this resulted in much higher costs (2).
- **Failure to emphasize reporting, including custom report development.** The use of report generators and user training in reporting applications is critical to project implementation success. One of the 'lessons learned' by the military manufacturer was that insufficient end-user training can generate resistance to using the system, largely because people are ill-prepared to use it effectively (2).
- **Insufficient discipline and standardization.** Another 'risk factor' which is closely associated with the software itself is insufficient adherence with the standardized specifications that the software supports. It is important to avoid compromising the system and its specifications. In terms of 'lessons learned', the pharmaceutical manufacturer's experience demonstrated the importance of using SAP's built-in 'best practices' (2).
- **Ineffective communications.** It is critical to communicate what is happening, including the scope, objectives, and activities of the ERP project (2).
- **Avoid technological bottlenecks.** Lack of an integrated technology strategy to support client–server implementation causes further risks and bottlenecks in project success. The different 'technology' environments within one organization created delays in establishing consistency and coordination in platforms, database management systems, and operating system environments for the PeopleSoft application. Technology bottlenecks can occur when designers try to implement bridges between ERP modules and

Table 6.5 Summary of risk factors in enterprise-wide/ERP projects

Risk category	Risk factor	Unique to ERP (*)
Organizational fit	Failure to re-design business processes	*
	Failure to follow an enterprise-wide design which supports data integration	*
Skill mix	Insufficient training and re-skilling	*
	Insufficient internal expertise	*
	Lack of business analysts with business and technology knowledge	*
	Failure to effectively mix internal and external expertise	
	Lack of ability to recruit and retain qualified ERP system developers	*
Management structure and strategy	Lack of senior management support	
	Lack of proper management control structure	
	Lack of a champion	
	Ineffective communications	
Software system design	Failure to adhere to standardized specifications which the software supports	*
	Lack of integration	*
User involvement and training	Insufficient training of end-users	
	Ineffective communications	
	Lack of full-time commitment of customers to project management and project activities	
	Lack of sensitivity to user resistance	
	Failure to emphasize reporting	
Technology planning/integration	Inability to avoid technological bottlenecks	*
	Attempting to build bridges to legacy applications	

legacy applications. At the military aircraft company, building a bridge between the PeopleSoft human resources module and a legacy payroll application contributed to significant time and cost overruns (2).

A summary of the risk factors affecting the management of enterprise-wide/ERP projects, and a description of which factors are unique to ERP projects described in these case studies, is shown in Table 6.5.

The Unique Risks of Enterprise-wide/ERP Projects

The third question, 'What risks are there inherent in enterprise-wide/ERP projects that are not found in non-ERP projects?' revealed factors dealing with organizational fit, skill mix, software system design, and technology integration.

The first challenge which was universally supported by the respondents was the risk of failing to re-design business processes and of following an enterprise-wide design that supports data integration across the organization. This makes ERP projects unique because of their size, scope, and organizational impact. The integration of business functions, elimination of redundant databases, and streamlining of organizational processes are all essential for project justification.

A unique challenge involved in ERP system implementation is acquiring the necessary skills. Insufficient training and re-skilling of the IT workforce in new ERP technology, insufficient 'internal' expertise, failure to effectively mix internal and external expertise, and lack of 'business' analysts were all risks associated with the recruitment and retention of IT professionals. The unique challenge here is aggravated by the scarcity of ERP-trained system developers and the high market demand for their skills. The investment in recruiting, re-skilling and re-training IT professionals was considered very high. The problem of retention was further exacerbated by the tendency of highly trained ERP analysts to move to consulting firms where the salaries were even higher.

Traditional strategies for software systems design and construction were also devalued within the context of ERP projects. System analysts quickly learned that failure to adhere to standardized specifications which the software supports created risks. Data integration became a significant design issue and often entailed a top–down system integration strategy. When legacy systems were involved, many organizations found that attempts to integrate ERP systems with legacy applications could bring about significant cost and time overruns because of lack of integration and duplication of business processes.

Implications for Managing ERP Projects

By organizing these risk factors (see Table 6.6) within the context of the stages of an ERP project, and by identifying individuals responsible for managing risk factors at each phase, management can assign responsibility for managing each of these risk factors.

6.9 Recommendations

Some of the unique challenges to managing enterprise-wide/ERP projects which are highlighted through these findings include: the re-design of business processes, the investment in recruiting and re-skilling ERP system developers,

Table 6.6 Risk factors in ERP systems projects

Project phase	Responsibility	Risk factor to be addressed
Planning	User management IT management	Lack of top management support Lack of a proper management structure for the project Lack of a 'champion'
Requirements analysis	User management IT management Business analysts	Failure to re-design business processes Failure to follow an enterprise-wide design that supports data integration
System design	User management IT management IT designers	Lack of 'business' analysts Failure to adhere to standardized specifications which the software supports Lack of data integration
System implementation/ maintenance	IT management	Insufficient training and re-skilling of the IT workforce in new technology Insufficient 'internal' expertise Failure to effectively mix internal and external expertise
Technology integration and implementation	IT management User management	Unsuccessful attempts to integrate ERP with legacy applications

the challenge of using external consultants and integrating their application-specific knowledge and technical expertise with existing teams, and the challenge of recruiting and retaining business analysts who combine technology and business skills. Many of the strategies which can be used to minimize these risk factors were contributed by the ERP project managers (see Table 6.7).

Implications for Practitioners

Enterprise-wide/ERP projects pose new opportunities and significant challenges. Some of the 'summary' ideas which are re-iterated throughout the case studies are:

- Justify the enterprise-wide projects based upon cost justification and economies of scale.
- Re-engineer business processes to 'fit' the package, rather than trying to modify the software to 'fit' the organization's current business processes.
- Identify and implement strategies to re-skill the existing IT workforce and acquire external expertise through vendors and consultants when needed.
- Utilize 'business analysts' with both business knowledge and technology knowledge.

Table 6.7 Strategies for controlling risk factors in enterprise-wide/ERP projects

Type of risk	Strategies for minimizing risk
Organizational fit	Commitment to re-designing business processes
	Top management commitment to re-structuring and following an enterprise-wide design which supports data integration
Skill mix	Effective use of strategies for recruiting and retaining specialized technical personnel
	Effective re-skilling of the existing IT workforce
	Obtaining 'business analysts' with knowledge of application-specific modules
	Effective use of external consultants on project teams
Management structure and strategy	Obtaining top management support
	Establishing a centralized project management structure
	Assigning a 'champion'
Software system design	Commitment to using project management methodology and 'best practices' specified by vendor
	Adherence with software specifications
User involvement and training	Effective user training
	Full-time commitment of users to project management roles
	Effective communications
Technology planning/ integration	Acquiring technical expertise
	Acquiring vendor support for capacity planning and upgrading
	Planning for client–server implementation, including client workstations

- Obtain top management support for the project and establish strong project leadership.
- Make a commitment to training end-users in custom report development.
- Manage change through leadership, effective communications, and the role of a champion.

Future Research

Without question, the effective management of these large projects is a new and unique challenge which requires the use of project management and control methods that have not been used extensively in the past. The sheer size of these projects requires centralized control, strict discipline, and extensive monitoring of project outcomes. Several research issues which can be explored in the future include conducting an assessment of the relative criticality of each of these risk factors and contrasting the risk factors which occur in large vs. small ERP projects. One of the greatest challenges is recruiting and retaining highly sought IT professionals with the specialized technical and application-specific skills. Further research could analyse factors contributing to effective recruitment and retention of IT professionals with these specialized skills.

REFERENCES

Adam, F. and O'Doherty, P. (2000) Lessons from Enterprise Resource Planning Implementation in Ireland – Toward Smaller and Shorter ERP Projects. *Journal of Information Technology*, **15**, 305–316.

Ang, J. S. K, Sum, C. C., and Yang, K. K. (1994) MRP II Company Profile and Implementation Problems: A Singapore Experience. *International Journal of Production Economics*, February, Amsterdam.

Bancroft, N., Seip, H., and Sprengel, A. (1998) *Implementing SAP R/3*, 2nd edn. Greenwich, CT: Manning Publications.

Barki, H., Rivard, S., and Talbot, J. (1993) Toward an Assessment of Software Development Risk. *Journal of Management Information Systems*, **10**(2), 203–225.

Beath, C. (1991) Supporting the Information Technology Champion. *MIS Quarterly*, **15**(3), 355–373.

Bingi, P., Sharma, M., and Godla, J. (1999) Critical Issues Affecting an ERP Implementation. *Information Systems Management*, Summer, pp. 7–14.

Block, R. (1983) *The Politics of Projects*. Yourdon Press, Prentice-Hall.

Boehm, B. W. (1991) Software Risk Management: Principles and Practices. *IEEE Software*, **8**(1), 3241.

Caldwell, B. (1996) Client–Server: Can It Be Saved? *Information Week*, **584**, 36–44.

Cash, J., McFarlan, F. W., McKenney, J., and Applegate, L. (1992) A Portfolio Approach to IT Development, *Corporate Information Systems Management*, 3rd edn. Irwin Publishing.

Charette, R. N. (1989) *Software Engineering Risk Analysis and Management*. New York: Intertext.

Davenport, Thomas H. (1998) Putting the Enterprise into the Enterprise System. *Harvard Business Review*, July–August, 121–131.

Duchessi, P., Schaninger, C., and Hobbs, D. (1989) Implementing a Manufacturing Planning and Control Information System. *California Management Review*, Spring, 75–90.

Ewusi-Mensah, Kweku (1997) Critical Issues in Abandoned Information Systems Development Projects. *Communications of the ACM*, **40**(9), 74–80.

Ginzberg, M. I. (1981) Early Diagnosis of MIS Implementation Failure: Promising Results and Unanswered Questions. *Management Science*, **27**(4), 459–478.

Hammer, M. and Champy, J. (1993) *Re-engineering the Corporation: A Manifesto for Business Revolution*. London: Nicholas Brearley Publishing.

Holland, Christopher and Light, Ben (1999) A Critical Success Factors Model for ERP Implementation. *IEEE Software*, May/June, 30–35.

Holland, C., Light, B., and Gibson, N. (1999) A Critical Success Model for Enterprise Resource Planning Implementation. Proceedings of the 7th European Conference on Information Systems, Copenhagen, Denmark, June, 273–287.

Keil, M., Cule, P., Lyytinen, K., and Schmidt, R. (1998) A Framework for Identifying Software Project Risks. *Communications of the ACM*, **41**(11), 76–83.

Keil, M. and Montealegre, R. (2000) Cutting Your Losses: Extricating Your Organization when a Big Project Goes Awry. *Sloan Management Review*, **41**(3), 55–68.

Kremers, M. and Van Dissel, H. (2000) ERP System Migrations. *Communications of the ACM*, **43**(4), 53–56.

Lucas, H., Walton, E., and Ginzberg, M. (1988) Implementing Packaged Software. *MIS Quarterly*, December, 537–549.

Markus, L. M., Axline, S., Petrie, D., and Tanis, C. (2000) Learning from Adopter's Experiences with ERP: Problems Encountered and Success Achieved. *Journal of Information Technology*, **15**, 245–265.

McFarlan, F. W. (1981) Portfolio Approach to Information Systems. *Harvard Business Review*, **59**(5), 142–150.

Mumford, E. (1981) Participative Systems Design: Structure and Method. *Systems, Objectives, Solutions*, **1**(1), 5–19.

Parr, A. N., Shanks, G. and Darke, P. (1999) Identification of Necessary Factors for Successful Implementation of ERP Systems. In *New Information Technologies in Organizational Processes: Field Studies and Theoretical Reflections on the Future of Work*, Ngwerryama, O., Introna, L., Myers, M. and DeGross, J. (eds) IFIP TC8 WGB8.2 International Working Conference on New Information Technology in Organizational Processes: Field Studies and Theoretical Reflections on the Future of Work. 21–22 August. St. Louis, Missouri, USA.

Parr, A. and Shanks, G. (2000) A Model of ERP Project Implementation. *Journal of Information Technology*, **15**, 289–303.

Soh, C., Kien Sia Siew, and Tay-Yap, J. (2000) Cultural Fits and Misfits: Is ERP a Universal Solution? *Communications of the ACM*, **41**(4), 47–51.

Subramanian, A. and Lacity, M. C. (1997) Managing Client/Server Implementations: Today's Technology, Yesterday's Lessons. *Journal of Information Technology*, **12**(3), 169–186.

Wiegers, Karl (1998) Know Your Enemy: Software Risk Management. *Software Development*, October.

Willcocks, L. and Margetts, H. (1994) Risk Assessment and Information Systems.
European Journal of Information Systems, **3**(2), 127–138.

Willcocks, L. and Sykes, R. (2000) The Role of the IT Function. *Communications of the ACM*, **41**(4), 32–38.

7 A Framework for Understanding Success and Failure in Enterprise Resource Planning System Implementation

Christopher P. Holland and Ben Light

7.1 Introduction

Companies continue to move away from developing IT systems in-house and purchase instead standard package software. PriceWaterhouse (1995)[1] predicted that by the year 2000, two-thirds of all business software would be bought off the shelf. Following this Deloitte and Touche (1997) stated that enterprise resource planning (ERP) systems, standard package-based software that supports core business processes, were the preferred method by which businesses replaced 'legacy systems'[2]. In 2000 a Consultant's Advisory survey indicated that 61% of respondents planned to invest in existing or new ERP implementations. The use of standard systems has been the predominant strategy for several years in the area of desktop computing and it is clearly now being adopted elsewhere in organizations.

Even in the light of the phenomenal uptake of this form of strategy and subsequent potential for learning, implementation is still problematic for many organizations. There are mixed reports concerning the outcome of ERP projects. Successful ERP implementations are certainly publicised, such as Monsanto (Edmondson, Baker, and Cortese, 1997) and Guilbert-Niceday (Gibson, Holland, and Light, 1999), but less-successful attempts, such as FoxMeyer Drug (Bicknell, 1998; James, 1997), have received limited attention. It is estimated that at least 90% of ERP implementations end up late or over budget (Martin, 1998). However, this may be due to poor cost and time estimation rather than a failure in project management. Changes in the scope of ERP projects may also contribute to this figure (Holland and Light, 1999). The integrated nature of ERP software may also explain some of the

[1] Now PriceWaterhouseCoopers

[2] The notion that legacy systems can be replaced does not concur with the authors' definition of legacy information systems discussed later in the paper. The authors argue that ERP systems are the dominant method for trying to deal with legacy information system problems.

problems. Enterprise consensus is required to reengineer an organization's core business processes and to take advantage of the software (Davenport, 1998; Knowles, 1997). If the system is to be implemented globally, then global consensus is required. In contrast to custom approaches, designed specifically for an individual company, standard package software is generic and requires extensive configuration to match the business processes of a specific organization. The associated 'reengineering' activity, acknowledged as a significant part of ERP projects, is widely reported as requiring a focus upon organizational and IT contexts (Grover, Seung, and Teng, 1998; Avison and Fitzgerald, 1995; Martinez, 1995; Hall, Rosenthal, and Wade, 1993), a potential factor contributing to increased complexity, diversity, and difficulty. Finally, according to Martin (1998) there are as many different methodologies for implementing ERP as there are consultants who will partner an organization through an ERP project. All of the points cited here have great resonance with the authors' research into nearly 40 ERP implementations.

In order to try and improve implementation experiences a number of strategic approaches have evolved which differ in terms of the technical and business scope of the project. The main technical options are the implementation of the standard package with minimum deviation from the standard settings provided by the supplier and custom adaptation of the system to suit particular local requirements. The main business options revolve around the issue of compromise over fitting the system to the organization or vice versa (KPMG, 1998). Several authors have written on success and failure in ERP implementation from a number of perspectives, including Parr and Shanks (2000) and Bingi, Sharma, and Godla (1999). In this paper the authors offer a contribution by adopting an applied management perspective and the use Critical Success Factors (CSFs) theory to explain differences in project outcomes. CSF models have been applied to general project management problems (Slevin and Pinto, 1987), manufacturing system implementation (Lockett, Barrar, and Polding, 1991; Roberts and Barrar, 1992) and the area of reengineering (Bashein, Markus, and Riley, 1994). The approach is particularly suitable for the analysis of ERP projects because it provides a framework for including the influence of tactical factors such as technical software configuration and project management variables together with broader strategic influences such as the overall implementation strategy. The analysis of the CSFs identifies the critical role of organization legacy on the implementation process and the importance of business process change and software configuration.

ERP implementation is a complex and difficult process that can potentially reap enormous benefits for successful companies and be disastrous for those organizations that fail to manage the implementation process. The following section is a brief review of the critical success factors literature applied to information systems, before introducing a critical success factors framework for understanding ERP implementation and applying it to five case study examples of ERP implementation.

7.2 Critical Success Factors

Slevin and Pinto (1987) offer a project implementation profile that consists of ten critical success factors organized in a strategic/tactical framework that enables project assessment. It is argued that to manage projects successfully, project managers must be capable in the strategic and tactical aspects of project management. The critical success factors can be divided between the planning (strategic) phase and the action (tactical) phase of the project. Strategic issues specify the need for a project mission, for top management support, and a project schedule outlining individual action steps for project implementation. These issues are most important at the beginning of the project. Tactical issues increase in importance towards the end of the project and include communication with all affected parties, recruitment of necessary personnel for the project team, and obtaining the required technology and expertise for the technical action steps. User acceptance, monitoring, and feedback at each stage, communication to key project people, and trouble shooting are also classified as tactical issues. Strategy and tactics are not independent of each other. Projects that exhibit a high quality in both strategy and tactics are likely to be successful. Further work has highlighted CSFs in an IT context. Benjamin and Levinson (1993) identify the need to manage organization, business process and technology changes in an integrative manner. Kotter (1995) emphasizes the need to create an atmosphere that is open to change and provides the necessary resources to effect those changes. CSFs can therefore be seen as a useful way for examining success and failure in a variety of business and IT projects. The authors now discuss their approach to the research including the CSF framework that was calibrated for ERP implementation.

7.3 Research Method

A series of case studies were conducted across a range of industries looking at companies implementing ERP systems as shown in Table 7.1. A case study

Table 7.1 The cases in the study

Case	Industry	ERP Project
Threads	Textile	Global SAP System
Chemical	Chemical	Global SAP System
Bell	Manufacturing – Retail	Global SAP System
StatCo.	Office Supplies	European SAP System
CompCo.	Information Technology	National Masterpack System
Pump	Manufacturing – Industrial	Global SAP System
PlasCo. PharmCo.	Plastics	National Movex System
	Pharmaceuticals	Global SAP System

research strategy was employed as ERP implementation is a contemporary area and understanding the area poses content, context, and process questions which deal with operational links over time (Pettigrew, 1985; Miles and Huberman, 1994; Yin, 1994). Case study research to build theory (Eisenhardt, 1989) was the method used in order to understand the implementation process of ERP software in these companies. Theoretical sampling was used to choose cases, with the intention being to select cases representing different aspects of ERP project success and failure, thereby enabling the development of a richer theoretical framework. The initial framework developed based on literature and previous research and was used to give focus to interviews. Interview questions were related to the framework and included investigation of the IT and business legacy of the company, the approach taken for implementing the ERP software, and the outcomes from the implementation (where projects had reached their conclusion). More general background information about the company was also obtained. The framework was considered tentative and therefore subject to change during the iterative process of case study research. Two researchers collected data in the organizations over a period of two years. They conducted interviews lasting two to three hours, with key company business and IT personnel including managers, users, and consultants. Further documentary evidence was collected in the form of project documents and company literature, such as annual reports, and from the Internet. Interviews were held every six months starting from the formation of the strategy through to the 'go live' and the bedding in of the IT system and the reengineered business processes. This approach facilitated the triangulation of data and strengthened both construct and internal validity. The data were analysed within individual cases and across cases to generate and refine the theoretical framework. The resulting framework is shown in Figure 7.1 in the next section.

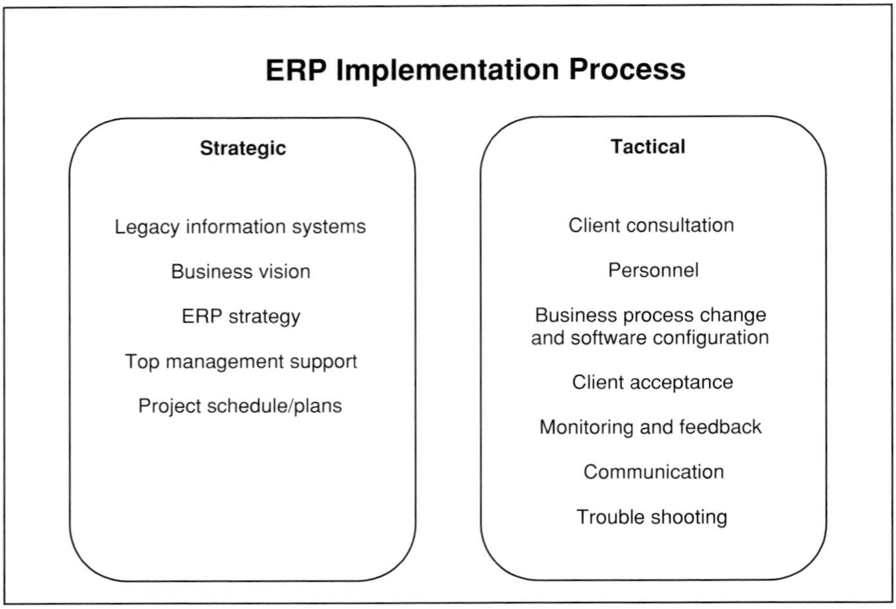

Figure 7.1 A critical success factors framework for ERP implementation

7.4 The Research Framework

The structure of the framework shown in Figure 7.1 is based on grouping the CSFs into strategic and tactical factors. Each group of factors is now discussed with the emphasis on those factors specific to ERP projects.

Strategic Factors

The strategic factors business vision, top management support, and project schedules/plans are based on Slevin and Pinto (1987). In addition, we have identified the importance of ERP strategy, and the role of legacy information systems, similar to what Roberts and Barrar (1992) refer to as 'antecedents'. A brief definition of Slevin and Pinto's factors are given before describing the role of legacy information systems and ERP strategy in more detail:

- Business vision is the clarity of the business model behind the implementation of the project. Is there a clear model of how the organization should work? Are there goals/benefits that can be identified and tracked?

- Top management support is the level of commitment by the senior management in the organization to the project in terms of their own involvement and the willingness to allocate valuable organizational resources.
- Project schedule/plans is the formal definition of the project in terms of milestones, critical paths, and a clear view of the boundary of the project.

Legacy information systems

Legacy information systems encapsulate the existing business processes, organization structure, culture, and information technology (Young Gul, 1997; Adolph, 1996; Bennett, 1994; Johnson, 1992). They cannot be controlled by a company to the same extent as may be other variables in the framework. Inevitably, they determine the amount of organizational change required to successfully implement an ERP system and will dictate the starting point for implementation. By evaluating legacy information systems, it is possible to gain an appreciation of the nature and scale of problems that may be encountered. This should influence ERP strategy selection. For example, if legacy information systems are extremely complex, with multiple technology platforms and a variety of procedures to manage common business processes, then the amount of technical and organizational change required may be high. If an organization already has common business processes and a simple technical architecture, change requirements may be low. Legacy information systems can therefore not be viewed as business or technical problems since their design and operation bind so many components of a business, such as organizational culture and processes.

ERP strategy

The ERP strategy is concerned with how the software package is to be implemented. For example, a skeleton version of the software package can be implemented initially, and extra functionality can then be added gradually once the system is operating and the users are familiar with it. A much more ambitious strategy is to implement a system that offers all the functionality that the organization requires in a single effort. Independent of the level of functionality chosen, there are different approaches to linking with the existing system ranging from implementing one ERP module at a time and interfacing with the legacy system or going for a big bang approach. The single module approach can be done in parallel with the existing system or on its own. International projects add further complexity regarding the choice of country-by-country rollout of the ERP system or parallel teams operating in different regions. It is clear that an organization's propensity for change should influence the choice of ERP strategy. A further technical choice is whether to carry out custom development on the package software and how this will affect the organization when upgrading the system in the future. The amount of custom development depends on whether an organization is willing to change their business to fit the software, or whether they prefer to change the software to fit their business. However, modifying the software to fit the business means that it is possible that any potential benefits from reengineering business processes will not be achieved. Once a decision has been made on the ERP strategy, issues surrounding how the project should be managed can be considered.

Tactical Factors

Client consultation, personnel, client acceptance, monitoring and feedback and communication are based on Slevin and Pinto (1987).

- Client consultation is the involvement of the users in the design and implementation of business process that includes formal education and training.
- Client acceptance is the user acceptance of the system and represents 'buy in' from the owners of the business processes.
- Monitoring and feedback is the exchange of information amongst members of the project team and the analysis of feedback from organization users.
- Communication is the formal promotion and advertisement of the project's progress from the project management team to the rest of the organization.

Business process design and software configuration

The additional factor Business process change and software configuration, recognizes the critical role of aligning business process to the ERP software during implementation. Although the standard project management factors are still important, they play a supporting role. Organizations need to understand their current business structure and business processes associated with their existing IT systems, and relate this to the business processes contained within the ERP system. ERP software configuration is different than building a customized system, because the focus of the development effort shifts from systems analysis and design to software configuration. The majority of the systems analysis and design effort has already been captured within the software and, consequently, much of the systems development effort focuses on enabling the required functionality embedded within the ERP system's business model.

7.5 Case Data

This section applies the research framework to two of the cases in the study. This is followed by an analysis of the interplay of the factors, particularly those identified as different in ERP implementation. A table summarizing and contrasting CSFs for the two cases (Table 7.2) is presented at the end of this section.

Threads

Threads is an international textile firm that had fragmented legacy information systems. There were over 40 separate accounting systems, and the information systems were a mixture of custom software packages that had Year 2000 compliance problems. The fragmentation inhibited the company's strategic vision, a coordinated approach to the European market. The senior management recognized this and developed a business that incorporated a new organizational structure in Europe based on a new business model. The strategic objectives were to improve the customer interface by linking sales and marketing with production and distribution systems across Europe, hopefully reducing overhead costs by at least 10%. The ERP strategy was to roll out the SAP R/3 package over Europe country by country; full functionality of the system was exploited immediately and the system was run in parallel with the existing systems. Threads aimed to have 90% of its business processes in each country the same. Threads' top managers supported the project; this gave them control. Furthermore, board pressure to reduce overhead costs in Europe gave the project a high profile within the company, and at least two senior directors were actively involved in the day-to-day execution of the project. There was a clear project schedule that aimed to implement a fully functional ERP system and reengineer the business process. The implementation process took three years though the idea for the project and changes to the project's scope can be traced over six years. The delays are primarily due to the changes in the project scope. Originally, the ERP system was to support the business processes of a particular product market within the business, but between 1993 and 1996 Threads determined that the whole business needed to be supported by the same system. The geographic complexity associated with designing common systems across Europe and a high turnover of external consultants also contributed to the delay. We estimate that the combination of these factors have extended the life of the project by around 30–50%. The project team, including consultants, top internal staff from functional business areas, and a change manager, managed the business process change and software configuration process. The senior project group and project directors involved in the day-to-day implementation met regularly. Numerous workshops, involving approximately 150 staff, were held to facilitate client consultation and to examine business processes. Thirty main business processes were identified and then defined in detail, providing the basis for configuring SAP. In isolated instances, some local systems were retained but the objective was still to achieve 90% commonality across all countries. The team aimed to secure client acceptance by involving users in the system testing process at pilot sites and requesting feedback. It was found however, that additional training was required as staff initially tried to use the system in the way that they had used their old one. The high levels of communication throughout the project facilitated troubleshooting although some problems were difficult to resolve, such as differences in national business processes and cultures. The change manager was responsible for ensuring that users were aware of the current state of the project and managing human resources issues. The Threads' change manager and the external consultants also had different approaches to managing the change process; the consultants preferred a radical approach, whereas the company preferred a more incremental change process.

StatCo.

StatCo. is a European stationery supplier. Its legacy systems were the result of a history of mergers and acquisitions, and its business was comprised of autonomous companies, each with its own IT system. The IT systems were not Year 2000 compliant and were not capable of running an integrated business. There had been problems in the past when some of the autonomous companies had tried to integrate, but the senior management still wanted to create a unified business and chose an ERP system to support this strategy. Their ERP Strategy provided each site with a system that matched or exceeded the functionality of the existing legacy system. Once all of the sites were on the common platform, they aimed to implement the remaining functionality of SAP. StatCo. chose not to customize the systems, and they staggered the implementation. They did not run the old and new systems in parallel. The project had top management support and the managing director was actively involved on a day-to-day basis. The two-year project schedule implemented SAP quickly across all sites to establish a basic common system and then built up functionality across the whole business. The methodology was to fast track SAP implementation in the sites needing only the

Table 7.2 Summary of the case data. Details an analysis of the Threads and StatCo. cases by CSF

CSF \ Case	Threads	StatCo.
Legacy information systems	Fragmented IT systems (e.g. over 40 separate accounting systems in Europe) that were either custom or heavily tailored packages, not year 2000 compliant and not considered a good fit for business. Business is geographically dispersed and considered uncoordinated in Europe. IT legacy was impeding the integration of the various companies and viewing Europe as one market.	Organization comprised of autonomous companies, each with their own IT system. These systems were not year 2000 compliant and none was capable of running an integrated business. There had been problems when some of these autonomous companies had tried to integrate.
Business vision	New organization structure in Europe that creates a pan European business with links between all national sales units with production and distribution sites across Europe. Reduce overhead costs by at least 10%.	Create one UK business out of several autonomous businesses arising from merger and acquisition activities.
ERP strategy	Implementation of standard SAP package with rollout over Europe country by country. System exploited full functionality of SAP and was run in parallel with the existing system. The aim was for a 90% common core.	Provide each site with system that matches/exceeds functionality of legacy system. Once all sites on common platform exploit functionality of SAP. No custom development. Staggered implementation throughout the UK and Europe. No parallel running of old and new system.
Top management support	Board approval for project enabled top managers from throughout the company to be seconded on to the project. Pressure from board to reduce overhead costs in Europe gave the project a high profile throughout the company with active participation of at least two directors.	Managing director is actively involved in the project because it is viewed as the enabler of creating a single, integrated European organization from a group of what are now separate businesses.
Project schedule/plans	Clear plan divided into phases organized around the design of common business processes. The growth in the scale of the project (to include all parts of the business) and the geographic complexity of designing common systems across Europe has led to delays in the schedule.	The project schedule is to implement SAP quickly across all sites to establish commonality and then build up the functionality across the whole business. The methodology is fast track SAP implementation where only the minimum of SAP functionality is adopted.

(*cont.*)

Table 7.2 *(cont.)*

CSF \ Case	Threads	StatCo.
Client consultation	Numerous workshops to examine business process involving approximately 150 staff from throughout the businesses.	Communication with users via a project newsletter and consultation with 'business champions' at each site.
Personnel	Project team included consultants, internal staff ('top' people from functional areas of business) and a change manager (the Human Resource Director).	Managers taken out of the business to work full time on project, team is cross functional and has a 'team charter'. Did not use external management consultants.
Business process change and software configuration	Numerous workshops held to examine the generic business processes. Identified 30 main processes and then defined them in detail. This provided the basis for configuring SAP. In isolated instances, some local systems were retained but the objective was to achieve 90% commonality in all countries.	Technical expertise for software configuration was sought from experienced SAP consultants. The philosophy of the business process change was to align the business processes to the software and simplify business processes to eliminate redundant activities.
Client acceptance	System testing involved users from some of the pilot sites. Further training was seen as required as management felt enough had not been provided.	User acceptance was obtained through user testing trials and extensive training on system and new business processes. Training continues after 'go live' to maintain standards.
Monitoring and feedback	Regular meetings held by senior project group and project director involved in day-to-day implementation on site.	Identified issues relating to data quality, training and change management that were weak in their first implementation have improved in these areas in subsequent implementations.
Communication	Communication between change manager and users.	Had weekly meetings so that decisions could be made rapidly. Communications with users via project newsletter.
Troubleshooting	High turnover of consultants delayed project. Change manager and external consultants had different opinions over the change process with the consultants preferring a more radical approach to the company's chosen incremental strategy.	Adequate testing of the system and a trial run of 'go live' for each site avoided potential problems.

minimum ERP functionality. Project representatives at each site obtained client acceptance through user testing trials and extensive training on the system and new business processes. Users also received a project newsletter. To maintain standards, the training continued after the system went live. The project team, including managers taken out of the business to work full time on the project, was cross functional and had a charter detailing the project philosophy. The business process change and software configuration activities were split between in-house staff and consultants, but the team also sought technical expertise for software configuration from experienced SAP consultants. The philosophy of the business process change was to align the business processes to the software, simplifying the processes to eliminate redundant activities. The main forum of communication through out the project was a weekly meeting so that decisions could be made rapidly. Careful testing and trial runs of the system before the final delivery date avoided additional problems.

Case Discussion

The following discussion demonstrates the interaction of the strategic and tactical aspects of the CSF's framework and highlights the importance of legacy information systems, ERP strategy and business process change and software configuration. Threads set a clear business vision to overcome extremely complex legacy information systems but the implementation process has been very slow. The scope of the project was changed to include all parts of the business and this coupled with the geographic complexity of the organization slowed the implementation process. The difficulties of implementation were exacerbated by the ERP strategy of attempting to implement the full functionality of SAP immediately on each new site. The combination of these factors meant that even with top management support and a clearly articulated business vision, the ERP implementation process was very slow and difficult. Tactically, the client consultation and user acceptance was exhaustive and the philosophy of the human resources director was to involve users at all stages in the business process design activities. This approach reflected the paternalistic culture of the firm and their non-adversarial approach to change. Although the implementation is viewed as a success overall, the time scales have been much longer than was first anticipated and the associated implementation costs are much higher than the original estimates. Threads contrasts sharply with the StatCo. case. The legacy of StatCo. was much simpler than Threads and, although it is international, the actual business is simpler from an information management perspective. The lack of a dominant culture and accepted ways of working also meant that the managers in the separate business units were more open to change. The inertia deriving from legacy information systems was much lower than in Threads. The ERP strategy was a fast track implementation in which the minimum functionality was implemented across all the sites that will provide the basis for further development. The project schedules and plans were therefore simpler to manage and the testing of basic SAP systems was simpler than that which occurred in Threads where full functionality systems were tested in parallel with the legacy information systems. Figure 7.2 goes some way to explaining the different experiences of Threads and StatCo. It shows the CSFs framework with the interplay of legacy information systems, ERP strategy and business process change and software configuration being shown supported by the other factors. This demonstrates ERP strategy should be selected in the light of legacy information system conditions due to the consequences for business process change and software

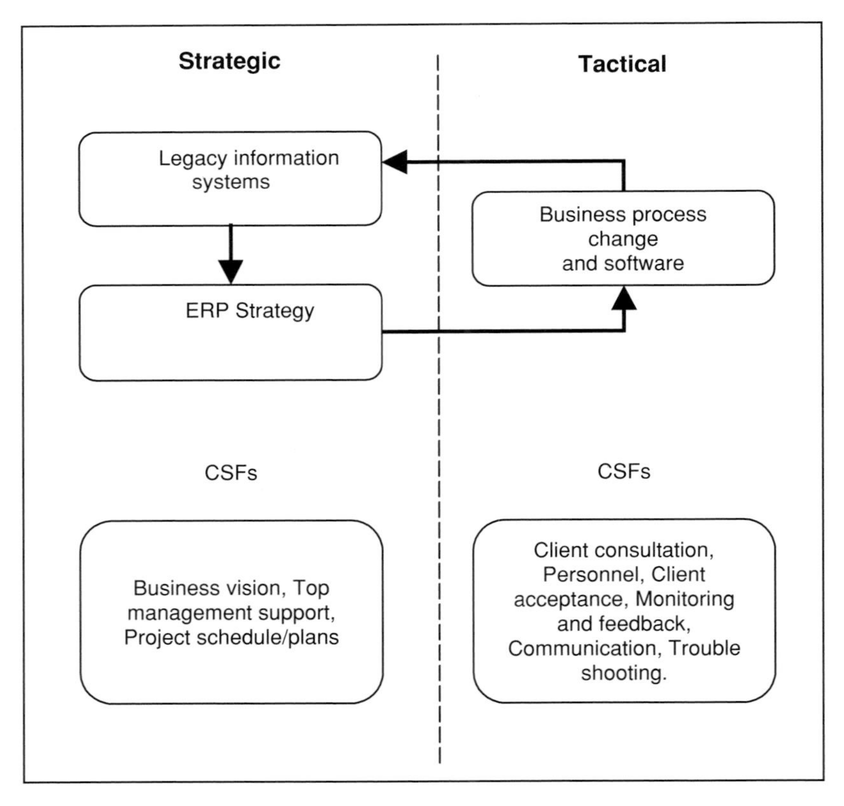

Figure 7.2 The interplay of the CSFs

configuration. It also completes the loop, indicating that the results of the process feed back into the legacy information systems conditions.

7.6 Conclusions

ERP systems link together an organization's strategy, structure, and business processes with the IT system. Although the technical risk from developing software is considerably reduced, risk during implementation is associated with aligning the processes with those of the software package and the corresponding change management and software configuration issues. The analysis of the cases reveals that in addition to standard project management critical success factors, there are other factors that need to be taken into account in ERP implementation. These are legacy information systems, ERP strategy and business process change and software configuration. Even though Threads' and StatCo.'s legacy information systems were similar, different ERP strategies were adopted. This demonstrates that choosing to implement ERP software is not a generic strategy since there are alternative ways those implementations can be approached. The key difference lies with the choice of whether to implement a skeleton or full functionality version of the system. The legacy information systems affected the success of the projects as each had some form of legacy to contend with since none was a 'green field' site. StatCo. attempted to minimize the negative impact their legacy could have on the project by implementing a skeleton strategy. Threads, even though they had considered change management issues, underestimated the impact that change would have on their organization and opted for a full functionality strategy. The multiple custom software packages and huge national differences throughout the business subsequently had a significant impact on the implementation making it much longer than that of StatCo.

The pervasive nature of the ERP platform means that it will be central to the IT infrastructures of many organizations for some time to come. This paper presents a CSFs framework that can aid in understanding this complex area. This paper is situated within a spectrum of research in the area of ERP; much of this focussed upon implementation. Further work should focus upon exploring this in greater depth and from different perspectives. However, additional branches of research could focus upon pre- and post-implementation issues. Interesting areas, where work is emerging, include the impact of ERP systems upon competition, web-enabled ERP, ERP maturity frameworks, social aspects of ERP usage, ERP selection, and the integration of ERP systems on an intra- and inter-organizational basis.

REFERENCES

Adolph, W. S. (1996) Cash Cow in the Tar Pit: Reengineering a Legacy System. *IEEE Software*, **13**(3), 41–47.

Avison, D. E. and Fitzgerald, G. (1995) *Information Systems Development: Methodologies, Techniques and Tools*, 2nd edn. London: McGrawHill.

Bashein, B. J., Markus, M. L., and Riley, P. (1994) Preconditions for BPR Success and How to Prevent Failures. *Information Systems Management*, **11**(2), 7–13.

Benjamin, R. I. and Levinson, E. (1993) A Framework for Managing IT Enabled Change. *Sloan Management Review*, **34**(4), 23–33.

Bennett, K. (1994) Legacy Systems: Coping with Success. *IEEE Software*, **11**(1), 19–23.

Bicknell, D. (1998) SAP to Fight Drug Firm's $500m. Suit over R/3 Collapse. *Computer Weekly*, 3 September, 3.

Bingi, P. Sharma, M. K., and Godla, J. K. (1999) Critical Issues Affecting an ERP Implementation. *Information Systems Management*, **16**(3), 7–14.

Davenport, T. H. (1998) Putting the Enterprise into the Enterprise System. *Harvard Business Review*, **76**(4), 121–131.

Deloitte & Touche (1997) *1996 CIO Survey: Major Packages*. Deloitte & Touche LLP.

Edmondson, G., Baker, S., and Cortese, A. (1997) Silicon Valley on the Rhine. *Business Week*, 3 November, 40–47.

Eisenhardt, K. M. (1989) Building Theories from Case Study Research. *Academy of Management Review*, **14**(4), 532–550.

Gibson, N., Holland, C., and Light, B. (1999) A Case Study of a Fast Track SAP R/3 Implementation at Guilbert Niceday. *Electronic Markets*, **9**(3), 190–193.

Grover, V. Seung Ryul, J., and Teng, J. T. C. (1998) Survey of Reengineering Challenges. *Information Systems Management*, **15**(2), 53–59.

Hall, G. Rosenthal, J., and Wade, J. (1993) How to Make Reengineering Really Work. *Harvard Business Review*, **71**(6), 119–131.

Holland, C. P. and Light, B. (1999) Global Enterprise Resource Planning Implementation. *Proceedings of the 32nd Hawaii International Conference On System Sciences*, Los Alamitos, California: IEEE Computer Society Press (CD-ROM).

James, G. (1997) IT Fiascoes and How to Avoid Them. *Datamation*, November, 84–88.

Johnson, G. (1992) Managing Strategic Change: Strategy, Culture and Action. *Long Range Planning*, **25**(1), 28–36.

Knowles, J. (1997) Competing Effectively: Buy Versus Build in Six Words. *Datamation*, January, 31.

Kotter, J. P. (1995) Leading Change: Why Transformation Efforts Fail. *Harvard Business Review*, **73**(2), pp. 59–67.

KPMG (1998) *Exploiting Packaged Software*, London: KPMG.

Lockett, A. G., Barrar, P. R. N., and Polding, M. E. (1991) MRPII Systems: Success Factors in the Process of Implementation. In *Production Research: Approaching the 21st Century*, Pridham, M. and O'Brien, C. (eds.), London: Taylor & Francis.

Martin, M. H. (1998) An ERP Strategy. *Fortune*, 2 February, 49–51.

Martinez, E. V. (1995) Successful Reengineering Demands IS/Business Partnerships. *Sloan Management Review*, **36**(4), 51–60.

Miles, M. B. and Huberman, A. M. (1994) *Qualitative Data Analysis: An Expanded Sourcebook*, 2nd edn. London: Sage.

Parr, A. and Shanks, G. (2000) A Model of ERP Implementation. *Journal of Information Technology*, **15**(4), 289–303.

Pettigrew, A. M. (1985) Contextualist Research: A Natural Way to Link Theory and Practice. In *Doing Research that is Useful in Theory and Practice*, Lawler, E. E. Mohrman, A. M. Mohrman, S. A. Ledford, G. E. Cummings, T. G. And Associates (eds), San Fransisco: Jossely-Bass.

Price Waterhouse (1995) *Information Technology Survey*. London: Price Waterhouse.

Roberts, H. J. and Barrar, P. R. N. (1992) MRPII Implementation: Key Factors for Success. *Computer Integrated Manufacturing Systems*, **5**(1), 31–38.

Slevin, D. P. and Pinto, J. K. (1987) Balancing Strategy and Tactics in Project Implementation. *Sloan Management Review*, **29**(1), 33–44.

Yin, R. K. (1994) *Case Study Research: Design and Methods*, 2nd edn. London: Sage.

YoungGul, K. (1997) Improving Legacy Systems Maintainability. *Information Systems Management*, **14**(1), 7–11.

8 Critical Success Factors Revisited: A Model for ERP Project Implementation

Anne Parr and Graeme Shanks

8.1 Introduction

Implementation of ERP systems is widely recognized (Parr, Shanks, and Darke, 1999; Ambrosio, 1997; Fine, 1995; Gartner Group, 1998; Horwitt, 1998; Stedman, 1998a, 1998b; Martin, 1998; Piszczalski, 1997; Tebbe, 1997) as both problematic, and likely to overrun time and budget allocations for the actual implementation project. Of course IT projects do have a history of having such problems but ERP implementations are more intractable than most. Martin (1998) reported that over 90% of ERP implementations were late and/or over budget; some have terminated in expensive legal actions (James, 1997; Shanks et al., 2000); and the popular literature (*Infoweek*, *Computerworld*, etc.) contains numerous stories which describe ERP implementations in colourful terms, such as 'endurance' tests, 'fiascos', 'living to tell about it', and 'war stories' (see, for example, Tebbe, 1997; Horwitt, 1998).

In response to these problems, the academic literature has taken a two-pronged approach. First, it has focused on CSFs (Parr, Shanks, and Darke, 1999; Holland, Light, and Gibson, 1999; Sumner, 2000; Shanks et al., 2000; Parr and Shanks, 2000b) for implementation of ERP systems. Secondly, it has developed process models of successful implementation (Bancroft, 1996; Bancroft, Seip, and Sprengel, 1998; Markus and Tanis, 2000; Holland, Light, and Gibson, 1999; Parr, Shanks, and Darke, 1999; Ross, 1998). Both approaches aim to achieve more successful implementation and deeper insight into the process.

In this paper we first characterize CSFs which were elicited from an earlier stage of the research. These CSFs were founded on interviews with experienced ERP implementation practitioners. In the section on CSFs we describe the elicitation process and define 'critical success factor'. Secondly, we review other process models of ERP implementation and we argue that there is justification to create a project phase model (PPM) of ERP implementation. This model differs from those in the literature in two respects:

- Its focal point is the individual, discrete phases of the implementation *project* itself. This differs from other models which treat the project as a single phase in the whole implementation effort.
- It clarifies which CSFs complement which phase of the implementation project. It demarcates the individual, discrete phases of the implementation *project* itself and, during each phase, it accentuates the relevant CSFs.

The aim of the PPM is to combine phases of the project with the relevant CSFs. Finally, we report the results of empirical testing of this combination of CSFs and the PPM model by describing and analysing two case studies. The first of these was unsuccessful in project terms, that is, it was substantially over time and over budget. The second company, in these terms, achieved a highly successful implementation. The companies were carefully selected. The two companies are closely related, and the unsuccessful case occurred earlier which meant that the later one had access to the learning gained in the first case. This combination of CSFs and the PPM proved to be a useful lens for understanding ERP implementation projects. It extended previous research which focused either on the CSFs or the process model for the entire implementation process. Findings from the case studies show the way to a best-practice process model.

The combination of CSFs and the PPM provides researchers with a foundation for additional empirical research, and offers direction for practitioners when considering future ERP projects. There are several valuable inferences that may be drawn from the research. Firstly, the unsuccessful project facilitated organizational learning. The major outcome of this learning was that a designated full-time senior 'champion' with clearly defined responsibilities *must* drive a successful ERP implementation. Secondly, large projects are not likely to succeed: they need to be broken up into smaller projects involving less business process reengineering (BPR). These simpler projects are referred to here as 'Vanilla' projects. Finally, the combination of CSFs and the PPM underlines the CSFs that are necessary for successful ERP implementation, the phases and when they will be critical when planning ERP implementation projects.

The proposed model consists of two concepts: CSFs and the ERP process models. Each of these is discussed in turn below.

8.2 CSFs

The CSFs used in the model described above were identified in an earlier stage of this research (Parr, Shanks, and Darke, 1999). The purpose of identifying

CSFs is to provide guidance to practitioners in planning and monitoring an ERP implementation. Critical success factors are defined as 'those few critical areas where things must go right for the business to flourish' (Rockhart, 1979). They have been applied to many aspects of information systems, including project management, manufacturing system implementation, reengineering, and, more recently, ERP system implementation (Parr, Shanks, and Darke, 1999; Holland, Light, and Gibson, 1999).

Williams and Ramaprasad (1996) noted that although CSFs are widely used by academic researchers and practitioners, it is important to discriminate between different levels of criticality. They distinguish four types of criticality:

> factors linked to success by a known causal mechanism; factors necessary and sufficient for success; factors necessary for success; and factors associated with success. (p. 250)

These factors are in descending order of power: for example, a causal link between a factor and an outcome is empirically and logically stronger than a mere association between a factor and an outcome. Similarly, a necessary and sufficient link is stronger than a link that is only necessary. The CSFs used in this paper were generated in a previous study (Parr, Shanks, and Darke, 1999). In the taxonomy provided by Williams and Ramaprasad (1996) the level of criticality is the third one: they define those factors which, while not sufficient to ensure a successful outcome, are necessary to achieve success. Both the concepts of causality and necessary *and* sufficient conditions are concepts so rigorous that they were regarded by the authors as unachievable in the analysis of complex social, organizational, and technical interactions, such as an ERP implementation. Association is too weak a notion. The concept of necessity, while not optimum, does however provide guidelines for practitioners on those factors that must be present if successful implementation is to be achieved.

The CSFs for ERP implementation were identified using a review of the literature on system implementation followed by an extensive set of interviews with ERP implementation. The interviewees had been involved in a total of 42 ERP implementation projects in Australia and the USA (Parr, Shanks, and Darke, 1999). The CSFs identified in the earlier study are shown in Table 8.1.

8.3 ERP Implementation Process Models

Several researchers have developed process models of ERP implementation. In this section we review three of those models. The implementation of an ERP system implies that a company must focus on, evaluate and define in precise

Table 8.1 Critical success factors for ERP implementation

Factors		Description
1	Management support	Top management advocacy, provision of adequate resources, and commitment to project
2	Best people fulltime	Release full-time on to the project of relevant business experts
3	Empowered decision makers	The members of the project team/s must be empowered to make quick decisions.
4	Deliverable dates	At planning stage set realistic milestones and end date
5	Champion	Advocate for system who is unswerving in promoting the benefits of the new system
6	Vanilla ERP	Minimal customization and uncomplicated option selection
7	Smaller scope	Fewer modules and less functionality implemented, smaller user group; single/fewer site/s
8	Definition of scope and goals	The steering committee determines the scope and objectives of the project in advance and then adheres to it
9	Balanced team	Right mix of business analysts, technical experts and users from within the implementation company and consultants from external companies.
10	Commitment to change	Perseverance and determination in the face of inevitable problems with implementation

detail relevant company processes. Implementing the ERP is itself a process that begins with planning for the system. After planning is complete, a project team embarks on and then moves through a number of discrete project phases. After the system is up and running, there may be a post-implementation review and later a stabilization phase. As several authors (Shanks et al., 2000; Markus and Tanis, 2000) have stated, the implementation process of an ERP is best conceptualized as a business project, rather than the installation of a new technology.

Bancroft, Seip, and Sprengel (1998) presented a view of the implementation process which was derived from discussions with 20 practitioners and from studies of three multi-national corporation implementation projects. The Bancroft model has five phases: Focus, As Is, To Be, Construction and Testing, and Actual Implementation. The Focus phase is essentially a planning phase, in which the key activities are the set-up of the steering committee, selection and structuring of the project team, development of the project's guiding principles, and creation of a project plan. The As Is phase involves analysis of current business processes, installation of the ERP, mapping of the business processes on to the ERP functions, and training of the project team. The To Be phase entails high-level design and then detailed design subject to user acceptance, followed by interactive prototyping accompanied by constant communication with users. The key activities of the Construction and Testing phase are the development of a comprehensive configuration, the population of the test instance with real data, building and testing interfaces, writing and testing reports, and finally system and user testing. The final phase, Actual Implementation, covers building networks, installing desktops, and managing user training and support. In summary, the model of implementation extends from the beginning ('Focus') of the project proper to the cutover to the live system.

Ross (1998) developed a five-phase model based on 15 case studies of ERP implementation. The phases are Design, Implementation, Stabilization, Continuous Improvement, and Transformation. The Design phase is essentially a planning phase in which critical guidelines and decision making for the implementation is determined. Ross' implementation covers several of Bancroft's phases: As Is, To Be, Construction and Testing, and Actual Implementation. For Ross, Stabilization occurs after cutover, and is a period of time in which system problems are fixed, and organizational performance consequently improves. This is followed by a continuous period of steady improvement in which functionality is added. Finally firms expect to reach the stage of transformation in which organizational boundaries and systems are maximally flexible.

Markus and Tanis (2000) have developed a four-phase model of ERP implementation: Chartering, Project, ShakeDown, and Onwards and Upwards. The Chartering phase begins before Bancroft's Focus and Ross's Design phases. It includes development of the business case for the ERP, package selection, identification of the project manager, and budget and schedule approval. The description of their Project phase is similar to Ross' Project phase, and it covers all of Bancroft's As Is, To Be, Construction and Testing, and Actual Implementation phases. The main activities of this phase are 'software configuration, system integration, testing, data conversion, training, and rollout' (Markus and Tanis, 2000: 20). Markus and Tanis's Onward and Upwards phase is essentially a synthesis of Ross's (1998) Stabilization and Continuous Improvement phases.

Several points need to be made about these three models. Firstly, Markus and Tanis, and Ross include a planning phase which occurs prior to the actual implementation project. Secondly, these two models collapse the actual implementation project into one discrete unit. By contrast, Bancroft (1996) categorizes the stages of the actual project into four project sub-phases (As Is, To Be, Construction and Testing, and Actual Implementation). Thirdly, two of the models (Ross, and Markus and Tanis) include a post-project phase (variously called Continuous Improvement, Transformation, and Onward and Upwards) in the model of the whole ERP implementation enterprise. Finally, none of them relates CSFs to phases of implementation.

The model on which this research is based, the PPM, synthesizes previous models, in that it recognizes the importance of the planning and post-implementation stages. However the focus of the model is on the implementation project, and the factors which influence successful outcome at each of the phases of the implementation. There are three justifications for this focus:

1 Many problems documented in the literature relate to the actual implementation project (Parr, Shanks, and Darke, 1999; Ambrosio, 1997; Fine, 1995; Gartner Group, 1998; Horwitt, 1998; Stedman, 1998a, 1998b; Martin, 1998; Piszczalski, 1997; Tebbe, 1997). A strength of the PPM is that it identifies the discrete sub-phases of the project, while also recognizing the importance of the planning and post-implementation phases. Although organizations are primarily interested in business success (Markus and Tanis, 2000), understanding how to also achieve project success is critical in ERP implementation due to the high costs involved.
2 The PPM model relates success factors to the phases of the ERP implementation process. This builds on earlier work, which identified those factors which experts in ERP implementation believe lead to successful implementation. It also augments the model in that it links factors leading to success with implementation stages.
3 The PPM is concerned with the concept of project 'success'. Project success simply means bringing the project in on time and within budget. This concept of success is recognized by Markus and Tanis (2000), and is the concept used by the implementation experts discussed below. This differs from the Ross and Markus models which are aimed at 'success' in terms of contribution to company performance. It is recognized that the notion of 'success' changes over the life of the implementation enterprise (Shanks et al., 2000). The purpose of a process model of implementation is to provide guidance for 'successful' project implementation. A process model which extends over the life of the implementation enterprise, extending beyond the project, and into refinement and organizational transformation, and which incorporates the project as only one of several phases, is aimed towards a concept of success that involves the contribution of the ERP to the performance of the implementing company.

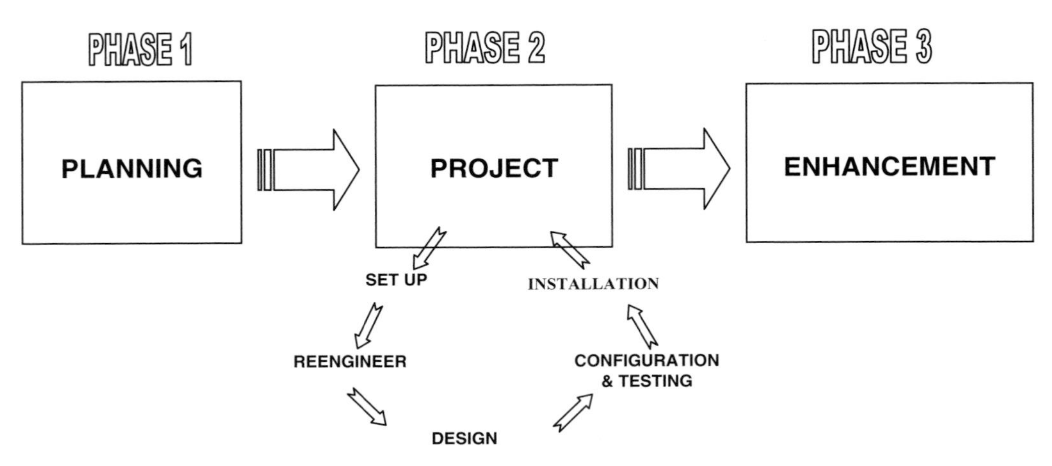

Figure 8.1 PPM Model of ERP Implementation

Project Phase Model (PPM) of ERP Implementation

The PPM consists of two concepts. CSFs have been discussed above. In this section the implementation phases are discussed.

PPM phases

The PPM (see Figure 8.1 above) has three major phases: Planning, Project, and Enhancement. The Planning phase includes selection of an ERP, assembly of a steering committee, determination of high-level project scope and broad implementation approach, selection of a project team manager, and resource determination. The Project phase extends from the identification of ERP modules, through to installation and cutover. The Enhancement phase may extend over several years, and includes the stages of system repair, extension, and transformation; that is, it encapsulates Ross's Stabilization and Continuous Improvement phases, and Markus and Tanis's Onwards and Upwards phases.

Additionally, because the focus of the model is on the implementation project itself, the Project phase is divided into five sub-phases: set-up, reengineer, design, configuration and testing, and installation. In the set-up phase the project team/s are selected and structured with an appropriate mix of technical and business expertise, the team/s integration and reporting processes are established, and guiding principles are developed and/or re-affirmed. The reengineering phase involves analysis of current business processes, often to determine the level of BPR required, installation of the ERP, mapping of the business processes on to the ERP functions, and training of the project team/s. The design sub-phase entails high-level design and then detailed design subject to user acceptance. This is followed by interactive prototyping and is accompanied by constant communication with users. The major activities of the configuration and testing sub-phase are the development of a comprehensive configuration, population of the test instance with real data, building and testing interfaces, writing and testing reports, and finally system and user testing. Finally the installation sub-phase includes building networks, installing desktops, and managing user training and support. These last four sub-phases are comparable to those described above for the Bancroft model. Figure 8.1 provides a graphical representation of the PPM.

Combining the PPM with the CSFs

The CSFs elicited from the first stage of the research may be combined with the PPM in order to examine the relationship between the CSFs and the PPM. Several questions about the relationship are investigated. First, which CSFs are necessary in which phase? Second, when comparing a successful case and an unsuccessful case, what are the differences in the relationship between the phases of the PPM and CSFs? Thirdly, how are these differences explained?

8.4 Research Method

The case study research method was used in this study to investigate the relationship between phases in the PPM and necessary CSFs in a real-world context. Case studies are useful to study a contemporary phenomenon in a real-world context and to create and refine theory (Yin, 1989; Eisenhardt, 1989). Two case studies are reported in order to allow cross case comparison. The first case (Oilco) concerns the implementation of an ERP into the Australian and New Zealand subsidiaries of a multinational oil company. The second (Exploreco) is an Australian implementation of an ERP into the exploration arm of the same company, some years later. Both cases were perceived within each company to have been 'successful' in that they have brought considerable benefits to the companies. However they differed in terms of *the project concept of success* defined above. The Oilco implementation went substantially over budget and over time. Eventually it cost A$70m and extended over a seven year period. Exploreco on the other hand came in under budget, and within two weeks of its time allocation. This case, particularly, was thought by all participants to have been a model of successful project implementation that had incorporated the knowledge gained from the parent company's earlier implementation.

Data Collection

Data were collected in 1999 from semi-structured interviews, general company documentation, and implementation-specific documentation, such as project plans and post-implementation reports (Darke, Shanks, and Broadbent, 1998). The PPM provided a framework from which the interview protocol was developed. Interviews were conducted with *at least* five stakeholders in each of the companies. These were selected to cover a range of possible viewpoints, and between the two companies they included system managers, project managers, project team members, business analysts, consultants, and system users (expert and non-expert).

8.5 The Case Studies

In this section an overview of each case is presented, followed by a detailed comparison of the two cases. Subsequently a tabular summary of the importance of each CSF in the PPM phases is presented.

The Two ERP Implementations

The first case study involved Oilco, a refiner and marketer of a broad range of petroleum products in Australia and 11 countries in the Pacific. As one of Australia's major industrial companies, Oilco directly employs over 2000 people and owns assets valued at approximately A$2 billion. Oilco is the Australian subsidiary of one of the world's largest multinational oil companies. It has a nationwide network of 1800 locations, is one of the four major oil companies in Australia, and enjoys a substantial market share. In the late eighties the global oil industry underwent significant restructuring and increasing competition. As a consequence, Oilco wished to implement a new information system to achieve fully integrated process automation, improved levels of customer service, and to facilitate planned business restructuring. To meet these business requirements the company selected, in 1989, a mainframe-based ERP solution. With 1600 users in Australia, New Zealand and the Pacific Islands, this ERP system is now one of the largest and most complex mainframe implementations in the world. It processes 25 000–35 000 transactions per hour, and handles over one thousand orders per day across the country. The implementation of the system involved major change to the company's business processes so that they matched the ERP's processing methods. While recognizing that some existing business process changes were necessary, Oilco aimed to maximize the integration benefits of the ERP and simultaneously to streamline the company's existing processes. The implementation also involved development of an oil-industry specific module. The ERP (referred to here as ERP1) has now been implemented for over four years, and the business benefits are substantial. They include better sales forecasting, fully automated ordering and delivery processes, real-time financial data, improved data quality and streamlined business processes. Although eventually significant benefits have been achieved, the project itself went significantly over budget and over time.

The second case study involved Exploreco, an oil and gas exploration and production company in the southwest of Australia. Exploreco is a major affiliate of Oilco. The company is involved in off-shore gas and oil exploration and production. In 1997 Oilco acquired another oil exploration company which had an operational system that became the Exploreco operational system. However there was substantial dissatisfaction with this system within Exploreco, and it was not Y2K compliant. So in 1998 the company had to decide either to rework and upgrade the existing system, or to replace it. They chose a new system, and conducted a feasibility analysis of several ERP systems. They decided for budgetary reasons, and because it suited their exploration business, to implement an ERP system (referred to here as ERP2). Although Oilco had ERP1, the company already had experience with ERP2 systems in Canada, Nigeria, and Singapore. The budget and project scope was considerably more modest than the Oilco implementation, so they planned to implement the system within 12 months, from January 1998 to November 1998. Documentation on the existing system indicated that understanding of requirements was already advanced, but they took the opportunity to refine and reengineer the system, particularly given the level of dissatisfaction with the old system. Moreover, they needed to align the new system (ERP2) with Oilco's existing ERP (ERP1). The implementation project was driven by Oilco's head office, which performed costings, set the scope, made recommendations, and provided leadership on the steering committee. System goals were set via performance indicators. For example, the indicators included the number of cheque runs in a given period; a measured reduction in off-system payments, and a reduction in suppliers from 6000 to 600. Given the lessons learned in the Oilco implementation, the steering committee insisted that the best people be released full time for the life of the project, and a 'Project Champion' (that was his official title) was placed on the steering committee. The project was completed on time and on budget; and was described by the highly experienced project manager as the 'easiest implementation' he had 'ever been involved in'

(Interview, December 1999). It has now been in operation for 15 months, and the business benefits are significant. These include a measured reduction in manual processes, manual transactions and the number of suppliers which has led to improved procurement and inventory systems; stream-lined, real-time accounting systems; a reengineering of processes which involved a devolution of responsibility back into the hands of the operators; and improved time accounting (to 15 minute intervals). This last benefit has been particularly important since this company has many joint ventures.

Case Study Data

Table 8.2 below provides a detailed description and comparison of the two cases. It is clear that there is significant variation in the CSFs for each implementation in each phase of the PPM. Only those CSFs that were believed by most participants to be important to very important are identified in Table 8.2. They are shown in descending order of importance.

A tabular summary of the importance of each CSF in the PPM phases is then presented in Tables 8.3 and 8.4 below. Table 8.3 represents the Oilco case study findings; Table 8.4 the Exploreco findings. The tables show the PPM, its phases, and the CSFs. Each cell represents a particular CSF in a particular phase. The number of ticks in each cell represents the strength of the participants' consensus that the particular CSF was necessary in that phase. Four ticks indicate that the particular CSF was considered to be of major importance in that phase of the PPM. Three ticks indicate that the CSFs were considered very important. Two ticks indicate that the CSFs were considered important. One tick indicates that the CSFs were considered to be of minor importance and no ticks indicate that the CSFs were considered to be unimportant. We have not included 'smaller scope' as a CSF in Table 8.3 as one implementation was clearly large in scope and the other smaller in scope. We draw on this difference later in the case study analysis.

Discussion and Analysis

The two cases were selected for two prime reasons. Firstly, while both cases were perceived to have brought considerable business benefits to the organizations, the Oilco implementation was considered a failure in a project sense and the Exploreco implementation was considered successful in a project sense. Secondly, the Exploreco implementation represented an evolution of ERP implementation: Oilco is now a 'dinosaur' of the ERP implementation process,

Table 8.2 Case study findings using the PPM

Phase tasks	Oilco	Exploreco
1 Planning		
Clarification of the system rationale, selection of an ERP, determination of high-level project scope and broad implementation approach, and resource determination.	Prior to the Asia-Pacific implementation, in 1988 the international parent of Oilco began a pilot ERP project in Europe. Oilco conducted a 6-month evaluation of the ERP. The evaluation included a technical evaluation, high-level evaluation by user personnel of the ERP modules to determine their fit with business requirements for financial and operational systems. The evaluation, while generally positive, highlighted some risks in adoption of the ERP. However, senior management insisted on a global commitment to the ERP. The level of management involvement focused the organization at senior levels, both locally and in the parent corporation. This led to the commitment and resources required to implement a common integrated systems. It was recommended that the ERP be used both for pricing and sales and to support system integration, and to reduce costs by minimizing the number of application technical platforms that were being supported. In 1990 it was decided to implement the ERP in multiple stages. Stage 1 was implementation of financials and purchasing for Australia and NZ, and sales and pricing for Australia. Stage 2 covered logistics and plant maintenance for Australia. Stage 3 extended this functionality to the Pacific Islands. Stage 4 was intended as a rollout of the Australian design to NZ but incorporated some local changes.	In 1997, Exploreco was dissatisfied with its existing system which had been inherited as a result of a takeover. The system was deficient particularly in the accounting and inventory areas, had poor levels of user acceptance and was not Y2K compliant. The parent company (Oilco) considered an upgrade of the existing system, or an ERP. The company had global experience of both ERP1 and ERP2 systems, and decided on the ERP2 system. Oilco did the costings and project scope. The single site, 12 month project was not expected to be problematic because there was company experience of the ERP, requirements were known because of the existing system, and the users' dissatisfaction with the existing system meant they were keen to obtain a superior system. CSFs: *Management support* *Champion* *Commitment to change* *Vanilla ERP* *Best people full time* *Deliverable dates* *Definition of scope and goals*

(*cont.*)

Table 8.2 *(cont.)*

Phase tasks	Oilco	Exploreco
	CSFs: *Management support* *Champion* *Commitment to change* *Vanilla ERP*	
2a Set-up		
Steering committee/team selection, resourcing, back-fill staff, project structures and reporting mechanisms.	The ERP implementation project was set up, under the leadership of a venture manager who came from the UK. The venture manager reported directly to the CEO Australia. The team consisted of business and IT personnel from both Australia and NZ, and consisted of 90 full-time people plus another 20 who developed documentation, plus the local user experts (LUEs). Project managers for each application area were responsible for a number of application teams. These managers reported to the venture manager. A steering committee had overall responsibility for the whole project. Separate teams were used to develop each application, and they consisted of business experts and IT staff, approximately 50% of each. CSFs: *Management support* *Balanced team* *Definition of scope and goals*	A project champion was appointed. He was also a member of the Leadership Council (Board). His task was to create acceptance within the organization, and head off problems, select a project manager, then select the project team (8 members + consultants), and communicate with the Leadership Council. Measurable goals were set via key performance indicators. It was decided to walk through all milestones. The end date was 'set in concrete'. The best staff from accounting were released full time. The team represented a mix of IT and business expertise. It was determined to do minimal customization of the ERP. CSFs: *Management support* *Champion* *Balanced team* *Vanilla ERP* *Deliverable dates* *Definition of scope and goals*
2b Reengineer		
Analysis of current business processes, installation of the ERP, mapping of the business processes on to the ERP functions, identification of data and system interfaces	The decision had been taken to have minimal customization of the software. However, it was found that an oil-industry specific module needed to be commissioned. Moreover, although some business processes did not fit well with the software, those processes were not reengineered as they would have	Although the decision had been taken to have minimal customization, it was agreed that this was an opportunity to improve business processes. The data quality was 'awful', and there was a proliferation of manual systems. There needed to be changes so that all transactions

Table 8.2 *(cont.)*

Phase tasks	Oilco	Exploreco
and training of the project team.	meant job re-structuring and it was considered that line operators would be resistant. CSFs: *Balanced team* *Definition of scope and goals*	were electronic. This required communication with and persuasion of users to accept change. Such change was more acceptable when proposed by an expert colleague. CSFs: *Empowered decision makers* *Management support* *Balanced team* *Definition of scope and goals*
2c Design High-level design followed by detailed design, and interactive prototyping. This is accompanied by constant communication with users.	This phase accompanied a company restructure, which meant that project personnel changed. Also there was a scarcity of expertise available at that time on this ERP. User commitment waned. The company had determined to have a policy of minimal customization, but encountered a push from those who wanted 'a gold-plated solution' (a wish list). So management energy was spent on conflict resolution. CSFs: *Best people full time* *Vanilla ERP* *Management support* *Commitment to change*	Decisions taken in the Planning phase by management resulted in very little 'pushback' from users in this stage. The decision to do minimal customization meant that there were no significant changes to the software apart from customizing it to a mainframe interface. The adherence to 'rock-solid' project milestones acted as a curb on the impulse to generate numerous reports of doubtful use. CSFs: *Vanilla ERP* *Management support* *Deliverable dates*
2d Configuration and testing Development of a comprehensive configuration, the population of the test instance with real data, building and testing interfaces, writing and testing reports, and finally system and user testing.	This phase was also concurrent with extensive company re-structuring. As a result personnel changed and the required people were not always released on to the project. This underlined the need for the best human resources. A huge amount of programming was done to generate reports. The decision to adhere to minimal customization precluded an easier, more user friendly option.	During this phase it was decided that the user interface was unacceptable, and that there should be a Lotus Notes interface to the ERP and to the users. It was also decided that the system should support EFT to the bank. Finally it was decided that the system should interface with an offshore company's maintenance system. These late changes meant

(cont.)

Table 8.2 (*cont.*)

Phase tasks	Oilco	Exploreco
	CSFs: *Best people full time* *Vanilla ERP*	that the system end date was (marginally) later than expected. These changes required management to authorize and support them, the business experts on the project team were valuable in gaining user acceptance, and the changes further highlighted the effects of not adhering to the principle of minimal customization. CSFs: *Management support* *Balanced team* *Vanilla ERP*
2e Installation		
Building networks, installing desktops, and managing user training and support.	Since this had been a long (7 years) and complex implementation, and because there had been extensive re-structuring during the time of the implementation, over the life of the project there had been surges of enthusiasm, followed by indifference and a low project profile. Training of users was an immense task: in the early stage 1525 people were trained in three months. This involved 1600 training sessions at 12 locations by professional trainers. In the second stage a different approach to training was taken: local user experts (LUEs) conducted training. These were often drawn from the business members of the project teams. CSFs: *Management support* *Commitment to change* *Balanced team* *Best people full-time*	Given the level of planning, the 'Go Live date was a non-event' (Project Manager). Management support was needed to engender enthusiasm. The technical membership of the project team managed the actual installation, while the user experts from the project team were responsible for user training and support. CSFs: *Management support* *Balanced team* *Commitment to change*
3 Enhancement		
System repair, extension, and	It is at best unclear whether transformation has occurred (see discussion below)	Since the system went live in 1998, there have been no significant repairs or changes.

Table 8.3 Oilco – Project Phase Model (PPM) of ERP implementation incorporating CSFs

Factor	Phase 1 Planning	Phase 2 Project: 2a Set up	2b Reengineer	2c Design	2d Configuration and testing	2e Installation	Phase 3 Enhancement
Management support	✓✓✓	✓✓✓	✓	✓	✓	✓✓✓	
Champion	✓✓✓	✓		✓		✓	
Balanced team		✓✓✓	✓✓	✓✓		✓✓	
Commitment to change	✓✓✓	✓		✓		✓✓	
Vanilla ERP	✓✓✓			✓✓	✓✓		
Empowered decision makers		✓		✓			
Best people F/T		✓	✓	✓✓	✓✓	✓✓	
Deliverable dates		✓				✓	
Definition of scope and goals	✓	✓✓	✓✓	✓		✓	

Table 8.4 Exploreco – Project Phase Model (PPM) of ERP implementation incorporating CSFs

Factor	Phase 1 Planning	Phase 2 Project: 2a Set up	2b Reengineer	2c Design	2d Configuration and testing	2e Installation	Phase 3 Enhancement
Management support	✓✓✓✓	✓✓	✓✓✓	✓✓✓	✓✓✓	✓✓✓	✓✓
Champion	✓✓✓	✓✓	✓	✓	✓	✓	✓
Balanced team		✓✓	✓✓✓	✓✓	✓✓	✓✓	✓
Commitment to change	✓✓✓	✓	✓	✓✓	✓	✓	✓
Vanilla ERP	✓✓✓	✓✓	✓✓		✓✓		
Empowered decision makers	✓	✓	✓✓✓	✓✓	✓	✓	
Best people F/T	✓✓✓	✓	✓	✓	✓	✓	✓
Deliverable dates	✓✓✓	✓✓	✓	✓	✓	✓	
Definition of scope and goals	✓✓✓	✓✓	✓✓	✓	✓	✓	✓✓

and the implementation at Exploreco represented a unique opportunity to pinpoint the next evolutionary stage using the PPM framework.

In discussing the two cases it is important to note that they differ in three important aspects. Firstly, the scope of the Oilco case is much larger than the Exploreco case. Secondly the Oilco case involved an earlier version of the ERP system that was more limited in functionality than the Exploreco system and required additional programming to generate required reports. Thirdly, although Oilco and Exploreco are companies in different geographical locations, the implementation project at Exploreco was driven from Oilco's head office, and the project 'champion' had access to extensive experience from the Oilco implementation. Our discussion highlights the similarities and differences in CSFs within each phase of the PPM between the two cases. We then offer some explanations for these differences, noting particularly the features of the successful case. Finally we consider the implications of the PPM for research and practice.

Case findings – similarities

A number of important similarities between the two cases are evident.

Both companies indicated the importance of the Planning phase. ERP implementation is clearly understood to be a major project, requiring substantial resources, commitment and change management. The CSFs identified by both companies during the Planning phase were management support, a champion for the project, commitment to the change, and a Vanilla ERP approach.

CSFs for installation were similar for both companies. Management support and having a balanced team were highlighted in both cases as the crucial CSFs in the installation phase. Commitment to change and having the best people full time were also considered important by both organizations for installation.

The Enhancement phase was insignificant in both companies. Although some CSFs were nominated in the Exploreco case, neither organization experienced significant enhancement after the project phase. The Enhancement phase was defined as including system repair, continuous improvement, and organizational transformation. The Exploreco ERP has now been in place for 15 months, and there has been no post-implementation review. All interviewees were delighted with the system: 'Never seen this done so well' (project champion); 'Cutover was a non-event' (project manager). Perhaps this explains why there has been no significant repair of the system (although they have made minor adjustments to the accounting and purchasing systems), nor has the business been transformed subsequent to a continuous period of steady improvement. The Oilco ERP has been in operation for three years. In that time, and during the seven-year implementation, there was extensive company re-structuring. Although it was claimed by one person that there has been continuous improvement in system operation and performance, 'no quantitative post-implementation audit has been done' (Interview with MIS Manager of the Oilco ERP). Given the complexities of the re-structuring and the lack of a post-implementation audit, it is at best unclear whether transformation has occurred.

Case findings – differences

A number of important differences between the two cases are evident.

Many CSFs were considered important over *all* phases in the Exploreco case. Exploreco considered seven CSFs to be necessary in *all* phases of the PPM. These were management support, a champion, commitment to change, Vanilla ERP, empowered decision makers, the best people full time, deliverable dates, and the definition of scope and goals. However, in Oilco, only one CSF, management support, was considered necessary overall. For Oilco most of the CSFs related to the Planning phase. For Exploreco, the CSFs are more evenly distributed across all phases.

Overall, CSFs within Exploreco were considered more important than in Oilco. There was much more enthusiasm for CSFs in the Exploreco case than in the Oilco case, generally over all phases of the PPM. For example, Exploreco interviewees thought nine CSFs were important/crucial in the reengineering phase; Oilco interviewees nominated, somewhat unenthusiastically, four CSFs as important in this phase.

Importance of people and firm deliverable dates in the Planning phase. Despite many similarities in the planning phase, the two organizations exhibited an important strategic difference. Exploreco, who had had the benefit of learning from the parent company's earlier implementation, added two more CSFs to the Planning phase: a 'rock-solid' decision to release the best business user experts full time on to the project, and the setting and delivery of project milestones, and a final end-date. Oilco had come to recognize the value of the business user experts particularly for training and user support, and they adopted an implementation methodology with clear processes and targets.

The scope of the two cases differed in several key dimensions: physical scope, level of BPR, technical scope, module implementation strategy, and resource allocation. The CSF 'definition of scope and goals' was considered very important in the Planning phase at Exploreco but not at Oilco.

Facilitating the achievement of CSFs. Even when both companies identified what appeared to be the same CSF, they differed in that Exploreco devised *a process* and *structures* to facilitate its achievement. The starkest example of this concerns their recognition that a project champion was crucial. In Exploreco the champion was actually known by that title, was allocated to the project for the life of the project, had defined responsibilities, and, most importantly, was a member of the board (called the Leadership Council) of the company. This level of seniority plus the daily hands-on approach proved invaluable. By contrast, in Oilco, this person was not officially recognized, and the person in the role changed over time. Initially the drive for the system came from a USA MD, who promoted the ERP as a global strategy. Subsequently the Venture Manager (brought in from the UK) became the *de facto* champion, and then later there was an in-house senior ERP 'convert'. There was neither a defined role, nor processes or structures via which his influence could be conveyed.

Focus in set-up phase for Exploreco. In Exploreco there was formal recognition of the 'champion' role throughout the set-up phase. Although in both cases a balanced team and clear definition of scope and goals were recognized as important in set-up, in Exploreco the strong focus on implementing a Vanilla ERP and setting clear deliverable dates were evident.

Importance of the reengineering and configuration phases. There is a marked difference in the reengineering and configuration and testing phases for the two cases. Exploreco recognized and ensured that all the CSFs were present. This indicates that they placed equal weight on each stage of the project. For Oilco, little emphasis was placed on these phases. Also decisions made by Exploreco at the Planning phase influenced other phases of the project. For example, during reengineering and design, management at Exploreco were actively involved at the 'coalface' in, for example, promoting the benefits of process changes to employees who would be directly affected by them, and participating in project group and departmental walkthroughs at milestones. In Oilco, during these phases, management activity can be seen as more remote – 'storming' around the country, making presentations. This remoteness was evident very early in the Planning phase. The evaluation team, who were principally locals, recommended against the ERP. They were overruled.

Consistency and persistence of CSFs throughout all phases. There is also a difference simply in the patterns of individual CSFs, which is underscored by the PPM. As it shows, commitment waxed and waned at Oilco. This is hardly surprising given the length of the project. At Exploreco, there was a steady, persistent commitment to the project. They did have advantages: it was a smaller, self-contained project and they were replacing an unpopular system. They were able to obtain persistent commitment because at the planning stage they carefully scoped the project, and because they set a immovable end date.

This variation in pattern can also be seen with other CSFs. Both companies adopted a policy of minimal customization and deliverable dates. However, Oilco was forced to commission an oil industry-specific module, and they generated endless reports often because it was possible rather than desirable (MIS Manager). These changes were accompanied by extensive company restructuring and it is unclear which of these caused them to go years over their projected end date. Exploreco adhered to the principles of minimal customization and fixed deliverable dates until their project was well advanced in the configuration and testing phase. At this stage it became clear that the interfaces were unacceptable to the users. They then bought in Lotus Notes, and wrote the necessary interfaces. This meant they ran two weeks over their 'rock-solid end date'.

Explanations for similarities and differences

Two explanations for the similarities and differences between the successful and unsuccessful cases are offered. These concern organizational learning and the scope and complexity of each project.

The organizational learning that took place at Oilco during the unsuccessful project led to a clear articulation of the project plan for ERP implementation at Exploreco. The project was driven from Oilco's head office and led to Exploreco's decision at the early Planning phase to adhere to a rock-solid deadline, to insist on minimal customization, and to appoint a senior member of the board on the steering committee as a champion, one of whose tasks was to provide the authority to release the best business experts on to the team. Also, they learned from prior experience that system training, and *de facto* 'selling', had to be done by the in-house experts who were committed to the new ERP, not by the consultants, or by senior managers. Previous experience of the Exploreco project team in ERP system implementation allowed them to readily identify both the CSFs required, and the right process for ensuring the achievement of each CSF. Processual details include what phase it should be present in, at what level, and in what form.

The scope and complexity of the Oilco implementation was clearly greater than the Exploreco implementation. We have argued elsewhere (Parr and Shanks, 2000a) that there are three archetypal categories of ERP implementations. These are 'comprehensive', 'middle-level', and 'vanilla'. Essentially, these categories are a grading in project scope from most extensive to simplest and are based on a set of ERP implementation characteristics. These include physical scope (multi-site, multinational boundaries versus single site, for example); technical scope (involves decisions either to modify or accept the ERP as is); module implementation strategy (essentially a modular or 'big-bang' approach); the level and type of reengineering involved, and in resource scope. In this categorization scheme, Oilco is a 'comprehensive' and Exploreco is a 'vanilla' implementation. 'Comprehensive' implementations are inherently large and complex, and IT projects with these characteristics are high risk with a significant probability of failure (Willcocks and Sykes, 2000). In contrast Exploreco is a 'vanilla' implementation with manageable scope. The Oilco implementation was also more complex and involved an earlier version of the ERP software and consequently involved development of a specific module and extensive programming for reports.

Implications of the PPM for practice and research

1 Organizations need to be aware that large-scale ERP implementation projects are high risk. They are highly unlikely to be implemented on time and within budget. It is important to consider partitioning large-scale ERP implementation projects into several simpler and smaller implementation projects. Further case studies of ERP implementations should investigate this proposition.

2 In both cases there was no marked 'enhancement' phase. This is in contrast to the findings of Ross (1998) and Markus and Tanis (2000), where significant post-implementation phases were identified. Further case studies of ERP systems implementation are required to investigate the nature of activities within the enhancement phase of the PPM.

3 The appointment of an experienced ERP implementer early in the project as 'champion' and empowered to make decisions about the project is important to the success of the project. This kind of role is highly relevant in the smaller-scale 'vanilla' category of ERP implementation projects. Further studies of ERP implementation projects are needed to better understand the nature

of the champion role and the importance of previous experience to project success.

4 The PPM model together with associated CSFs from the Exploreco case provides practitioners with a useful example of successful 'vanilla' ERP implementation, and may be considered a best-practice ERP implementation process model. Three amendments need to be made to the CSFs in Table 8.4 from the case study analysis. The 'smaller scope' CSF should be added to the set of CSFs, with particular emphasis in the planning phase. The 'vanilla ERP' CSF should include the assumption that the ERP selected is a recent version of software with suitable functionality. The 'Champion' and 'Best People Full-time' CSFs should emphasize the value of experience with ERP implementations. Further case studies of ERP system implementation are required to validate and extend the particular CSFs that are important in each PPM phase.

8.6 Conclusions and Future Research

This paper draws upon CSFs which were developed from an earlier stage of the research. We then review current process models of ERP implementation and argue that there is valid reason to develop a model (the PPM) which represents a combination of the CSFs and project phases. This model was then empirically tested using two case studies of ERP implementation projects: one successful and the other unsuccessful. The case study results suggest that a model that focuses solely on the project and the factors that are critical at each phase is valuable both to researchers and to practitioners. For practitioners, it provides a guide to the CSFs required during each project phase, and underlines the importance of the Planning phase. For researchers, it is a useful model of ERP system implementation and provides a foundation for further empirical research.

Nonetheless the research has certain limitations which give rise to further research questions. Care was taken internally to improve validity by using multiple sources of data and multiple viewpoints, and the cases were selected to demonstrate a transition from failure to success. However only two case studies are reported, and it is unclear whether other cases that are markedly dissimilar would generate the same results. It is recognized that ERP implementations vary markedly in terms of key factors (Parr and Shanks, 2000b; Soh et al., 2000; Shanks et al., 2000; Kremers and Dissel, 2000), such as multi-site versus single-site implementations, cultural differences, simple versus complex legacy systems, and decisions regarding the approach to Business Process Reengineering (BPR). Further the results of this study suggest that partitioning large projects enhances the probability of meeting budgetary and time constraints. Together, these factors imply that a refinement and classification of ERP systems is needed, and then it will be possible to strengthen the findings by application of the PPM to further cases, and to test it against cases which appear to be significantly similar and dissimilar.

REFERENCES

Ambrosio, J. (1997) Experienced SAP Users Share Ideas with Newbies. Online News Story, 27 August.

Appleton, E. L. (1997) How to Survive ERP. *Datamation* March, **43**(3), 50–3.

Bancroft, N. (1996) *Implementing SAP R/3.* Greenwich: Manning Publications.

Bancroft, N., Seip, H., and Sprengel, A. (1998) *Implementing SAP R/3,* 2nd ed. Greenwich: Manning Publications.

Darke, P., Shanks, G., and Broadbent, M. (1998) Successfully Completing Case Study Research: Combining Rigor, Relevance and Pragmatism. *Information Systems Journal,* **8**, 273–289.

Deloitte Consulting (1998) ERP's Second Wave: Maximizing the Value of ERP-enabled Processes. www.dc.com.

Eisenhardt, K. M. (1989) Building Theories from Case Study Research. *Academy of Management Review,* **14**(4), 532–550.

Fine, D. (1995) Managing the Cost of Client/Server. *Infoworld,* 17 November.

Gartner Group (1998) Implementing SAP R/3: Avoiding Becoming a Statistic. http://www.gartner.com/.

Holland, Light and Gibson (1999b) A Critical Success Factors Model for Enterprise Resource Planning Implementation. In Pries-Heje, Ciborra, C., Kautz, K., Valor, J. Christiaanse, E., Avison, D., and Heje, C. (eds), Proceedings of the 7th European Conference on Information Systems, Copenhagen Business School. Copenhagen, pp. 273–287.

Horwitt, E. (1998) Enduring a global Rollout – and Living to Tell About It. *Computerworld,* Framingham, **32**(14), 8–12.

James, G. (1997) IT Fiascos and How to Avoid Them. *Datamation,* November.

Kremers, M. and H.V. Dissel (2000) ERP Systems Migration. *Communications of the ACM,* **43**(4), 53–56.

Markus, M. L. and Tanis, C. (2000) The Enterprise Systems Experience – From Adoption to Success. In *Framing the Domains of IT Research: Glimpsing the Future Through the Past,* R.W. Zmud (ed.), Cincinnati, OH: Pinnaflex Educational Resources.

Martin, M., (1998) An electronics firm will save big money by replacing six people with one and lose all the paperwork, using Enterprise Resource Planning software. But not every company has been so lucky. *Fortune,* **137**(2), 149–151.

Nandhakumar, J. (1996) Design for Success?: Critical Success Factors In Executive Information Systems Development. *European Journal of Information Systems,* **5**, 62–72.

Parr, A. and Shanks, G. (2000a) A Taxonomy of ERP Implementation Approaches. Proceedings of the 33rd Hawaii International Conference on System Sciences, IEEE Computer Society.

Parr, A. and Shanks, G. (2000b) A Model of ERP Project Implementation. *Journal of Information Technology*, **15**(4), 289–303.

Parr, A., Shanks, G., and Darke, P. (1999) Identification of necessary factors for successful implementation of ERP systems. In *New Information Technologies in Organisational Processes* Ngwenyama, O., Introna, L.D., Myers, M.D. and DeCross, J.I. (eds.), Boston, MA: Kluwer Academic Publishers, pp. 99–119.

Piszczalski, Martin (1997) Lessons Learned from Europe's SAP Users. *Production*, **109**(1), 54–56.

Rockhart, J. F. (1979) Critical Success Factors. *Harvard Business Review*, March–April, 81–91.

Ross, J. W. (1998) The ERP Revolution: Surviving Versus Thriving. Research Paper, Centre for Information Systems Research, Research Paper, Sloan School of Management, MIT.

Shanks, G., Parr, A., Hu, B., Corbitt, B., Thanasankit, T., and Seddon, P. (2000) Differences in Critical Success Factors in ERP Systems Implementation in Australia and China: A Cultural Analysis. Proc. European Conference on Information Systems, Vienna (July). IEEE

Soh, C. et al. (2000) Cultural Fits and Misfits: Is ERP a Universal Solution? *Communications of the ACM*, **43**(4), 47–51.

Stedman, C. (1998a) Business Application Rollouts still Difficult. *Computerworld*, Framingham, **32**(28), 53–56.

Stedman, C. (1998b) ERP User Interfaces Drive Workers Nuts. *Computerworld*, Framingham, **32**(44), 1–24.

Sumner, Mary (2000) Risk Factors in Enterprise-wide/ERP Projects. *Journal of Information Technology*, **15**(4), 317–327.

Tebbe, M. (1997) War Stories Outnumber Successes when It Comes to Implementing SAP. *Infoworld*, **19**(27), 120.

Willcocks, L. and Sykes, R. (2000) The Role of the CIO and IT Function in ERP, *Communications of the ACM*, **43**(4), 32–38.

Williams, J. J. and Ramaprasad, A. (1996) A Taxonomy of Critical Success Factors, *European Journal of Information Systems*, **5**, 250–260.

Yin, R.K. (1989) *Case Study Research: Design and Methods*. San Francisco: Sage.

9 Offsetting ERP Risk through Maintaining Standardized Application Software[1]

Guy G. Gable, Taizan Chan, and Wui-Gee Tan

9.1 Introduction

Enterprise Resource Planning (ERP) systems purport to integrate the complete range of business processes and functions into a single information and technology infrastructure. By 2003 ERP software was highly configurable, and, while noting the qualifications made in earlier chapters, technically it went a long way towards accommodating the diverse needs of users across most sectors of the economy. This development has, in fact, happened over a long period. Though the breadth and tight integration of ERP has only become available in recent years, ERP have a pedigree in large, packaged application software that has been in widespread use since the 1970s (Klaus, Rosemann, and Gable, 2000).

The market for ERP products and services sustains a significant segment of the software and services industry, consisting of several of the largest software firms as well as the world's largest management consulting companies. Related annual revenues were expected to exceed $20 billion by 2002, approximately half vendor service revenue and half vendor license revenue (Gartner Group, 1999). Most multinational companies as well as many public sector organizations, have adopted ERP and increasingly small and medium-sized enterprises have been following suit. Indeed, by 2003 ERP formed the backbone of many application architectures and were being extended into developments such as e-business and e-government (see also Chapters 4 and 18).

As Chapters 1, 5, and 6 made clear, however, the potential benefits of ERP can often be realized only with substantial investment. These implementations have turned out to be a major challenge for information systems practice. Ninety percent of ERP implementations end up late or over budget,

[1] An earlier version of this paper first appeared in a special issue of *Journal of Software Maintenance and Evolution: Research and Practice* (2001, **13**(1):1–20) the theme of which is 'Large Packaged Application Software Maintenance'.

while reportedly, 67% of enterprise application initiatives could be considered negative or unsuccessful (Davenport, 2000). The magnitude of the costs and problems associated with ERP implementations have prompted a vein of research, much of it represented in this book, that seeks to identify implementation issues in order to ameliorate these issues in the future (see, for example, Chang et al., 2000).

Overemphasis on systems implementation can tend, however, to obscure the increasingly apparent reality that large-scale commercial software, such as ERP, is subject to special dynamics. Early examination of ERP installations reveals that their life cycle involves major iterations (Dailey, 1998; see also Chapters 1 and 4), very dissimilar to the traditional view, of the succession of development, implementation, and maintenance. Following initial implementation, there are subsequent revisions, re-implementations, and upgrades, going beyond what commonly has been considered system maintenance (see Chapters 7 and 8). The extensive and growing installed ERP base has thus precipitated further significant efforts for user organizations in terms of maintenance.

In the past, packaged software, though pervasive, have generally been under-researched and under-represented in curricula (Gable et al., 1997). This has changed only recently, due to the heed given to ERP implementations. Unlike implementation issues, however, issues related to the ongoing support, modification, and enhancement of ERP after its implementation have received little attention. Only a relatively small amount of mostly practitioner and some academic literature exists in the area. As mentioned above, the system life cycle of ERP is distinct from conceptions underlying earlier software maintenance research, suggesting that we question the generalizability of results from past research on software maintenance across all situations and their transferability into the ERP context. We believe that the significance of the installed ERP base poses new challenges for practice and opens new opportunities for research. We suggest that, to take up the research agenda, it is imperative to direct effort towards identifying key factors that impact maintenance costs and benefits, their incidence, problems, and strategy across diverse new software and support scenarios. We will further expand on these themes by highlighting more specific areas of related research need and neglect, presenting a tentative framework for research in the area, and instantiating partially that framework with examples. Ultimately, the tentative framework will stress the importance of software maintenance research by (a) focusing on the incidence of maintenance costs and benefits, (b) concentrating on benefits instead of costs, and (c) taking a lifecycle-wide view.

The discussion is interspersed with questions. Our list of questions does not purport to be complete, nor do all of these purport to be 'research' questions. Rather, they are simply sample questions intended to stimulate thinking on new areas of research for important packaged software.

9.2 The Incidence of Maintenance Cost and Benefits

In Chapter 3 Shang and Seddon sensitised us to the issues in implementation costs and benefits. But maintenance also incurs costs and yields benefits. Herein, the words 'costs' and 'benefits' are used in their broadest sense. Costs include (1) costs of doing maintenance (for example search costs, correction costs, testing costs) and (2) costs of not doing maintenance (for example opportunity costs of absent functionality). Benefits include improved system responsiveness, new functionality, and reduced risks (for example cost containment through preventative maintenance).

Costs are incurred and benefits reaped by different stakeholders. The organization using the software (user-organization) is always a stakeholder in these costs and benefits. Where the software is purchased, the vendor too is normally a stakeholder (unless source code is sold outright). There may be a third party 'service provider' who too experiences the incidence of maintenance costs and benefits. Finally, 'society' may be impacted (e.g. by the effective or ineffective incidence of maintenance responsibilities across the key stakeholders). With packaged software, in general, and ERP in particular, the vendor is now assuming much responsibility for maintenance, and also realizing substantial revenues from this activity.

From the point of view of business economics, the absolute and relative magnitude of costs and benefits is clearly of interest. From the point of view of applied economics, the value of better understanding the incidence of these costs and benefits across stakeholders is also obvious. There are significant economies of scale and related savings that are shared amongst the stakeholders which need to be determined, assessed, and apportioned.

A Benefits View of Maintenance

Maintenance generates benefits, and lack of maintenance can entail lost benefits (i.e. opportunity costs of not doing maintenance). The substantial cost and complexity of ERP implementation, and the long implementation periods often needed, increase the costs of switching out of ERP, and reduce the likelihood that organizations will replace their ERP in the short or even medium term.[2] Users are, therefore, more likely to explore means of generating further value from their existing investment – a premiss on which this book is based. Benefits from ERP can be sustained and increased through software upgrades (from the vendor), and software maintenance (improvements specific to the using organization's business environment).

Therefore, it is important that increased attention is given to possible benefits from maintenance. Recent emphasis on 'benefits realization' from ERP (for example Gable, Chan, and Tan, 2001) suggests that awareness of this potential is on the rise. There appears to be a range of explanatory frameworks that allow for exploring and accounting for the phenomenon of benefits from enterprise software, which put the matter of maintenance into intra- and inter-organizational perspectives. Examples of intra-organizational perspectives include: 'Resource-based Theory of the Firm' (Barney, 1991) and 'Barriers to Intra-firm Transfer of Best Practice' (Szulanski, 1995); while examples of inter-organizational frameworks are: 'Vertical Integration, Appropriable Rents and the Competitive Contracting Process' (Klein et al., 1978); and 'Organizational Virtualness' (Sieber and Griese, 1999).

Up until now however, maintenance has been viewed excessively negatively, with insufficient regard to related benefits. It could be argued that maintenance has been rated only as a source of cost. The problem with a negative, cost-focused view is less severe where opportunity costs of not implementing beneficial change are considered. Yet, this still does not render the full picture, since it does not account for benefits realized (only accounts for benefits not realized). A balanced view of maintenance should emphasize overall the net difference between costs and benefits.

Since maintenance is generally, and inappropriately, viewed negatively, we propose to amplify attention on maintenance benefits. We thus do not use the term 'maintenance burden' nor 'maintenance problem' in this discussion. Rather, we use the more neutral terms, 'maintenance activity' and 'maintenance strategy'. Adopting this view gives rise to the following questions:

Q1: How can maintenance realize benefits?

Q2: How do maintenance benefits relate to initially sought implementation benefits?

Q3: What are the key drivers of software maintenance benefits?

[2] Of course changing or getting rid of an ERP is not unheard of, and is often a central issue in mergers and acquisitions.

Q4: Do existing maintenance taxonomies sufficiently characterize packaged software related maintenance activities?

As regards the last listed question, recent work by Hirt and Swanson (2001), Nah, Faja, and Cata (2001) and Ng (2001) suggests that the existing taxonomies of maintenance as corrective, adaptive, and perfective activities are inadequate in characterizing ERP maintenance.

9.3 A Lifecycle-wide View

Implementing packaged application software is typically a long-term investment, having long-term maintenance implications. Thus, decisions concerning the source of the application software, as well as the source of support for its maintenance implicitly reflect long-term maintenance strategy (Gable, Scott, and Davenport, 1998).

The user-organization, vendor, and service providers are not only distinctive stakeholders, but important actors, and ostensibly partners, in a relationship that may span the life of the software. One recent development, Organizational Virtualness (Sieber and Griese, 1999), recognizes that modern organizations have moved to explicit and implicit models of shared services and have outsourced non-core activities (Bennett and Timbrell, 2000). Across the ERP lifecycle, user-organizations, consultants, and vendors work together to realize ERP benefits in a way also suggestive of an extended virtual organization (Sieber and Griese, 1999). Strategic conflict between 'members' may arise, threatening ERP benefits. An examination of the barriers to ERP benefits realization, and identification of relevant theoretical foundations for their explanation, has the potential to provide a unique, innovative perspective on strategic management of large-scale packaged software maintenance and evolution, and the extended virtual enterprise explicitly or implicitly deployed across the ERP lifecycle (see also Chapters 1, 2, and 3).

Broad research questions implicit in the preceding discussion include:

Q5: Are lifecycle-wide maintenance requirements anticipated sufficiently when making software and related support sourcing decisions

Q6: Have key players involved with ERP lifecycle support implicitly entered into an extended, virtual organization? Is this view on the relationship revealing?

In the next section we present a research framework for better understanding the scope of important package software related research need. We conclude with a brief summary of key messages suggested from discussion on the framework.

9.4 A Research Framework

We start by proposing three dimensions along which the scope of software maintenance research could be delineated: (1) Stakeholders, (2) Organizational size, and (3) Source of software, as shown in Figure 9.1.

The purpose of this a priori framework is to illustrate important dimensions and perspectives along which maintenance scenarios, strategies, and related research will vary. While software maintenance research to date has largely addressed the maintenance of custom, in-house developed applications that involve mainly the user-organization (see Figure 9.1), we seek to shift focus to issues with 'large packaged application software maintenance'.

We thus advocate in principle research that assumes alternative, less-conventional perspectives on the maintenance activity (for example the vendor view). Yet, for the purposes of scoping the discussion at the moment, it is useful to continue to assume a primarily user-organization centric perspective (thus a 'user-organization' centric context), the simple reason being that

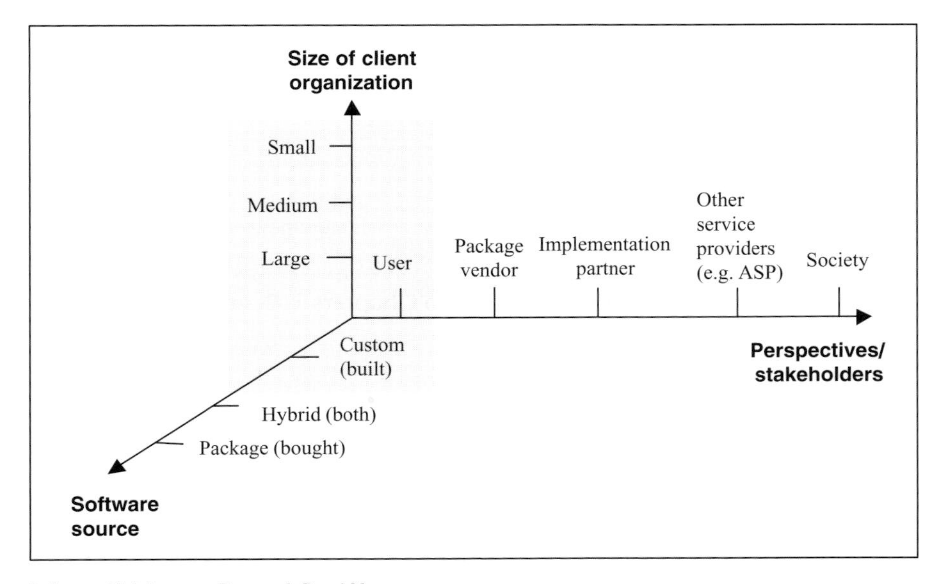

Figure 9.1 Software Maintenance Research Road Map

maintenance deals with software in use and all software of interest is ultimately used by organizations. Vendors licence the software to organizations. Third parties (and vendors) sell software support services to these organizations. In other words, it is the user-organization where software becomes effective, in concert with the economic transactions of purchasing licenses and services.

We proceed by deriving from the above a priori framework a user-organization centric research framework that captures many relevant and interacting factors that could impact ERP maintenance. The user-organization centric research framework provides the following extensions: (1) broader user-organization context (than simply size), (2) explicit recognition of environmental context (outside the organization), (3) consideration of software support-source as well as software-source, and (4) integration of perspectives and interplay of the key stakeholders – user-organization, vendor, third-party service providers, and society – as well as (5) the implicit 'virtual organization' of user-organization, vendor, and third parties (Figure 9.2). Note that examples only are indicated in the matrix in Figure 9.2 (for example ASP, Outsourcing, etc.). The potential combinations of software-source/support-source are many and include: (a) External custom/In-house supported; (b) External custom

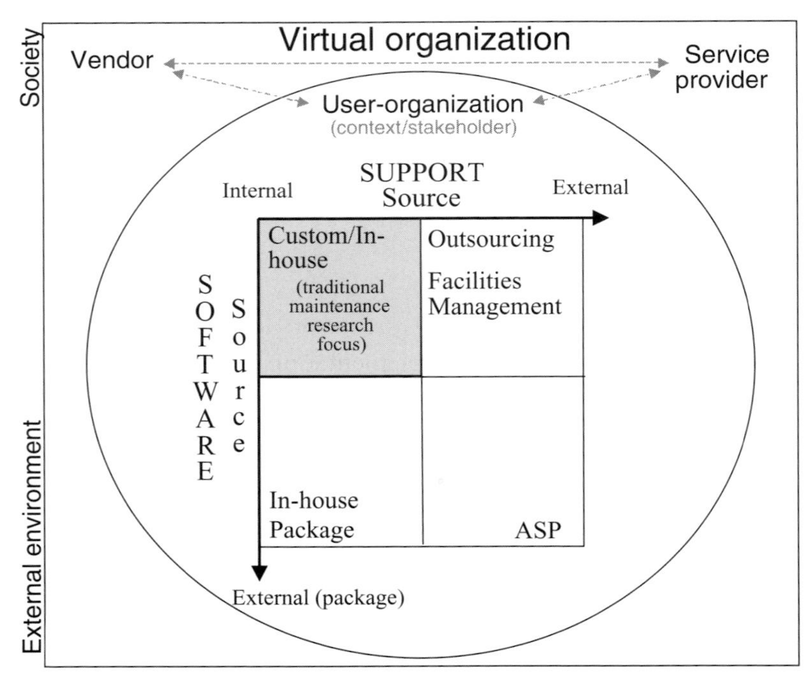

Figure 9.2 The software maintenance research framework

(by vendor (a)/External supported (by vendor (b)); (c) Internally developed package (sold to other organizations)/Supported internally (rare); (d) Bought in package/Supported by user-organization (user-organization acquires source, or links with package vendor are broken). These variants are implied further in subsequent discussion on support-source, software-source, and stakeholders.

The framework does not purport to reflect all aspects of software maintenance and related research need. For example, neither the a priori framework (Figure 9.1) nor the Software maintenance research framework (Figure 9.2) addresses the need to maintain personal software or software for the home; these are areas of growing importance and increasing societal investment and impact (for example all those PCs sitting idle due to lack of hardware and software maintenance). Our main concern is the maintenance of *application* software *in organizations.*

Organizational and Environmental Contexts

Maintenance problems and strategies will vary with organizational and environmental context. In the a priori model (Figure 9.1) we included the dimension, 'organizational size' as a key determinant of maintenance strategy. It should be noted that organizational size is a simplistic proxy for many potentially important contextual factors. While many of these factors do vary with organizational size, others do not. Organization size in the a priori model subsumes other organizational factors that may be important antecedents (or mediators or moderators of antecedents) of maintenance success. In the Software maintenance research framework (Figure 9.2) we include the more inclusive term 'user-organization context'[3] (rather than simply size).

A phenomenon of the environmental context would be, for example, the labour market situation. We know of one organization that when implementing an ERP made the implementation partner and vendor commit to 'not transferring knowledge into their organization'. In a climate of serious skill shortage, they were legitimately concerned that staff would leave once they gained ERP-related skills. Thus, environmental contingencies (possibly temporary) may result in enduring constraints regarding the maintenance strategy.

Q7: What are the key environmental factors impacting maintenance strategy? How are these changing over time?

[3] Note that 'user-organization' represents both a context of maintenance in the user-centric framework, as well as a stakeholder in the virtual organization.

Q8: What key organizational factors moderate or mediate causal relations between independent variables and maintenance success?

Q9: How can maintenance success be measured?

Q10: What other dependent variables should we seek to predict/explain?

Q11: What maintenance skills are required in which contexts?

Software Source

The organization's sourcing strategy for software and related support is a primary determinant of their maintenance strategy. The source of software utilized by the organization substantively impacts the incidence of maintenance. As suggested in our framework (Figure 9.2), software can be sourced internally or externally (see also Chapter 4, on developments in sourcing strategies). In the extremes, software may be sourced almost entirely internally, or externally. An example of the former would be an in-house 'custom-built' application, while an example of the latter would be a software package implemented without modification. In between these extremes is a range of possible scenarios, each involving both purchased and in-house software (see also Kern, Lacity, and Willcocks, 2002).

Q12: What factors determine the amount of in-house (non-vendor supplied) maintenance required of large packaged software and hybrid solutions?

Q13: To what extent are package maintenance concepts generic and extensible beyond a particular vendor's product?

Q14: How does package maintenance differ from custom software maintenance?

Q15: What differences in the roles and expertise of the maintenance team do these dissimilarities suggest?

Q16: What changes to tertiary curriculums might these differing roles and expertise suggest?

Q17: What factors determine the amount of in-house maintenance required of packaged software and hybrid solutions?

Q18: To what extent can maintenance be avoided through packaged software and hybrid solutions (to avoid costs) and what maintenance should be encouraged, in order to produce benefits?

Q19: What are the maintenance implications of implementing 'Vanilla' ERP versus customizing (for example, see Light, 2001)?

Q20: What are the maintenance implications of customization at implementation time?

Q21: What are the drivers behind the upgrade decision?

Support Source

It is useful to make a distinction between 'software-source' and 'support-source', as it is possible to acquire the software (a package) externally yet assume full responsibility for its ongoing maintenance (for example, purchase the source code). Alternatively, with application service provision (ASP), a relatively recent development, in the extreme one can contract the service provider to not only operate the software on their hardware, but to own the software (or its licenses) and rent it to the user-organization on a per-use basis. In this scenario, the application service provider is fully responsible for virtually all maintenance activities. Application service provision dramatically impacts the incidence of maintenance responsibility and activity and maintenance strategy (see Kern, Lacity, and Willcocks, 2002).

Q22: In what ways and to what extent can maintenance be outsourced to third parties? What aspects remain?

Stakeholders

Key stakeholders reflected in the framework (Figure 9.2) are user–organization, vendor, service providers (including implementation partners), and society. All stakeholders impact each other in the framework, as well as having differing perspectives on maintenance.

The user-organization

While the user-organization perspective has received most attention in the literature and in past research, it is clear that where software is sourced externally, the incidence of maintenance may change and relations between the user-organization and other stakeholders become important. Thus, several questions arise:

Q23: What is the relative incidence of costs and benefits across these stakeholder groups?

Q24: What factors impact the relative incidence?

In addition to the overall organization, individuals and groups of interest within the user-organisation may include: the maintainers, IT management, the IT function, end-users, business management, and others. For example, within the user-organization, maintenance benefits may accrue to the end-users (for example, improved functionality) or to the maintenance function (for example, reduced code complexity). Costs of this maintenance may be billable to the end-users or not. The incidence of these costs and benefits will undoubtedly influence the decisions of these stakeholders, and should influence their maintenance strategy.

Q25: Who are the key stakeholders within the user-organization?

Q26: What is the relative incidence of costs and benefits across these internal stakeholders?

Q27: How does the influence of costs and benefits influence stakeholder behaviour? How are these influences manifested in practice?

Q28: How might the incidence of costs and benefits across the stakeholders be influenced in order to benefit the overall user-organization?

The vendor

Software vendors may include the ERP vendor (for example, SAP, Baan, PeopleSoft, JD Edwards, Oracle Applications), vendors of systems management software (for example realTech, Computer Associates), the database vendor (for example, Oracle, Microsoft), or others. These stakeholders can potentially play an important role not only during the implementation but also in the ongoing maintenance of an ERP.

Q29: Who are the key vendors at implementation time? Who are the key vendors post-implementation? What roles do the various vendors play (if any) as regards software and system maintenance?

Q30: What user-organization maintenance activities are outsourced to the vendor through purchasing a package? Which remain?

Maintenance done by user-organizations can be usefully categorized as either internally generated (for example, user requests) or generated externally by the software package vendor. Vendor generated maintenance is usually referred to as 'patches', which are distributed to user-organizations by the vendor and implemented by the user-organization.

Q31: What factors influence the cost to user-organizations of implementing patches?

Q32: Would it be more cost effective (in total) for vendors to implement patches (rather than user-organizations)?

Q33: Could patches be implemented remotely? What would be the cost to the vendor? What would be the savings to the user-organization? Would user-organizations be willing to pay for this service?

Q34: Is this a cost that would become too visible for many user-organizations to accept (in the form of increased annual maintenance charges from the vendor or another stakeholder), though they implicitly incur it regardless?

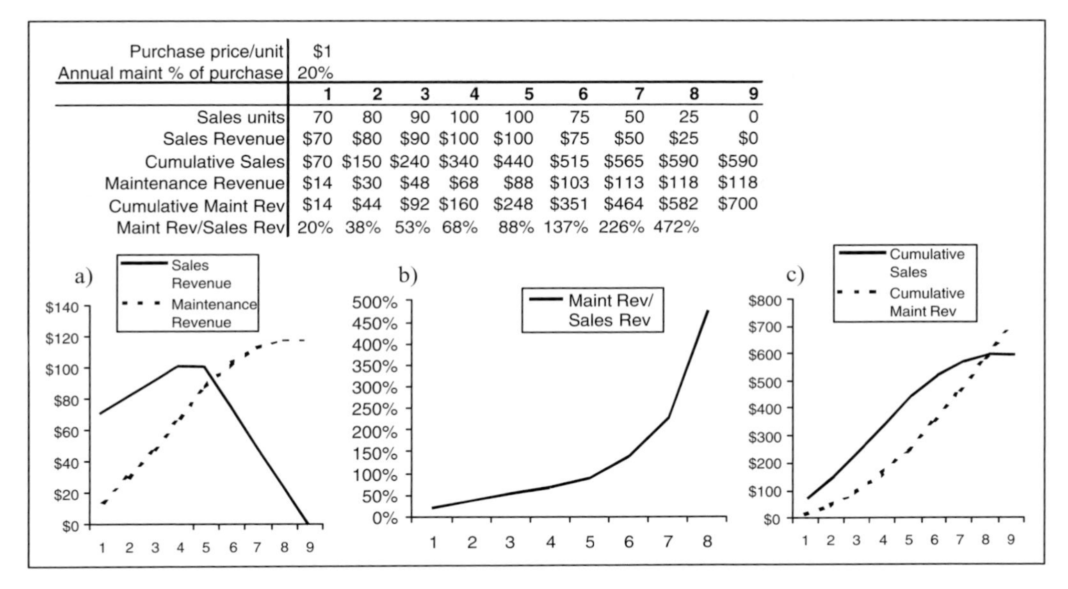

Purchase price/unit	$1								
Annual maint % of purchase	20%								
	1	2	3	4	5	6	7	8	9
Sales units	70	80	90	100	100	75	50	25	0
Sales Revenue	$70	$80	$90	$100	$100	$75	$50	$25	$0
Cumulative Sales	$70	$150	$240	$340	$440	$515	$565	$590	$590
Maintenance Revenue	$14	$30	$48	$68	$88	$103	$113	$118	$118
Cumulative Maint Rev	$14	$44	$92	$160	$248	$351	$464	$582	$700
Maint Rev/Sales Rev	20%	38%	53%	68%	88%	137%	226%	472%	

Figure 9.3 Sales vs maintenance revenue

Maintenance is an important revenue stream for the vendor. As the size of their installed base grows, toward theoretical market saturation, maintenance revenue will increase as a proportion of total revenue. Figure 9.3 presents a simple, contrived example in which hypothetical software sells for $1/unit, the vendor's annual maintenance fees on the software are 20% of the purchase price, annual sales grow from $70 in year 1 to $100 in years 4 and 5, then drop off during years 6 through 9 to zero as the market place is saturated. From the table and graphs we note: (a) annual maintenance revenue continues to grow with growth in the installed base of the software even while sales are declining, (b) annual maintenance revenue as a percentage of annual sales revenue grows exponentially as sales decline, and (c) somewhere around year 8, cumulative maintenance revenue (over the life of the software product) surpasses cumulative sales revenue.

While this simple example demonstrates the potential value of the maintenance activity to the software vendor as a revenue stream, it should be clear that maintaining multiple versions of a software package installed across numerous user-organizations, sometimes geographically dispersed, and possibly spanning multiple industries involving multiple integrated software modules, is a business in its own right (beyond software development, sales and implementation), entailing much cost and complexity. If not managed well, maintenance costs to the vendor could exceed related revenue (for example, see Sahin and Zahedi, 2001).

Q35: How can packaged software be best designed to minimize ongoing maintenance costs?

Q36: What is the relationship between package flexibility, package complexity and package maintainability?

Q37: How does the vendor's maintenance management strategy potentially impact user-organization maintenance strategy (zero-sum game?)?

Q38: When and on what basis should the vendor introduce a new version of their package?

Q39: Do maintenance strategies of vendors vary? In what ways? With what implications for user-organizations?

Q40: What are package vendors doing to reduce maintenance costs? For themselves? For user-organizations?

Service providers

Service-providers may be distinguished as those who are primarily involved at package implementation/upgrade time (for example, Big5 implementation partners, training and user documentation specialists) and others who may only become involved post-implementation (for example, facilities management).

Q41: Who are the main service providers to ERP-using organizations (and to other stakeholders in the ERP ecosystem) during implementation? Post-implementation?

Packages are developed outside the user-organization. Their depth and breadth, as well as their flexibility, make them complex. Consequently, the skills and knowledge required to customize and modify large packaged applications are often specialized and must often be sourced externally. It is therefore not surprising that with large packaged applications, hundreds of firms populate the large packaged application services ecosystem comprising tens of thousands of professionals that derive much, or their entire livelihood from large packaged system-related services. These external service providers range from the packaged application vendors themselves, to integration service partners of ERP vendors, to hardware vendors, to application service providers.

Large, regional, and international consulting firms, most often selected as 'implementation partner' by user-organizations, play a particularly prominent role in the implementation and maintenance of large packaged applications. As just one example, Table 9.1 lists numbers of consultants employed by large (>900 staff worldwide) and smaller SAP R/3 consulting firms (systems

Table 9.1 The SAP ecosystem

	Americas		Europe		The Rest		Total	
	#	%	#	%	#	%	#	%
Large firms	9610	41%	10200	43%	3780	16%	23590	66%
Small firms	3190	37%	4100	48%	1320	15%	8610	24%
SAP	1300	34%	2000	53%	500	13%	3800	11%
Grand total	14100	39%	16300	45%	5600	16%	36000	100%

Note: Adapted from Gartner Group 1999.

integrators) and SAP as at 1998 (these numbers have changed substantively since that time). Approximately two-thirds were employed by the 11 largest consulting firms with the remaining one-third employed by approximately 500 smaller consulting firms around the world. Forty-five per cent were employed in Europe, 39% in the Americas, and 16% elsewhere. SAP was then somewhat more concentrated in Europe than its partners. Although SAP has increased its service staff, it can be seen that SAP's consultant partners (90%) yet provided the bulk of user-organization service and support. It is thus interesting to ask:

Q42: How can service providers impact the incidence of package maintenance costs, benefits, and responsibilities?

Q43: What alternative maintenance strategies for user-organizations does the existence of these service providers make possible?

Q44: What further maintenance activities does moving to an ASP (for example, patches) outsource? Which if any remain?

Q45: What are the various forms and modes of outsourcing and how does each impact maintenance required in house? What are the key variables of interest?

Society

The shift of maintenance cost incidence from within user-organizations to external stakeholders has implications beyond the immediate stakeholders (user, developer, and service provider) involved. The significance of maintenance for value generation from the computer and services industry, needs to be identified. Trans-organizational division of labour issues abound, with consequent impacts on the dynamics of the IT labour market, changing the composition of services and the roles of IT professionals.

Q46: What are the total costs and benefits of package maintenance across all stakeholders? What is the national investment each year in package software maintenance?

Q47: From a holistic, societal perspective, is this most effective? How might it vary? Is there some optimal distribution of costs and benefits across the stakeholders?

Q48: Is the societal incidence of package maintenance costs similar across national cultures?

Q49: Can these costs be better rationalized and allocated through some form of central control or influence (for example, laws, regulations, associations, professions)?

Q50: To what extent are ASPs a response to the inefficiency of having maintenance expertise reside in each user-organization?

Q51: What are the key technological developments coming in packaged software and what implications might these have for maintenance costs, maintenance skills required, the incidence of maintenance costs?

The Virtual Organization

The view that maintenance of a user-organization's ERP installation involves the stakeholders in a virtual organization is supported by research on software maintenance by Swanson and his colleagues (Swanson and Beath, 1989; Hirt and Swanson, 2001). Based on their case studies of organizations using mainly in-house software Swanson and Beath (1989) proposed a relational foundations model for understanding the issues related to the maintenance of application systems within organizations. They proposed that a maintenance organization consists of three group of entities: the users, the IT staff, and the application systems, and that maintenance problems and issues could be understood in terms of the inter-relationships between the entities within each group and between entities from different groups (across the groups). In a recent study of an ERP user-organization, however, Hirt and Swanson (2001) found that the original relational foundations model involving only the entities, within the user-organization is not adequate for understanding ERP maintenance issues. They proposed that the relational model be extended to also include external entities, such as the ERP vendor and service providers. That is, to understand issues related to ERP maintenance, the maintenance organization should be expanded to include these external parties. We propose that the new relational model is best perceived as a virtual organization for ERP maintenance. Such perception raises the following questions:

Q52: Can ERP-related resources be better managed across the virtual organization?

Q53: Must ERP core capabilities exist somewhere within the virtual organization?

Q54: Is there value from the key players having a better understanding of their respective capabilities in the virtual organization, their respective goals, and how each can aid the other?

Q55: Should key players come together and recognize their place in the larger virtual organization, in order to effectively negotiate the incidence of costs and benefits across the partnership?

9.5 Conclusions

The standardization of business applications has changed the patterns of development and use of software, as well as many aspects of maintenance. We have proposed a framework for future research on large application packaged software maintenance that accounts for some of these changes, thereby revealing important new research questions. The commodification of software development and services, emerging alongside the pervasion of standard business software, has amplified issues of maintenance economics and related business strategy. The user-organization centric research framework presented herein, commences from the simple observations that maintenance generates benefits as well as costs and that maintenance strategy fundamentals are impacted on by a range of factors, including software source, support source, and organizational and environmental contexts. Within the new distributed maintenance arrangement, four key stakeholders participate: the user-organization, the software vendor, third-party service providers, and society (national economy). Important economic and business strategy issues arise from the fact that various software and related support sourcing alternatives have substantial maintenance incidence implications (incidence of costs, benefits, and responsibilities). In terms of organizational knowledge strategy, the complexity of standard business software and the concomitant commercial arrangements necessitate that maintenance knowledge sourcing decisions are made in light of lifecycle-wide maintenance knowledge requirements (see also Chapters 14 and 15 on this point). To optimize maintenance it appears necessary that all stakeholders have a lifecycle-wide view of maintenance costs and benefits, considering the four key factors, and underpinned by an understanding of the other stakeholders' perspectives.

From our framework and our questions, we hope it has become clear that business software maintenance can no longer be seen as the sole concern of managers of the IT-function, and that its economics go far beyond the effectiveness of maintenance work and the allocation of appropriate resources to that task. Maintenance has become subject to new commercial arrangements, and thus a new social division of labour. In our opinion, this brings to the foreground, issues of business economics and strategy, as well as issues of knowledge management. The complexity and significance of maintenance, at the same time, demand that all parties enter new partnerships and pay heed to the problems of inter-organizational collaboration.

REFERENCES

Barney, J. B. (1991) Firm Resources and Sustained Competitive advantage. *Journal of Management*, **17**, 99–120.

Bennett, C. and Timbrell, G. (2000) Application Service Providers: Will They Succeed? *Information Systems Frontiers*, **2**(2), 195–211.

Chang, S., Gable, G., Smythe, E., and Timbrell, G. (2000) A Delphi Examination of Public Sector ERP Implementation Issues. *21st International Conference on Information Systems.* Minneapolis, MI, pp. 494–500.

Dailey, A. (1998) SAP R/3: Managing the Life Cycle. *GartnerGroup Symposium/ITexpo*, 28–30 October, Brisbane.

Davenport, T. (2000) The Future of Enterprise-System Enabled Organizations, *Information Systems Frontiers*, **2**(2), 163–180.

Gable, G., van den Heever, R., Erlank, S., and Scott, J. (1997) Large Packaged Software: The Need for Research. 3rd Pacific Asia Conference on Information Systems, Brisbane, pp. 381–388.

Gable, G., Timbrell, G., Sauer, C., and Chan, T. (2002) An Examination of Barriers to Benefits Realisation from Enterprise Systems in the Public Service. Working paper submitted to European Conference on Information Systems.

Gable, G., Scott, J., and Davenport, T. (1998) Cooperative ERP Life-Cycle Knowledge Management. 9th Australasian Conference on Information Systems, Sydney, NSW.

Gable, G. G., Chan, T., and Tan, W. G. (2001) Large Packaged Application Software Maintenance: A research framework. Special issue of *Journal of Software Maintenance and Evolution*, **13**, 1–20.

Gartner Group (1999) The ERP Vendors Market. Symposium/ITexpo. Brisbane, 19–22 October.

Hirt, S. G. and Burton Swanson, E. (2001) Emergent Maintenance of ERP: New Roles and Relationships. Special issue of *Journal of Software Maintenance and Evolution*, **13**, 1–20.

Kern, T., Lacity, M., and Willcocks, L. (2002) *NetSourcing: Renting Your Applications and Services Over a Network.* New York: Prentice Hall.

Klaus, H., Rosemann, M., and Gable, G. (2000) What is ERP? *Information Systems Frontiers*, **2**(2), 155–176.

Klein, B. et al. (1978) Vertical Integration, Appropriable Rents and the Competitive Contracting Process. *Journal of Law and Economics*, **21**(2), 297–326.

Light, B. (2001) The Maintenance Implications of the Customisation of ERP Software. Special issue of *Journal of Software Maintenance and Evolution*, **13**, 1–20.

Nah, F. H., Faja, S., and Cata, T. (2001) Characteristics Of Erp Software Maintenance: A Multiple Case Study. Special issue of *Journal of Software Maintenance and Evolution*, **13**, 1–20.

Ng, C. (2001) A Decision Framework for Enterprise Resource Planning Maintenance and Upgrade: A Client Perspective. Special issue of *Journal of Software Maintenance and Evolution*, **13**, 1–20.

Sahin, I. and Zahedi, F. (2001) Policy Analysis for Warranty, Maintenance, and Upgrade of Software Systems. Special issue of *Journal of Software Maintenance and Evolution*, **13**, 1–20.

Sieber, P. and Griese, J. (eds) (1999) Organizational Virtualness and Electronic Commerce. 2nd International VoNet Workshop, Zurich, September. http:\\www.virtual-organization.net.

Swanson, E. B. and Beath, C. M. (1989) *Maintaining Information Systems in Organizations.* Chichester: John Wiley & Sons.

Szulanski, G. (1995) *Appropriating Rents from Existing Knowledge: Intra-Firm Transfer of Best Practice.* UMI, MI.

Part III

From Learning to Knowledge

10 Implementing Enterprise Resource Planning Systems: The Role of Learning from Failure

Judy E. Scott and Iris Vessey

10.1 Introduction

Problems associated with software implementations are not new, nor specific to enterprise resource planning (ERP) systems. Nevertheless, ERP systems have been blamed for the poor performance of several organizations (Osterland, 2000). A number of companies have reported negative impacts on earnings as the changeover to an ERP takes place, among them Hershey (Girard and Farmer, 1999), AeroGroup (Asbrand, 1999), and Snap-On (Hoffman, 1998). Hershey, for example, suffered a third quarter (1999) sales decrease of 12.4% and an earnings decrease of 18.6% compared with the preceding year (Osterland, 2000).

Implementing an ERP is a major undertaking and the Standish Group International estimates that 90% of SAP R/3 (the dominant ERP) projects run late (Williamson, 1997).[1] The trade press is now also replete with articles on ERP failures, cancellations, and cost/time overruns. Dell canceled its R/3 system after two years when it determined that it could not support the required processes (King, 1997). AeroGroup (Asbrand, 1999), Unisource (IW, 1998), and a number of garbage disposal companies (Bailey, 1999) also abandoned their ERP projects. The most dramatic example of an organization claiming to be damaged by an ERP implementation is that of FoxMeyer Drug Corporation. FoxMeyer claims that its SAP R/3 system sent the company into bankruptcy (Bulkeley, 1996). Note, however, that, at the same time, many companies have experienced benefits that far exceeded their expectations (Davenport, 1998; Deloitte Consulting, 1998).

Because of their ability to integrate all aspects of a business, ERP systems are of significant strategic importance to the companies that implement them.

[1] The AcceleratedSAP methodology, introduced by SAP in 1997, has had some success in decreasing R/3 project implementation times (Alschuler and Dorin, 1998). However, the fact that projects run late does not necessarily mean that they are failures.

From a business perspective, for example, they facilitate reengineering business processes, global operations, competitive agility, as well as data integration across the enterprise, while from a technical perspective they facilitate the installation of more flexible and scaleable architectures, and purchasing rather than building systems, thereby also promising to reduce costs. ERP systems therefore result in widespread changes in business processes and organizational structure, as well in the IT infrastructure, and the IT skills the company needs to implement and operate them. Implementing an ERP is, therefore, a very complex endeavor. That complexity is influenced by the number (and type) of modules implemented and by the implementation strategy chosen (Big Bang, phased, or roll-out; see Welti, 1999).

The implementation of an ERP may be viewed from a number of perspectives. In particular, research has investigated critical success factors (Holland and Light, 1999; Sumner, 1999) and risks (Scott and Vessey, 2000) (see, also, Markus and Tanis, 2000). Problems in ERP implementations may also be addressed from an organizational learning perspective. Not only are the software and the systems architecture new to the company, so, also, are the business practices that the software supports. Hence, an organization wishing to implement such a system must, of necessity, engage in organizational learning.

In this paper, we focus on how organizational learning relates to success and failure of ERP implementations. Organizational learning is often associated with organizational improvement and development and thus implicitly and indirectly with success (Fiol and Lyles, 1985; Huber, 1991; Nonaka, 1994; Stein and Zwass, 1995). While success and failure are usually thought of as outcome variables, organizational learning is a process (Dodgson, 1993; Pentland, 1995). We view large-scale ERP implementation projects as exercises in organizational learning. In particular, because of the complexity of such projects, we believe that failure, at some level, or in some areas of the project, is inevitable. The question then becomes: How we can turn those failures to our advantage? We present Sitkin's (1992) theory of intelligent failure as a way of thinking about the implementation of such systems. We then apply the adapted theory to two well-known ERP implementations to illustrate the utility of the concepts.

The paper proceeds as follows. In the next section, we present the theoretical background to organizational learning and the relevance of organizational learning to information technology (IT) implementations. Further, we examine the relationship between organizational learning and success and failure and introduce Sitkin's theory of intelligent failure. In the third section, we examine the applicability of Sitkin's theory to ERP implementations. In the fourth and fifth sections we apply Sitkin's modified theory to two SAP

R/3 implementations, that of Dow Corning, Inc. and FoxMeyer Drug Corporation. The fourth section presents the background to the companies and the implementations and outlines the similarities between the two, while the fifth section analyses the approaches that Dow Corning and FoxMeyer Drug Corporation took to implementing their ERP systems using our theoretical model. Finally, we present our conclusions, which include the implications of the concepts of intelligent failure to large, complex, software projects, particularly ERP projects, suggestions for further research, as well as the lessons learned.

10.2 Organizational Learning

In this section, we present the concepts associated with organizational learning and their application to IT implementations. We then focus on the role of success and failure in learning, and present Sitkin's theory of intelligent failure to support the notions of long-term success via short-term failure.

Organizational Learning and Its Application to IT Implementations

Organizations must learn and adapt if they are to prosper and survive in a business environment that is continually changing. Hence learning is an important component of organizational survival in times such as the present. Because some organizations are threatened by their ERP implementations, we examine the role of organizational learning in IT implementations.

How organizations learn

Organizational learning is a process (Dodgson, 1993; Pentland, 1995) during which information and knowledge are acquired, created, interpreted, distributed, stored, and retrieved (Huber, 1991; Nonaka, 1994; Stein and Zwass, 1995). Inputs to the process are data, information, knowledge, and organizational commitment (Senge, 1990).

The learning process may take place in two ways: (1) organizations may learn from their own experiences (experiential learning); or (2) they may learn from the experiences of other organizations (vicarious learning) (Huber, 1991). An important form of experiential learning is exploration with organizational experiments. To benefit from these experiments, organizations need to be able to determine cause and effect. The causal variables can be clarified by limiting the sources of change. Ambiguous feedback from too much noise hinders learning and outcome changes need to be of sufficient magnitude to rise above the background noise (Lounamaa and March, 1987).

Vicarious learning refers to the acquisition of experience second hand. Organizations may learn about what competitors or other firms are doing, and may then imitate them. Consulting firms, for example, use knowledge management to make second-hand experience available for future engagements. Another vicarious learning strategy is to obtain new knowledge and skills by hiring ('grafting') new employees (Huber, 1991). Engaging consultants is an example of 'grafting' on a temporary basis.

The organizational learning process results in one of two potential types of outputs, which vary by the extent of the learning, or change that occurs: exploitation and exploration (March, 1991). These outcomes result, most obviously, from the experiential learning process described above. On the one hand, researchers posit a change in the range of potential behaviors (Huber, 1991), such as improved actions from knowledge and understanding (Fiol and Lyles, 1985). This low-level, 'single-loop' learning results in adjustment of behavior relative to fixed goals, norms, and assumptions (Argyris and Schon, 1978) and reinforcement of routines, practices, and policies (Levitt and March, 1988). Moreover, this type of adaptive organizational learning (Fiol and Lyles, 1985) copes with problems via extrinsic motivation (Senge, 1990), and results in incremental improvement or exploitation (March, 1991).

On the other hand, a change in routines, practices, and policies (Levitt and March, 1988) challenges goals, norms, and assumptions, as well as behavior (Argyris and Schon, 1978). This type of high-level learning is known as 'double-loop' learning. It is strategic (Fiol and Lyles, 1985) and generative, requiring new insights and a systems view which seeks to eliminate underlying causes of problems. It is based on intrinsic motivation (Senge, 1990). This type of learning is also called exploration and requires unlearning (Hedberg, 1981). Exploration may take place via experiments or 'trial balloons'. While there is a risk that some experiments will fail, there is also a chance of breakthrough learning. According to March (1991), exploration involves creative organizational learning, which helps to avoid the problem of continuing to use traditional approaches even when change is warranted (see also, Levitt and March, 1988; Robey, Wishart, and Rodriguez-Diaz, 1995). This type of outcome can also be characterized as resilience (Sitkin, 1992).

Learning in IT implementations

A number of researchers have conceived of various aspects of IT from the perspective of organizational learning. Attewell (1992), for example, described the IT adoption process as a learning process. Pentland (1995) has argued that there is a close relationship between organizational learning and IT implementation.[2] Robey, Wishart, and Rodriguez-Diaz (1995) addressed business process reengineering (BPR) from the viewpoint of organizational learning. Process reengineering requires new organizational procedures, which may be institutionalized in the new systems that support the reengineered processes, and the 'data' that serves as the organizational memory. Ang, Thong, and Yap (1997) built on Robey et al.'s model of BPR as organizational learning to develop a model of IT implementation based on changing business processes.[3]

As we have seen, organizational learning is essential if an organization is to survive in a changing environment. Furthermore, IT is becoming increasingly important to a firm's survival. Organizational learning that involves IT may take two quite different forms, which involve different types of organizational personnel and are time dependent. First, organizational learning takes place during the implementation itself. This type of learning involves both IT and business personnel in two quite distinct roles, which highlight the fact that two types of organizational learning occur during the process of implementation. In the first role, business and systems personnel interact and determine the extent to which the new system differs from the system it is to replace. A system, for example, that automates an existing business system (be it manual or computerized) will simply embed existing routines, procedures, rules, and assumptions, in a single-loop learning environment. In March's (1991) terms, the governing variables will not change; from another perspective, although the form may change, the substance does not. A system, on the other hand, that makes explicit current assumptions, reevaluates and reformulates them, will involve double-loop learning. In the second role, systems personnel are responsible for implementing the system.

Second, organizational learning also takes place subsequent to the implementation when the users must adapt to the new system and perhaps also to new ways of working. The extent of the changes they experience will depend on whether the new system involves single- or double-loop learning.

Levitt and March (1988) refer to the type of learning that occurs during implementation as learning by doing and that occurring following implementation as learning by using. Figure 10.1 presents our model of IT-based organizational learning.

[2] Although Pentland used the term 'implementation' to refer to in-house developed systems, we use it generically to refer to the provision of a new information system, either developed or packaged.
[3] Note, also, that their model supports the notions of traditional systems analysis and design in addressing first the problem (that is, the business processes) prior to turning attention to the solution (that is, the physical requirements to solve the problem).

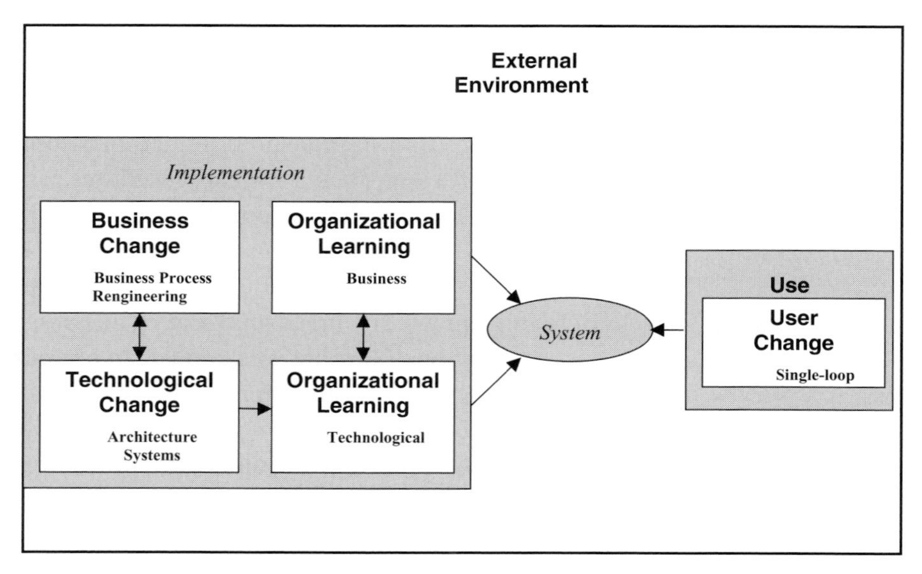

Figure 10.1 Relationship between IT implementations and organizational
Note: The arrows represent adaptations that must occur as a result of the changes and the learning that results.

The Effects of Success and Failure

Organizational learning may also be viewed as a series of processes interspersed with 'small' successes or failures. Short-term or small successes and failures are tactical rather than strategic in nature. Here we characterize the effects on organizations that seek to learn via small successes and those that seek to learn via small failures.

Learning from success

When an organization largely succeeds in its endeavors, it adopts a particular attitude toward change: an attitude that supports exploitation as the appropriate outcome of learning, rather than exploration. The organization, for example, becomes complacent, engages in less search behavior, and therefore pays much less attention to its environment. It is difficult to convince an organization to try new approaches when its current approaches are reasonably successful. Further, new approaches are risky because they are unproven. And, while an individual or group of individuals is unlikely to be held responsible for failing while using the traditional approach, they are likely to be blamed for failing using a new, and therefore untried approach, further reducing the likelihood that they will try new approaches.

Further, focusing on prior success means that the organization will pay less attention to information that does not support traditional approaches. Focusing on prior success will also result in homogeneous procedures, personnel, and organizational structures, further restricting the organization's ability to adapt when the environment changes. These types of behaviors discourage the double-loop learning that occurs via exploration and therefore experimentation (Argyris, 1985).

Learning from failure

According to Sitkin (1992), rather than seeking to avoid failure, organizations should seek to pursue the strategy of learning by experimentation. This means that organizations will sometimes fail, giving them the opportunity to learn from their failures. The experience of failure produces learning readiness. If the cause of the failure is determined, organizational learning takes place. New knowledge on cause and effect relationships enables organizations to choose more promising actions in the future.

Sitkin argues that 'small prior failures' will mean: (1) deeper processing of information about potential problems; (2) greater recognition of problems based on past experience; (3) deeper levels of search; (4) an organization that is more flexible and open to change; (5) a greater level of risk tolerance; (6) a greater variety of personnel and organizational procedures; (7) and a greater experience available to address future problems.

Learning by doing or by experience (with inevitable mistakes) was observed in aircraft production in the 1930s and in manufacturing during the second world war and accepted as a valid approach to learning (Levitt and March, 1988). Using this approach, short-term or small failures may pave the way to ultimate success.

Implications of learning from success and learning from failure

We have seen that learning from success encourages exploitation, which seeks to maintain the status quo (single-loop learning), and hence leads to reliable organizational responses. This situation is particularly efficient when the environment is stable. On the other hand, learning from failure encourages exploration (in the form of experimentation), which seeks to push the envelope (double-loop learning), which, in turn, leads to an organization that is resilient to change. This situation is particularly effective when the environment is changing.

In general, because of the benefits of resilience, Sitkin (1992) advocates seeking long-term success through short-term failure.

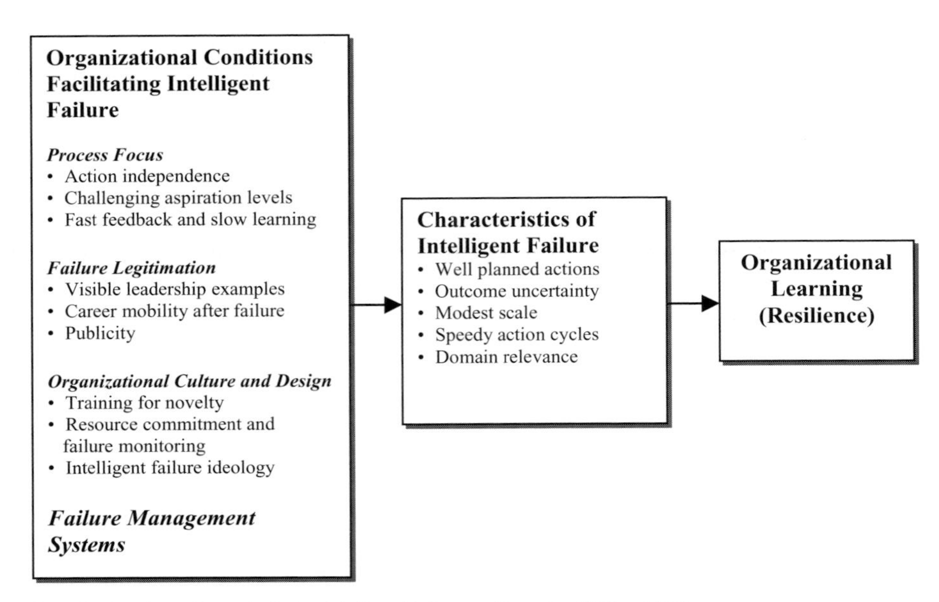

Figure 10.2 Organizational conditions fostering intelligent failure and learning (Sitkin, 1992)

The Role of Intelligent Failure in Long-Term Success

Here we focus on intelligent failures, those that 'are most effective at fostering learning' and that therefore provide intelligence relating to failures (Sitkin, 1992). Figure 10.2 presents Sitkin's (1992) model of intelligent failure. We first address the organizational conditions that must be in place to facilitate learning from (intelligent) failure, followed by what an organization needs to do to benefit from (intelligent) failure.

Organizational conditions facilitating intelligent failure

Sitkin's (1992) theory includes four categories of organizational conditions that should be in place to promote intelligent failure. He uses the term 'strategic failure' to refer to the systematic approach to organizational failure outlined here.

First, the focus should be on the process rather than outcomes of organizational learning. Many small, independent experiments provide learning opportunities through feedback. The goal should be to generate a range of outcomes some of which will involve failure. Second, the organization should legitimize intelligent failure. Legitimization of failure removes inhibitions that prevent learning (Argyris and Schon, 1978). This can be achieved via visible leadership examples, career mobility after failure, and publicity. There needs to be public commitment to supporting employees who are willing to engage in intelligent failure, organizational examples highlighting the failures of executives or other employees whose careers did not suffer as a result, as well as examples publicizing intelligent failures.

Third, the organizational culture and design must be such that individuals or groups can fail without suffering reprisals from their failures. An organization can prepare individuals for intelligent failure by providing training to help them overcome unusual experiences, and also by committing resources to, and monitoring, failure. Fourth, the focus should be on systems that will produce failure rather than on individuals because they do not fail in sufficiently systematic ways for the outcomes to be useful. This means that organizations need to implement strategic failure programs (failure management systems) to increase the number and diversity of failures (Sitkin, 1992).

Characteristics of intelligent failure

There are a number of characteristics that must be in place for an organization to implement intelligent failure. First, only actions that are well planned (that is, to succeed) can be used to determine why failure occurred. The actions and the reasons for failure then provide information on alternative possible future actions, and hence lead to organizational learning. It is difficult, however, to plan adequately or anticipate potential outcomes when the domain is unfamiliar (Sitkin, 1992; Cliffe, 1999).

Second, outcomes must be uncertain otherwise the potential for failure would not exist, and hence no information would be gained from which learning could occur. Third, problems must be of such a size and scope that they are noticed, but not so large as to be overwhelming. This balance allows the organization to be responsive rather than defensive. Fourth, the action and the associated feedback should occur quickly to ensure evaluation, which, in turn, can lead to adjustment, followed by further action and feedback. This type of experimental situation is particularly important when the organizational environment is changing rapidly.

Fifith, for a failure to be 'intelligent', it must occur in a familiar domain, and in one that is important to the organization, otherwise people may not recognize it and/or they may be unable to formulate a response (Sitkin, 1992). When failure challenges fundamental organizational assumptions (as may happen in double-loop learning), defensiveness may preclude even the acknowledgment of the problem because of the deep-seated changes that would most likely be necessary to address it.

10.3 Learning in ERP Implementations

In this section, we justify using an organizational learning lens to analyze ERP implementations, adapt Sitkin's theory of intelligent failure to ERP implementations, and highlight certain further characteristics that distinguish ERP implementations.

The ERP Implementation Process as Organizational Learning

As we have seen in Figure 10.1, the opportunities for organizational learning in the context of IT implementations, in general, are many and varied. ERP systems are much more complex and far-reaching from the viewpoint of both implementing them (because of the many different and novel types of knowledge required) and using them (because of the fact that they may change significantly the way people work) than the majority of in-house developed systems. Here, we focus exclusively on the implementation process (see Figure 10.2). Further, as in IT implementations, in general, learning takes place in both the business and technological aspects of the implementation. This fact becomes all the more apparent in ERP implementations where the integration of the software across functional areas renders the two aspects much more closely related and therefore interdependent.

There are numerous reasons why organizational learning is an appropriate theoretical lens for studying ERP implementations. Here we mention just a few to illustrate the concept; we elaborate further in the specific context of our case analyses. When implementing an ERP, the usual recommendation is that companies design their 'to be' processes with the target package in mind (Champy, 1997).[4] Adopting ERP best practices changes organizational routines (double-loop learning). At a more tactical level, ERP configuration tends to be trial-and-error. SAP R/3, for example, has almost 15 000 tables (version 4.0B), resulting in myriads of configuration options (Bancroft,

[4] Note that this advice is diametrically opposed to that given for systems development projects since the inception of computing. A major reason for this recommendation is that any changes to the package must be reflected in subsequent upgrades. Vendors have significant configuration management problems due to the numerous software versions they need to manage, including those for specific industries, and therefore typically support only three prior versions. Making substantial changes to the software therefore has major time and cost implications.

Seip, and Sprengel, 1998). Prototyping and extensive testing are necessary and widely used to find configuration errors. Experience with these 'small failures' provides feedback for organizational learning.

The complexity of ERP implementations means that, as projects evolve, they may be shaped by unanticipated circumstances, for example, in the external environment (Cliffe, 1999). Organizations need to adjust the parameters of their projects to reflect changes in the environment. Adhering to project plans that are no longer relevant might cause long-term failure. Learning from small failures, such as not meeting intermediate project deadlines, is preferable.

Adaptation of Sitkin's Theory

While a large part of Sitkin's intelligent failure theory fits an ERP implementation, some parts do not. Here we examine that fit, and adapt Sitkin's theory for use in evaluating ERP implementations from the perspective of what we call 'learning from failure', rather than intelligent failure. Figure 10.3 presents Sitkin's intelligent failure model adapted to ERP implementations.

The major differences between our model and Sitkin's model lie in the organizational conditions that facilitate intelligent failure, and all stem from the notions of strategic failure; that is, that an organization seeks to have failure occur so as to learn from it, to remain alert, and to gain the advantages of 'pushing the envelope'. Organizations seeking to implement ERPs are unlikely to choose to implement a system in such a way as to encourage systematic failure. First, such implementations are sufficiently complex that failure will inevitably occur in many of the necessary sub-tasks, thereby allowing analysis of cause

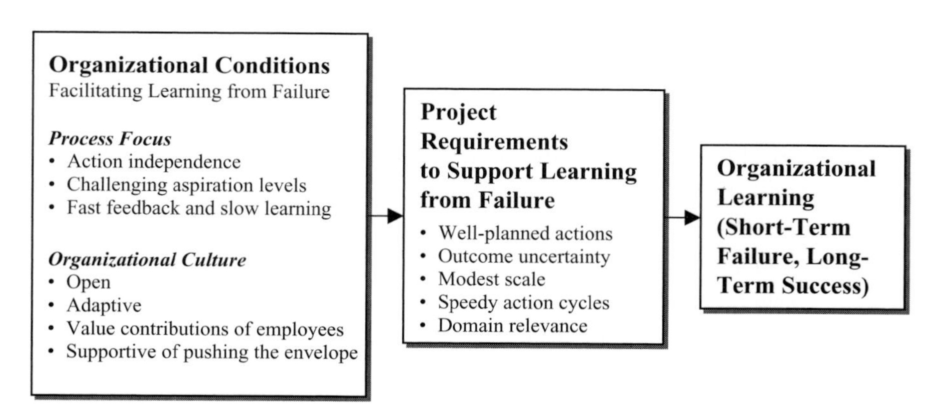

Figure 10.3 Organizational conditions fostering learning from failure in ERP implementations (Sitkin, 1992)

and effect relationships and identification of alternative approaches. Second, ERP implementations are usually one-of-a-kind projects and therefore only certain types of implementation strategies lend themselves to the systematic approach to failure that Sitkin proposes. Phased and roll-out strategies involve repetitive activities; a Big Bang strategy does not.[5]

Failure management systems are at the heart of Sitkin's suggestions. This notion is closely linked to the 'failure legitimation' and 'organizational culture and design' that Sitkin also envisages; that is, designing the organization and the reward systems to support strategic failure. Rather than purposefully establishing a strategic failure program and implementing an organizational design to support failure, we believe that learning from failure can be facilitated predominantly by the organizational culture alone. Hence, we exclude failure management systems from the conditions that facilitate intelligent failure in Sitkin's theory and replace 'failure legitimation' and 'organizational culture and design' with organizational culture (Figure 10.3).

Sitkin's characteristics of intelligent failure are appropriate to learning from failure and consistent with ERP implementations. Hence they are included unchanged in Figure 10.3.

10.4 The Dow Corning and Foxmeyer ERP Implementations

To illustrate the utility of an organizational learning approach to ERP implementations we chose two SAP R/3 implementations, those of Dow Corning, Inc. and FoxMeyer Drug Corporation. To facilitate comparison we chose two companies that were in similar (troubled) circumstances at the same time, companies that both chose to implement the same software to help them solve their business problems.

Research on failed implementations is difficult because organizations often adopt a conspiracy of silence on failures making it difficult for researchers to obtain data and any data obtained may be biased by participants' inclinations to rationalize their role in the fiasco (Sauer, 1999). The information on Dow Corning's implementation is based on three teaching cases (Ross, 1996a, 1996b, and 1999c), while the information on FoxMeyer's comes from several published (trade) sources (including its legal claims). Both cases therefore have rich data and high external validity based on the views of multiple stakeholders.

[5] These implementation strategies are discussed in more detail in the case analysis.

Table 10.1a Important milestones in Dow Corning's ERP implementation history

Phase	Year	Month	Milestone
Prior to initial implementations	1991		Global Order Entry System (GOES) Project
			Global process consensus difficult
	1994		Need to restructure around processes rather than functions
Pilot implementations	1995	Early	Decision to implement R/3 and Project PRIDE (Process Reengineering through Information Delivery Excellence)
			1 Established Process and Information Technology Board (PITB) – responsible for setting strategic directions for global information technology
			2 Combined IT function with business process reengineering function into a unit called Business Processes and Information Technology (BPIT)
			3 Appointed Charlie Lacefield, the then VP of Manufacturing and Engineering, to manage the new organizational unit
		June	Formation of global team of 40 operations and IT specialists from around world
	1996	February	Restructure sub-teams from 8 to 4 to match 4 key operational processes
			New project manager, Ralph Reed
		September 30	3 UK pilot implementations on time but cut scope and there were communication and change management problems
			Global Workstation Project – new distributed infrastructure
Follow-up	1997		Formed area teams in Americas, Europe, and Asia
			Established PRIDE academy in Midland, MI – global team to train area teams
			Formed 'think tanks' to determine future processes; aid in change management
	1998	December	Full implementation

Here we present the background to the two SAP R/3 implementations, followed by the similarities between them. Tables 10.1a and 10.1b present important milestones in the history of the Dow Corning and FoxMeyer implementations, respectively.

Table 10.1b Important milestones in FoxMeyer's ERP implementation history

Phase	Year	Month	Milestone
Pre-implementation	1993	May	Lost major customer, Phar-Mor Inc.
		July	Planning begins on Delta Project
			Testing scalability on Digital and HP servers
		September	Contract with SAP
Implementation	1994	January	Contract with Andersen Consulting
			HP chosen as hardware platform
			Implementation of Delta Project begins
		July	Contract with UHC Systems
	1995	January to February	UHC warehouses open interface problems result in millions of dollars in inventory losses
		January to April	Implementation proceeds
			FI and SD live for UHC
		May	Washington Courthouse warehouse scheduled to go live
		August	Washington Courthouse warehouse live; damaged inventory and losses from duplicate shipments cost $15.5 million
Aftermath	1996	February	COO Thomas Anderson asked to resign
		July	$34 million charge for inventory and order mix-ups
		August	Chapter 11
		October	McKesson agrees to pay $80 million for FoxMeyer
	1997	March	Chapter 7
		Fall	Asset liquidation
	1998	July	Sues Andersen Consulting
		August	Sues SAP
			Sues Deloitte
	1999	May	Case against Deloitte dismissed

Backgrounds

At the time of its R/3 implementation, beginning in 1995, Dow Corning Incorporated was a $2.5 billion producer of silicone products. The company was facing competitive pressures as well as lawsuits worth $2 billion due to well-publicized problems with silicone breast implants. Existing systems were focused on specific departments; they were therefore fragmented and lacked the ability to present a common face to the customer. The company decided that its survival depended on reengineering its business processes to become a truly global company, an objective that it believed could be met only with appropriate information systems. The decision to reengineer was supported by incorporating responsibility for business processes and information technology within the same organizational unit, Business Processes and Information Technology (BPIT).

During preparation for its initial implementations, three pilot sites in the UK, the company had problems training the project team appropriately, gave up some of its best business practices, did not use an implementation partner, and paid little attention to managing change within the organization. Yet the company and the system survived.

This scene can be compared with FoxMeyer Drug Corporation's R/3 implementation. FoxMeyer's business was wholesale drug distribution, an industry rendered very competitive and relatively unstable by health-care reforms of the early 1990s. Hoping to gain a competitive advantage, FoxMeyer initiated its $65 million Delta project, in 1993. The Delta project involved implementation of R/3 integrated with extremely ambitious warehouse automation software. FoxMeyer expected to save $40–$50 million dollars annually from the project, as well as to grow rapidly and gain market share (Jesitus, 1997).

FoxMeyer's plans did not work out. Its major customer, Phar-Mor, went bankrupt in May 1993, and FoxMeyer signed up a major new customer, University Healthsystem Consortium (UHC) in July 1994. However, the UHC contract required major changes to the already ongoing project. Costs soared to over $100 million; and in August 1996 FoxMeyer filed for Chapter 11 bankruptcy protection, after taking a charge of $34 million the previous month for inventory and order mix-ups. Following liquidation of the company's major assets in November 1997, FoxMeyer's trustee sued Andersen Consulting, SAP, and Deloitte for $500 million each in July and August 1998.[6] The case against Deloitte was dismissed in May 1999.

Similarities

Both companies implemented R/3 for their primary activities (that is, those that are on the value chain), which typically include procurement, production/manufacturing, and sales and distribution functions (Porter and Millar, 1985). Value-chain implementations are much more complex than those for

[6] Other, similar cases are referenced below under domain relevance.

secondary activities (that is, those that support the value chain), which typically involve financial and/or human resources functions (Brown and Vessey, 1999). Dow Corning implemented production planning, sales and distribution, materials management, and financials modules. Because FoxMeyer was a distributor rather than a producer, it did not implement production planning.

Both Dow Corning and FoxMeyer were operating in 'threatened' environments leading up to their ERP purchase decisions. Dow Corning was facing costly lawsuits and increased competitive pressures, while FoxMeyer was facing decreasing margins and high transaction volumes. From the viewpoint of organizational context, both companies had well-defined business strategies and viewed information technology as key to supporting them. Because both companies viewed integrated business processes as the key to their success, both decided to implement an ERP. Both companies restructured their organizations prior to or in the same timeframe as initiating their R/3 implementations.

From the viewpoint of implementing an ERP, both companies had a champion at high levels within their respective corporations. There was some doubt about the fit of the R/3 package to both companies' needs: Dow Corning was uncertain whether the software would ultimately support the number of users it foresaw for its global operations, while FoxMeyer was uncertain whether the system could process the required number of daily transactions. FoxMeyer ran simulations and determined that R/3 could, indeed, process the volumes needed; Dow Corning did not address the software's limitations in the implementations reported.

The R/3 system and enterprise resource planning, in general, were new to both companies and the knowledge and skills were therefore not available in-house. Acquiring the necessary skills was problematic for both companies. The only training Dow Corning's employees received was for R/2 (mainframe), instead of R/3 (client/server). Similarly, the training of members of FoxMeyer's implementation team appears to have been inadequate.

Both companies needed to convert from their existing centralized mainframe infrastructure to a distributed client/server infrastructure. Both companies implemented so-called 'Vanilla' versions of the software,[7] although both implemented less than expected due to time pressures: FoxMeyer had intended to make process changes by customizing the system; Dow Corning had intended to bolt-on other software (for example, to automate authorization notices for large purchases, and to read bar codes to track inventory).

[7] 'Vanilla' implementations are those that largely accept the processes embedded in the software.

From the viewpoint of the project itself, neither company appeared to have very good project planning in place. Further, neither company appeared to expend a great deal of effort on organizational change management. Dow Corning focused on reducing the length of the change period out of concern for the project team, while FoxMeyer had problems with personnel at its warehouses.

10.5 Model-Based Analysis of the Implementations

We now apply the theory on learning from failure (Figure 10.3) to the SAP R/3 implementations at Dow Corning and FoxMeyer. We first discuss the necessary organizational conditions that must be in place to support learning from failure followed by the requirements of the project that allow project teams to learn from failure. We conclude by evaluating the organizational learning that took place in each company. Table 10.2 summarizes the implementations from the viewpoint of the theoretical model.

Characteristics of the Organization

Here we compare the Dow Corning and FoxMeyer implementations based on the organizational characteristics needed to support learning from failure.

Process focus

We have seen that for Sitkin's theory of intelligent failure to apply, the focus must be on the process rather than on the product or outcome. At least two different types of process focus can be identified in ERP implementations. First, a process is inherent in any software implementation, systems development or package implementation, usually in the form of a life cycle, which outlines the activities that need to be completed and the general sequence of their completion. Many opportunities for learning will occur during the course of a specific implementation. Although there may be limited opportunities for the same problems to recur, learning will be evident as team members apply their experience to new situations.

Second, the implementation strategy chosen may have an in-built process focus. On the one hand, a Big Bang strategy, in which all modules are implemented throughout the company at one time, has an outcome focus. On the other hand, phased strategies, in which certain modules, usually support modules such as finance and accounting or human resources, are implemented

Table 10.2 Summary of the Dow Corning and FoxMeyer implementations from the viewpoint of learning from failure

Learning factors	Dow Corning Ross (1996b)		FoxMeyer (Bulkeley, 1996)	
Organizational characteristics				
Process focus	Used a process approach to implementation (i.e., roll-out)	+	Used a one-step approach to implementation (i.e., Big Bang)	–
Organizational culture	Open	+	Not sufficiently open	–
	Confidence in, and supportive of, employees → Employees were loyal	+	Did not heed employees' warnings → Employees lacked loyalty	–
Project characteristics				
Well-planned actions	Low-level planning poor	–	Low-level planning there, but suspect	–
	High-level planning excellent	+	High-level planning disastrous (when external environment changed)	–
	De-escalation when could not meet deadlines	+	Escalation when could not meet deadlines	–
Outcome uncertainty	New software	–	New software	–
	New hardware	–	New hardware	–
	Appropriate skills and knowledge	–	Appropriate skills and knowledge	–
	Handling required organizational change	–	Handling required organizational change	–
			Implementing new, ambitious warehouse automation software at the same time as R/3	–
			Interfacing warehouse software with R/3	–
			Using R/3 for large-scale distribution	–
Modest scale	Shelved business process reengineering plans to focus on developing global infrastructure	+	Over-optimism regarding role of technology	–
			Simultaneous implementation of R/3 and extremely ambitious automated warehouse interfaced	–

Table 10.2 *(cont.)*

Learning factors	Dow Corning Ross (1996b)		FoxMeyer (Bulkeley, 1996)	
	Did three pilot implementations	+		
	Implemented R/3 at remaining locations one by one (i.e., roll-out)		Expanded project scale for UHC requirements	
Speedy action cycles	Adjusted quickly when something 'did not work'	+	Showed no signs of adjusting based on feedback	–
Domain relevance	Task-Technology fit	?	Task technology fit	–
	Did not use an implementation partner	–	Used an implementation partner	+
	Knowledge and skills problematic	–	But knowledge and skills still problematic	–

Notes: '+' and '–' refer to the positive and negative influences of a particular action. '?' means that the effect is unknown.

across the company at the same time, and roll-outs, which implement the system in each business unit (or plant) sequentially and then in the rest of the company following a successful operational period (Welti, 1999), have a process focus. A roll-out strategy may be based on either a Big Bang or a phased implementation (Welti, 1999). Clearly, the Big Bang strategy is riskier than the other strategies because fewer learning opportunities are incorporated into the approach.

Dow Corning treated its ERP implementation as a process. Its objective was to introduce R/3 into 109 locations in 17 countries.[8] It implemented the system in each of the locations in a roll-out strategy based on a Big Bang (sometimes called a Small Bang; Bancroft, Seip, and Sprengel, 1998); that is, all the modules are implemented in one location at a time. The roll-out was initiated with the three pilot projects in the UK. and then moved to Wiesbaden in Germany, and so on. Apart from the process focus inherent in any IT implementation, this further emphasis on process gave Dow Corning the opportunity to learn

[8] Additional sites in Argentina and Brazil were postponed until a later version of the software was implemented.

from its failures and it did so quite effectively (see, below, under 'well-planned actions').

FoxMeyer Drug, on the other hand, implemented R/3 in a Big Bang implementation, demonstrating that the company's implementation strategy was not process focused. Further, the majority of the decisions FoxMeyer made were focused on the outcome rather than on how to achieve the outcome, or whether the outcome was achievable or not. This implies that FoxMeyer had little chance of learning through failure.

In summary, Dow Corning was well positioned to learn from failure, while FoxMeyer was not.

Organizational culture

The two companies differed markedly in their organizational culture. Dow Corning had an open culture that invited communication. In fact, Dow Corning's value statement:

> articulated its commitment to an open and creative culture that recognized employees as the primary source of ideas, actions, and delivery of performance. Dow Corning's consistent growth performance had resulted in a stable work environment that reinforced the loyalty of its employees. (Ross 1996a)

Dow Corning made the decision to implement R/3 without the help of an implementation partner because it had confidence in the ability of its employees.

> When Charlie Lacefield moved into his role what he had was a very good group of folks, and what he did was show a lot of confidence in those folks, and it was genuine. The proof that he felt that way was the fact that he didn't outsource this thing to Andersen or Price Waterhouse. And not just Charlie but other management said too, 'We're going to let our people do this and we're going to put them in charge of this and we'll use consultants when we need to but we have the expertise to do this.' This is proof that management really has confidence in us and that's a very exciting thing. (Information Architecture Department Manager (Ross, 1996b))

FoxMeyer's culture, on the other hand, did not invite open communication and therefore information sharing. When FoxMeyer brought forward its implementation 90 days as a result of its contract with UHC, employees apparently voiced their concerns to management; however, management ignored what they had to say.

> 'We were given an assignment to find any gaps in the SAP system', recalls one FoxMeyer information-systems manager. But systems people found they were encouraged to minimize problems.' It wasn't appropriate to criticize SAP', the manager says. Adds a

> consultant who worked on the project: 'Every time we showed something that didn't work, they'd say, "Is this a deal-breaker?" Well, no one was a deal-breaker. But if you put enough cinder blocks in a rowboat, it's going to sink.' (Bulkeley, 1996)

The lack of open communication meant that employees were not particularly loyal to FoxMeyer, which cost the company substantially. For example: (1) project personnel thought they could obtain much higher salaries with their newly gained R/3 knowledge; (2) employees at existing warehouses, knowing they were going to be out of work, left in droves prior to the implementation of the automated warehouse, which resulted in a $15.5 million loss due to order mix-ups (Bulkeley, 1996).

In summary, Dow Corning had a culture that facilitated organizational learning and that was supportive of employees who tried and failed (and therefore supportive of Sitkin's notions of 'intelligent failure'). FoxMeyer's culture, on the other hand, was not supportive of organizational learning and employees were discouraged from discussing the possibility of failure.

Characteristics of the Project

The characteristics of the implementation project must be set up in such a way as to support learning from failure.

Well-planned actions

ERPs were in their infancy at the time of the Dow Corning and FoxMeyer implementations. As we have seen, it is difficult to plan adequately when the domain is unfamiliar.

Dow Corning personnel were not well prepared for the project; they were trained on R/2 rather than R/3, had no consultants to guide them, and therefore did not know the software sufficiently well to plan in detail.

> Some folks on my team who come from a rigorous manufacturing background asked, 'Where's our project plan?' We're doing a lot of this on faith. I mean how do you put a plan out there and say, 'Learn SAP. OK, you've got two weeks to learn that and then you've got two more weeks to learn this.' You really can't plan learning or knowledge transfer the way you plan a production schedule. (Ross, 1996b)

Dow Corning's philosophy was 'to establish a limited set of strategic priorities for the IT unit' and then 'cut loose the IT people so that they could do things'.

This hands-on approach, although seemingly not well planned, did result in opportunities for exploration and valuable experiential learning. Moreover,

Dow Corning did a very good job of managing the project at a high-level: (1) it had extremely strong leadership in the head of BPIT, Charlie Lacefield; and (2) it maintained high-level control over the project by monitoring progress and adapting the project objectives.

FoxMeyer's consultants, Andersen Consulting, provided the implementation methodology that was used as the basis for planning. The methodology was based on experience with prior implementations. FoxMeyer claims that, in pre-sales discussions, Andersen Consulting led the company to believe that it had the methods in place to implement R/3 effectively.

> Our reengineering methodology (VDRE) and our systems development methodology, Method/1, provide the necessary tools for planning and estimating the project activities. For the core SAP components, we have several SAP-specific estimating tools which utilize tried-and-proven guidelines and have been developed over the course of several SAP implementations (e.g., resulting in Method/SAP). (LJX Files, 1997)

Such vicarious learning is often very effective when the new engagements are similar to previous ones. The FoxMeyer implementation was not amenable to a rigid plan, however, because many aspects of the project were novel and because the business environment changed during the project. Moreover, there were warnings that FoxMeyer's early plans would not be feasible. For example, an independent consultant considered their timeline too aggressive (Jesitus, 1997).

FoxMeyer's planning, however, failed at a high level as well as the low level. By far the most crucial change that either company experienced was FoxMeyer's loss of a substantial customer a few months prior to its R/3 implementation. Because the company was driven by the need to have a system in place so that it could regain some of its lost revenues, it signed up a new customer, which brought with it new challenges, including opening six new warehouses, which increased the workload substantially. To meet the contract deadline, which required deliveries to start in early 1995, the ERP needed to be operational three months sooner than planned. Meeting the schedule necessitated cutting testing, which resulted in disastrous data errors, and eventually the company sought bankruptcy protection.

These changes to FoxMeyer's ERP project were therefore not well thought out, and, worse still, went against traditional wisdom in responding to increased complexity by shortening the implementation time.

Hence, while low-level planning was suspect in both companies, high-level planning was particularly effective at Dow Corning and particularly ineffective at FoxMeyer.

Outcome uncertainty

Uncertain outcomes are expected in volatile environments and with new technology. There is considerable uncertainty about the performance of new technology from the viewpoint of response times, interfacing to other systems, tools available, and data storage requirements (for example, Barki, Rivard, and Talbot, 1993; Block, 1983; Lyytinen, Mathiassen, and Ropponen, 1998; Willcocks and Griffiths, 1997). Further, the introduction of ERP technology means that different knowledge and skills will be needed, which raises the question of how to obtain them. Further, again, the significant organizational changes associated with implementing an ERP will most likely cause resistance to that change until people become familiar with the technology and with new ways of working (Bancroft, Seip, and Sprengel, 1998). These issues form part of the business-focused organizational learning that must take place for an implementation to be effective, in particular change management (see Figure 10.1).

Because both Dow Corning and FoxMeyer, as noted above, were operating in threatened environments, organizational learning was essential for both to survive. Implementing an ERP had far-reaching implications for both companies, both from the viewpoint of the integration of all the areas of the companies' businesses, as well as from the viewpoint of introducing a new technological architecture (a client–server system that replaced existing mainframe systems) without the requisite skills and knowledge. Neither company appeared to have addressed organizational change management, though it gained Dow Corning's attention as time went by, via think tanks designed to define new business processes (see under 'speedy action cycles').

FoxMeyer, however, had additional areas in which the outcome was uncertain. The company was implementing extremely ambitious warehouse automation software at the same time as R/3, software that was designed to pick 80% of goods automatically when the industry norm was 20%. Further, they were integrating the software to R/3, a risky undertaking that had not been done previously. Further, again, R/3 had not been implemented for large-scale distribution.

Hence there is little doubt that the outcomes of both companies' implementations were uncertain. FoxMeyer, however, had considerably more outcome uncertainty than Dow Corning.

Modest scale

Complexity makes it difficult to discern cause and effect and so hinders organizational learning (Lounamaa and March, 1987). Interestingly, Dow Corning appears to have sought 'simplicity' while FoxMeyer accepted 'complexity'.

The two companies viewed technology in markedly different ways. In the first half of the 1990s, Dow Corning spent considerable time and effort reexamining its business processes with the aim of responding to its competition by restructuring around business processes rather than along functional lines. The company was prepared, however, to move forward gradually, foregoing its process reengineering vision to get the system up and running so as to reap the benefits of a common, global system prior to seeking more sophisticated processes. It therefore followed the notion of pursuing 'small wins' (Kotter, 1995). Further, Dow Corning first conducted three pilot implementations prior to rolling out R/3 across its global operations, a further sign of proceeding with a certain amount of caution.

On the other hand, FoxMeyer appeared to have unrealistic expectations of what R/3 could do for it (see, also, Markus and Benjamin, 1997):

> Insiders say top management was so overoptimistic about computerization that it recklessly underbid contracts, expecting electronic efficiencies to lower costs enough to make the deal profitable. (Bulkeley, 1996)

As we have seen, at the same time as implementing R/3, FoxMeyer installed ambitious warehouse automation software, which needed to be interfaced to R/3 at a new state-of-the-art warehouse. Further, FoxMeyer expanded the scope of its project to meet the needs of its new customer, UHC.

In summary, then, FoxMeyer did not seek to keep its activities to a modest scale in its ERP implementation, while Dow Corning took a number of steps to ensure that its implementation proceeded smoothly.

Speedy action cycles

Both companies were focused on implementing their ERP systems in a given time-frame, and therefore on meeting deadlines. Hence turnaround needed to be quick and effective. The notion of speedy action cycles relies on feedback, which, in turn, relies on communications.

Management at Dow Corning monitored progress constantly and made decisions to ensure that the deadlines would be met, decisions that depended on effective feedback. For example: (1) when the project was losing momentum in February 1996, it reduced the number of teams working on the project from eight to four to regain focus; (2) when it appeared unlikely that the three UK pilots would be implemented on time, the company cut scope to meet the deadline; (3) when it realized that local knowledge was needed for regional implementations, the company formed area teams; (4) when area teams could not address R/3 issues effectively, it established the PRIDE Academy to train area team members; and (5) the company eventually addressed change management as a side effect of the think tanks set up to address future business processes.

Feedback at FoxMeyer, on the other hand, was not effective because the communication and attention necessary for fast and effective feedback were missing (Bancroft et al., 1998; Holland and Light, 1999). For example (as quoted previously under 'organizational culture'):

> 'We were given an assignment to find any gaps in the SAP system', recalls one FoxMeyer information-systems manager. 'But systems people found they were encouraged to minimize problems. It wasn't appropriate to criticize SAP', the manager says. Adds a consultant who worked on the project: 'Every time we showed something that didn't work, they'd say, "Is this a deal-breaker?" Well, no one was a deal-breaker. But if you put enough cinder blocks in a rowboat, it's going to sink'. (Bulkeley, 1996)

Minimizing problems and discouraging criticism resulted in ambiguous feedback (Lounamaa and March 1987) and lack of action that might have resolved issues. By refusing to acknowledge problems, FoxMeyer hindered organizational learning (Argyris and Schon, 1978).

In summary, while Dow Corning implemented effective and speedy action-feedback cycles to aid it in recovering from its 'failures', FoxMeyer's ineffective communication precluded meaningful feedback, let alone the ability to react to it. Hence, while Dow Corning had an effective action-feedback cycle, FoxMeyer did not.

Domain relevance

There are two major aspects of domain relevance in this context: (1) choosing technology to fit the task at hand (task-technology fit); and (2) having the appropriate knowledge and skills to implement the system effectively (Lyytinen, Mathiassen, and Ropponen, 1998).

From the viewpoint of task-technology fit, as noted earlier, Dow Corning was uncertain whether the software would ultimately support the number of users it foresaw for its global operations, while FoxMeyer was uncertain whether the system could process the required number of daily transactions. FoxMeyer ran simulations and determined that R/3 could, indeed, process the volumes needed; Dow Corning did not address the software's limitations in the implementations reported. Hence domain relevance was suspect for both firms from the viewpoint of technology fit.

From the viewpoint of the knowledge and skills to implement an ERP, many aspects of an ERP system are different from the functionally oriented systems they replace, and so is the implementation of such complex systems.

Organizations, therefore, usually partner with consultants, who have expertise in the software and its implementation.

Dow Corning chose not to use an implementation partner because its culture encouraged employees to try new things. Pressure for organizational learning was therefore high because the company could not conveniently fall back on consultants.

> When Charlie Lacefield moved into his role what he had was a very good group of folks, and what he did was show a lot of confidence in those folks, and it was genuine. The proof that he felt that way was the fact that he didn't outsource this thing to Andersen or Price Waterhouse. (Information Architecture Department Manager (Ross, 1996b))

FoxMeyer was under less pressure to learn because it expected Andersen Consulting, as its implementation partner, to take care of the implementation. At the time of FoxMeyer's R/3 project, however, knowledgeable personnel were in short supply, and in-house staff and consultants were frequently hired away, disrupting projects. Further, consultants who had implemented R/3 in manufacturing were not familiar with R/3 in distribution, and integrating warehouse automation software with R/3 had not been done before. According to FoxMeyer, Andersen used trainees (Caldwell, 1998) and the Delta project became a 'training ground' for 'consultants who were very inexperienced' (Computergram International, 1998). Similarly, FoxMeyer claimed that SAP treated it like 'its own research and development guinea pig' (*Financial Times*, 1998). Note that similar claims have recently been made by both W.L. Gore and SunLite Casual Furniture in their lawsuits against Deloitte (Osterland, 2000).

In summary, then, because of its use of consultants, FoxMeyer appeared to be better positioned to implement R/3 from the viewpoint of domain relevance than Dow Corning was, although it appears that the advantage might not have been as strong in the long run as perhaps it might have been. Given the shortage of ERP knowledge and skills at that time, even the use of consultants did not guarantee sufficient domain relevance from prior knowledge.

Organizational Learning

What was the outcome of the two implementations? Dow Corning completed its R/3 implementation in December 1998, on time and within budget (Ross, 1999c). Because FoxMeyer needed to support six new warehouses in the west, the company implemented R/3 to support those warehouses and used the old mainframe system for the rest of the country. As noted previously, R/3 costs exceeded $100 million and the company took total charges of $34 million as a result of its inventory and order woes. FoxMeyer declared bankruptcy, claiming that its R/3 project had a lot to do with the company's problems.

It will be clear from the preceding descriptions of the conditions needed to learn from failure why Dow Corning fared much better than FoxMeyer. Dow Corning set up its project in such a way that it was well positioned to recover from any (small) failures along the way (in Sitkin's terms the organizational learning that took place was 'resilient'). In other words, Dow Corning's exploratory experiential learning without consultants appears to have been more effective than FoxMeyer's vicarious learning with consultants. This finding is surprising given that most organizations use consultants and many of these ERP implementations are ultimately satisfactory.

On deeper analysis there is an explanation for this seeming inconsistency: with fewer hands-on opportunities, FoxMeyer's employees used exploration less than the employees at Dow Corning and had minimal experiential learning. Whether FoxMeyer had sufficient knowledge transfer from its consultants to enable organizational learning is not known (Gable, 2000). However, it is known that at that time vicarious learning was limited by the shortage of expertise in consulting firms and the ERP vendors. FoxMeyer was the first ERP implementation to support large-scale distribution and the first needing interfaces to automated warehouses. FoxMeyer also attempted to do too many, extremely complex things all in the same timeframe. This hindered its organizational learning because it was difficult for FoxMeyer to isolate cause and effect relationships.

10.6 Discussion and Implications

In this paper we used Sitkin's (1992) theory for organizational learning through intelligent failure as the basis for examining ERP implementations. Sitkin proposes that firms that focus solely on success suffer the consequent 'liabilities of success', such as complacency, inertia, and, over time, a decrease in resilience and eventual failure. Failures, on the other hand, are opportunities for an organization to learn. Sitkin proposes that firms should actively seek to create situations in which small failures will occur on a (somewhat) regular basis so that employees are challenged, with consequent benefits to the firm. His model includes both organizational characteristics that facilitate intelligent failure as well as the characteristics that need to be in place to support it.

We adapted Sitkin's theory to apply to ERP implementations and then applied the adapted theory to SAP R/3 implementations at Dow Corning, Inc. and FoxMeyer Drug Corporation.

Utility of the Learning from Failure Model

We modified Sitkin's theory to exclude from its application to ERP implementations the notion of systematically seeking organizational failure in the form of failure management systems. ERP implementations represent a very complex endeavor and, because they are comprised of multitudes of tasks spread over a significant time period, they themselves inherently present sufficient opportunities for failure. Similarly, we also eliminated the related aspects of failure legitimation initiatives and designing the organization to support deliberate failures. We refer to our adapted model as describing how to 'learn from failure.'

We applied the 'learning from failure' model to two ERP implementations, one of which failed while the other succeeded. The two implementations were similar in many respects, including unstable business environments, software (SAP R/3), date of implementation (that is, knowledge of the software and of how to implement it), and the need to implement new technology infrastructures. These similarities allowed analysis of the organizational and project issues needed to learn from failure.

On the surface, both implementations appeared to be quite problematic, as illustrated by the problems they faced. Theory-based analysis reveals, however, that Dow Corning's implementation demonstrated many of the characteristics necessary to allow it to learn from failure. Its organizational learning was both single-loop adaptive and double-loop exploratory. The characteristics of FoxMeyer's implementation, on the other hand, were essentially diametrically opposed to what was needed to support learning from failure (see Table 10.2). FoxMeyer attempted to do too much, far too fast, and in one step. Furthermore, it was not responsive to either changes in the external business environment, or to the warnings of its employees. It neither adapted nor explored sufficiently to give it the resilience needed for ultimate success. Nonetheless we must take care to make clear that it is impossible to state, unequivocally, that FoxMeyer failed because it did not use the recommended approach. It is true, however, that Dow Corning followed many of the tenets of the theory and succeeded, while FoxMeyer did not and failed.

From the viewpoint of the model itself, we found that it did not account for one aspect of Dow Corning's implementation that was very beneficial.

The learning from failure model focuses on small failures and is therefore tactical in nature. Dow Corning's tactical (task-level) planning was essentially non-existent; yet its more strategic-level planning and leadership were excellent. Hence, although Dow Corning did not plan well tactically, its high-level planning was sufficient for its implementation to be judged a success in the long-term. This observation suggests that 'well-planned actions' in the learning from failure model might be further differentiated as: 'well planned actions, at either the strategic or tactical levels'.

This statement assumes that if tactical planning is effective, so too will be higher-level planning. Note that the level of planning is an issue only in an endeavor that involves many interdependent actions, such as an IT implementation, and is not an issue in the circumstances in which Sitkin developed his theory.

Further Insights

It is interesting to ponder the difference between the 'learning from failure' focus and a success focus, which views any event that is less than effective as a negative. For example, from the latter perspective, Dow Corning's approach would be considered reactive rather than proactive, while from the learning from failure perspective it is considered a success because the company recovered from numerous, small 'failures'. Despite the fact that Dow Corning's global R/3 project was completed on time and within budget (Ross, 1999c), an issue for the company is whether it would have been better off in the long-term not to have had to learn everything for itself, but instead had learned vicariously from consultants.

It is also interesting to ponder whether FoxMeyer would have succeeded in implementing R/3 using a roll-out rather than a Big Bang implementation strategy. Note, however, that successful Big Bang implementations have been reported (Radosevich, 1997; Stedman, 1998). The Quantum Corporation implementation, for example, although Big Bang, did involve organizational learning, adapting to changing conditions, and rethinking assumptions (Radosevich, 1997). Quantum had a well-planned execution and in contrast to FoxMeyer, Quantum extended the project time when it encountered setbacks. The extra window of time gave Quantum the chance to adapt and explore alternative actions. Nevertheless, learning from failure is much more feasible using a phased or a roll-out approach to implementing an ERP.

FoxMeyer's problems were largely associated with its responses to changes during the implementation in its extremely threatened business environment.

Sitkin (1992) notes that: 'In the face of large and potentially threatening losses, organizational responses are more likely to be protective rather than exploratory.' In other words, companies take a defensive stance and remain committed to prior plans. He cites a range of responses to extreme threat, ranging from hypervigilence, threat-rigidity responses, and the escalation of commitment to the status quo, in this case to prior plans. When FoxMeyer lost a major customer, it did not reexamine its alternatives. Instead, it continued to focus on its position that technology would solve all of its problems and escalated commitment to its original plan by declaring that the R/3 project would fulfill the needs of the new, and therefore changed, environment 90 days earlier than initially determined (see also, Scott, 1999). Hence, FoxMeyer may have found it difficult to react appropriately to substantial environmental changes because it was confronted with the possibility, not of a series of small failures, any of which could have been handled effectively, but of one gigantic failure. This analysis further emphasizes the benefits of actively seeking just small failures so as to learn during the process, improve the possibility for recovering from failure, and hence to avoid large failures.

Implications for Research and Practice

This study focused on developing a better understanding of the role of learning from small failures in the implementation of ERP systems. Further research could refine the organizational learning concepts and the concept of learning from failure that we have introduced in this paper. An issue that has arisen from our research is whether Sitkin's theory applies equally to independent and interdependent tasks. Sitkin does not address this issue explicitly. We assumed that the theory was applicable to interdependent tasks in the sense that if low-level tasks are addressed appropriately, the higher-level issues will also be addressed effectively. And our analyses largely support this assumption, with the exception of the efficacy of strategic as opposed to tactical planning. It is possible that a richer theory might be developed that incorporates notions of task interdependence.

An analysis of multiple case studies of ERP implementations might reveal whether learning from failure could be generalized as an effective implementation strategy. Furthermore, such a study might explore whether learning from failure differentiates among organizations that are satisfied with their ERP systems and those that are not. Organizational learning concepts could also be used to address issues surrounding organizational change management (see organizational learning from the business perspective in Figure 1) and also the issues surrounding system use once implemented.

We also believe that our study has yielded several insights that could be useful to practitioners who will be faced with implementing an ERP system in the future, as well as for those who may be undertaking any kind of far-reaching, integrated system implementation such as those that will be necessary to conduct meaningful e-business. Following are some of the lessons to be learned from our model and its application to these two cases.

- Recognize the importance of organizational culture; foster an open culture and encourage open communications to facilitate organizational learning.
- Consider using a phased or roll-out strategy to facilitate organizational learning and provide the opportunity for learning from (small) failures.
- Take a realistic view of the role technology can play in supporting your firm's strategy; engage in a strategy of 'small wins' or 'small failures' to leverage knowledge gained. (Note that this could be achieved even within the outcome-focused Big Bang strategy.)
- Plan and manage the project; employ a strong project leader and a well-defined methodology so that changes during the project are addressed appropriately; even if there is insufficient knowledge to manage at the tactical level, ensure the effectiveness of high-level planning and management of the project.
- Learn from unforeseen circumstances; be flexible in adapting to changes in the business environment by adjusting at the project level; defer the Go Live date, reduce the project scope, change the number/composition of teams, organize training, think tanks, etc.

Finally, we believe that ERP platforms provide appropriate infrastructure solutions for not only achieving a variety of current strategic business benefits, but also for embarking on competitive inter-organizational systems for e-business. We hope that this study will enable readers to better understand how to achieve this complex enterprise system phenomenon.

REFERENCES

Alschuler, D. and Dorin, R. (1998) AcceleratedSAPWorks! An Executive White Paper, Aberdeen Group, June. http://www.aberdeen.com.

Ang, K., Thong, J., and Yap, C. (1997) IT Implementation through the Lens of Organizational Learning: A Case of Insuror. Eighteenth International Conference on Information Systems, Charlotte.

Argyris, C. (1985) *Strategy, Change and Defensive Routines*. Boston, MA: Pitman.

Argyris, C. and Schon, D. A. (1978) *Organizational Learning*. Reading, MA: Addison-Wesley.

Asbrand, D. (1999) Peering Across the Abyss: Clothing and Shoe Companies Cross the ERP Chasm. *Datamation*, Executive Diary: Part 3, January.

Attewell, P. (1992) Technology Diffusion and Organizational Learning: The Case of Business Computing. *Organization Science*, **3**, 1–19.

Bailey, J. (1999) Trash Haulers Are Taking Fancy Software to the Dump: Allied Waste, Following Waste Management, to Shed SAP's Costly R/3. *The Wall Street Journal*, 9 June, p. B4.

Bancroft, N., Seip, H., and Sprengel, A. (1998) *Implementing SAP R/3: How to Introduce a Large System into a Large Organization*. 2nd edn, Manning Publications, Chapters 6, 7, and 11.

Barki, H., Rivard, S., and Talbot, J. (1993) Toward an Assessment of Software Development Risk. *Journal of Management Information Systems*, **10**(2), 203–225.

Block, R. (1983) *The Politics of Projects*. Yourdon Press, Prentice-Hall, Chapter 1.

Brown, C. and Vessey, I. (1999) ERP Implementation Approaches: Towards a Contingency Framework. Research-in-Progress, Twentieth International Conference on Information Systems, Charlotte.

Bulkeley, W. M. (1996) When Things Go Wrong. *The Wall Street Journal*, 18 November, R25.

Caldwell B. (1998) Andersen Sued On R/3. *InformationWeek*, 6 July.

Champy, J. (1997) Packaged Systems: One Way to Force Change. *Computerworld*, 22 December, 61.

Cliffe, S. (1999) ERP Implementation: How to Avoid $100 Million Write-Offs. *Harvard Business Review*, January–February, 16–17.

Computergram International (1998) FoxMeyer Plus Two Sue Andersen for SAP Snafus. *Computergram International*, 20 July.

Davenport, T. H. (1998) Putting the Enterprise into Enterprise Systems. *Harvard Business Review*, July–August, 121–131.

Deloitte Consulting (1998) ERP's Second Wave: Maximizing the Value of ERP-Enabled Processes.

Dodgson, M. (1993) Organizational Learning: A Review of Some Literatures. *Organization Studies*, **14**(3), 375–394.

Financial Times (1998) SAP in $500m US lawsuit. *Financial Times*, Surveys, 2 September, 2.

Fiol, C., and Lyles M. (1985) Organizational Learning. *Academy of Management Review*, **10**(4), 803–813.

Gable, G. G. (2000) The Expert's opinion – A Conversation with Tom Davenport. *Journal of Global Information Management*, **8**, 60–62, Idea Publishing Group.

Girard, K. and Farmer, M. A. (1999) Business software firms Sued over implementation. *CNET News.com*, 3 November. http://news.cnet.com/news/0-1008-202-1428800.html.

Hedberg, B. (1981) How Organizations Learn and Unlearn. In *Handbook of Organizational Design*, P. C. Nystrom and W. H. Starbuck (eds), Vol. 1, New York: Oxford University Press, pp. 3–27.

Hoffman, T. (1998) Software Snafu Triggers Order Delays, Loss: Snap-On Software Leads to $50m. Dip, *Computerworld*, News, 6 July, 6.

Holland, C. P. and Light, B. (1999) A Critical Success Factors Model for ERP Implementation. *IEEE Software*, March/June, 30–35.

Huber, G. (1991) Organizational Learning: The Contributing Processes and Literatures. *Organization Science*, **2**(1), 88–115.

Information Week (IW) (1998) SAP Installation Scuttled: Unisource Cites Internal Problems for $168m Write-off. *Information Week*, 26 January.

Jesitus, J. (1997) Broken Promises? FoxMeyer' s Project was a Disaster. Was the Company Too Aggressive or was it Misled? *Industry Week*, 3 November, pp. 31–37.

King, J. (1997) Dell Zaps SAP. *Computerworld*, 26 May, 2.

Kotter, J. P. (1995) Leading Change: Why Transformation Efforts Fail. *Harvard Business Review*, **73**(2), 59–67.

Levitt, B. and March, J. G. (1988) Organizational Learning. In Scott, W. R. and Blake, J. (eds) *Annual Review of Sociology*, **14**, 319–340.

LJX Files, The Complaint in Brown v. Andersen Consulting (1997), http://www.ljx.com/LJXfiles/bankruptcy/andersen.html.

Lounamaa, P. H. and March, J. G. (1987) Adaptive Coordination of a Learning Team. *Management Science*, **33**, 107–123.

Lyytinen, K., Mathiassen, L., and Ropponen, J. (1998) Attention Shaping and Software Risk – A Categorical Analysis of Four Classical Risk Management Approaches. *Information Systems Research*, **9**(3), 233–255.

March, J. G. (1991) Exploration and Exploitation in Organizational Learning. *Organization Science*, **2**(1), 71–87.

Markus, M. L. and Benjamin, R. I. (1997) The Magic Bullet Theory in IT-Enabled Transformation. *Sloan Management Review*, **38**(2), 55–68.

Markus, M. L. and Tanis, C. (2000) The Enterprise Systems Experience – From Adoption to Success. In *Framing the Domains of IT Research: Glimpsing the Future Through the Past*, R. W. Zmud (ed.). Cincinnati, OH: Pinnaflex Educational Resources.

Nonaka, I. (1994) A Dynamic Theory of Organizational Knowledge Creation. *Organization Science*, **5**(1), 14–37.

Osterland, A. (2000) Blaming ERP. *CFO Magazine*, January. (http://www.cfonet.com/Articles/CFO/2000/00Jablam.html)

Pentland, B. T. (1995) Information Systems and Organizational Learning: The Social Epistemology of Organizational Knowledge Systems. *Accounting, Management and Information Technologies*, **5**(1), 1–21.

Porter, M. and Millar, V. (1985) How Information Gives You Competitive Advantage. *Harvard Business Review*, **63**, 149–160.

Radosevich, L. (1997) Quantum's Leap. *CIO*, 15 February.

Robey, D., Wishart, N. A., and Rodriguez-Diaz, A. G. (1995) Merging the Metaphors for Organizational Improvement: Business Process Reengineering as a Component of Organizational Learning. *Accounting, Management and Information Technologies*, **5**(1), 23–39.

Ross, J. (1996a) Dow Corning Corporation (A): Business Processes and Information Technology. Center for Information Systems Research, Sloan School of Management, MIT.

Ross, J. (1996b) Dow Corning Corporation (B): Reengineering Global Processes. Center for Information Systems Research, Sloan School of Management, MIT.

Ross, J. (1999c) Dow Corning (C): Transforming the Organization. Center for Information Systems Research, Sloan School of Management, MIT.

Sauer, C. (1999) Deciding the Future for IS Failures: Not the Choice You Might Think. In *Rethinking Management Information Systems: An Interdisciplinary Perspective.* W. Currie and R. G. Galliers (eds), Oxford: Oxford University Press.

Scott, J. E. (1999) The FoxMeyer Drugs' Bankruptcy: Was it a Failure of ERP? *Proceedings of the Fifth Americas Conference on Information Systems*, pp. 223–225.

Scott, J. E. and Vessey, I. (2000) Toward a Multi-Level Theory of Risks in Enterprise Systems Implementations. Working Paper.

Senge, P. M. (1990) *The Fifth Discipline: The Art and Practice of the Learning Organization.* Doubleday.

Sitkin, S. B. (1992) Learning Through Failure: The Strategy of Small Losses. *Research in Organizational Behavior*, **14**, 231–266.

Stedman, C. (1998) Big-bang R/3 Rollout Forced Compromises with Business Goals. *Computerworld*, **32**(11), March 16, 1, 97.

Stein, E. W. and Zwass, V. (1995) Actualizing Organizational Memory with Information Systems. *Information Systems Research*, **6**(2), 85–117.

Sumner, M. (1999) Critical Success Factors in Enterprise Wide Information Management Systems Projects. Proceedings of the ACM-SIGCPR Conference, 297–303.

Welti, N. (1999) *Successful SAP R/3 Implementation: Practical Management of ERP Projects.* Addison-Wesley Longman, Chapter 2.

Willcocks, L. P. and Griffiths, C. (1997) Management and Risk in Major Information Technology Projects. In *Managing IT as a Strategic Resource*, Willcocks, L. P., Feeny, D., and Islel, G. (eds), Berkshire: McGraw-Hill, pp. 203–237.

Williamson, M. (1997) From SAP to 'nuts!'. *Computerworld*, 31, 45, 10 November, 68–69.

11 ERP Projects: Good or Bad for SMEs?

Frédéric Adam and Peter O'Doherty

11.1 Introduction

The enterprise resource planning (ERP) movement has been gathering momentum for the best part of ten years and has now reached a global dimension, with companies across the world and, more importantly, across very different industries jumping on the ERP band wagon. The pace of implementations has been such that SAP have posted on their web site the news that they have now implemented their software in 30,000 sites and have a user population of 10 million. This gives an idea of the scale and pace of the whole ERP movement.

ERP systems are integrated enterprise-wide software packages that use a modular structure to support a broad spectrum of key operational areas of the organization. They are widely acknowledged as having the potential to radically change existing businesses by bringing improvements in efficiency and the implementation of optimized business processes (Rowe, 1999). The key reasons why managers have sought to proceed with difficult ERP projects have been reported to be to end the fragmentation of current systems, to allow a process of standardization, to give more visibility on data across the entire corporation, and, in some cases, to obtain competitive advantage. Thus, ERP projects have been described as strategic projects whose success or failure will have a great impact on the organization (Rowe, 1999; Shakir, 2000; Wood and Caldas, 2000). Shakir (2000) concluded that ERP projects are expensive and time consuming with costs typically exceeding US$100,000 and a timeframe for evaluation, selection, and implementation of an ERP system between six months and two years.

However, many studies have revealed that the rationale pursued by managers in acquiring ERP packages is not always very robust or very well informed (Wood and Caldas, 2000). Moreover, the track record of ERP implementations has not been consistently good with many examples of much advertised failures. Dell and Unisource have both written off their large investments in aborted SAP implementations (Bingi, Sharma, and Godla, 1999) and the FoxMeyer Drugs case study (Kalakota and Robinson, 1999) has become a kind of a SABRE landmark case study for ERP (but, in a negative sense). The significance of such a case study cannot be understated because it shows a healthy company going bankrupt and losing its independence (it was bought by McKesson) after trying to implement a type of information system widely advertised as being 'the perfect solution' for companies who want to stay competitive (Wood and Caldas, 2000). Other case studies of organizations who ran into trouble as a direct results of implementing ERP confirm that the stakes in ERP projects are quite high. Hershey, a large distributor of candies implemented in July 1999 a $112m ERP only to find the shelves of hundreds of their regular stores empty during Halloween, costing the company enormous losses in turnover and market share. The software they implemented (SAP) turned out to be unable to understand exceptional variations in demand, such as those that characterize Halloween, and grossly underestimated the requirements (Miller, 2000).

In fact, very rigorous studies of the ERP market have shown that few ERP implementations are actually entirely successful, with approximately half failing to meet the implementing organizations expectations, due in the most part to an underestimation of the effort involved in change management (Stefanou, 2000; Appleton, 1997). Furthermore, it is estimated that approximately 90% of ERP implementations end up over-time and over-budget, due to poor cost and time estimation (Kelly, Holland, and Light, 1999; Shanks *et al.*, 2000), and changes in project scope (Shanks *et al.*, 2000).

On the other hand, Microsoft, Colgate (Kalakota and Robinson, 1999), Topps International (Adam and Doyle, 2001), and many other published success stories represent equally interesting case studies of successful implementations of ERP, which saw these organizations earning substantial benefits from their adoption of ERP packages. These success stories may explain that the level of interest in the concept of ERP systems amongst managers and the evolution of the worldwide market with growth rates of 35% to 40% (Bingi, Sharma, and Godla, 1999) makes the ERP movement one of the most dynamic items of the information systems agenda.

However, ERP systems have inherent strengths and weaknesses, and are therefore better suited to certain types of organizations and certain circumstances. Thus, managers interested in ERP systems must understand their trade-offs before they make any decision regarding the potential of the ERP concept in their case. As usual, consultants and media reports will be prone to emphasize the benefits of ERP implementations, but the key issues reside in understanding the specific needs of each organization and the business model best suited to its operations. The main difficulty in ERP projects is that few companies, if any, could possibly contemplate developing such vast applications in house. For the majority of companies, the decision to implement ERP functionalities will mean buying a software package from one of the major suppliers on the ERP market – for example, SAP, Baan, JD Edwards, PeopleSoft, Oracle, or MFG/PRO. The software selection phase is not straightforward and managers must understand what ERP packages are on offer, how they differ, and what is at stake in selecting one ERP over another. Each ERP package uses a business model as an underlying framework and they can differ quite a bit in terms of how they operate or the business processes they support. The problem for managers is that not all business models fit all organizations and the cost of failing to recognize the relationship between the nature of one's business and the ERP system to be purchased can be very high indeed. Quite literally, selecting the right software package, that is the right blueprint for one's organization is a critical failure factor in ERP projects.

This problem is particularly acute for SMEs because there is a perception that ERP packages are primarily designed for large corporations and not for smaller firms with simplified internal structure and exacerbated needs for rapid adaptation to market changes. Also, the amount of disruptions to everyday business reported in ERP implementations makes it difficult for managers in SMEs to commit to ERP projects without genuine fear of sabotaging currently sound businesses.

This paper investigates the potential of ERP packages for smaller organizations. First, the commonly cited arguments for and against ERP systems are reviewed, then a two-stage study of SME implementations of ERP in Ireland is presented and conclusions are proposed regarding the duration of ERP projects and about the potential of the ERP concept for SMEs.

11.2 Strengths and Weaknesses of ERP Systems

The Case for ERP

In a certain way, ERP systems represent the implementation of the old managerial dream of unifying and centralizing in one single system (or at least under one name) all the information systems required by the firm (Rowe, 1999). Most notably, they support the recording of all the business transactions from purchase orders to sales orders and the scheduling and monitoring of manufacturing activities. Most ERP systems are based on an inventory control module that records the movements of goods in and out of the company, which makes them particularly suited to organizations seeking rationalize their internal processes and obtain higher performance out of their operations.

ERP systems provide organizational actors with a common language and a common pool of data. At a practical level, ERP have very beneficial effects that remove the need for often disparate and unreliable end-user applications, standardize operating procedures and reporting, and optimize some of the key processes of the firm (for example, order acquisition and processing or inventory control). At the same time, these systems offer high levels of portability and reasonable flexibility in adapting to the requirements of specific organizations (Rowe, 1999; Forest, 1999).

One of the key strengths of ERP systems is that they are built on top of a relational database which enables a reliable and rapid circulation of the data between the modules and eliminates the need for multiple data entry and much of the error checking that data migrations between legacy systems traditionally brings. Thus, ERP systems simplify, accelerate and automate much of the data transfers that must take place in organizations to guarantee the proper execution of operational tasks. The relational database underlying an ERP can be quite large with some SAP applications implemented reported to have in excess of 8000 different tables.

At this point in time, the case for ERP systems seems compelling and the development of more powerful and user-friendly platforms makes it now possible to integrate many large systems in a way that was not possible a few years ago (Wood and Caldas, 2000). Thus, Microsoft spent ten months and $25m replacing 33 existing systems in 26 sites with SAP. Managers in this company claim to save $18m annually as a result and Gates reportedly expressed great satisfaction with the SAP software (White, Clark, and Ascarely, 1997). According to Kalakota and Robinson (1999), Microsoft had grown so fast that it could not keep up with itself and the number of applications developed to support the company's operation and their lack of integration meant that IS staff had lost control over the complexity of the systems they administered. Thus, as many as 90% of the 20,000 batch programs retrieving data and passing it between systems may have been redundant. The move to a single architecture enabled better linkages between business areas as well as with suppliers and customers.

The Case against ERP Systems

The strengths of ERP packages are matched by the high level of risk associated with ERP projects. ERP projects are complex and require reliance on many different types of expertise, often sourced outside the organization. Consultants often advise managers to undertake some degree of reengineering of key processes before acquiring ERP systems (Bancroft, 1996) and this adds to the complexity and political character of the projects. There is empirical evidence of the dangers inherent in such vast projects. The case of FoxMeyer who went bankrupt in 1996 after three years of unsuccessful implementation of SAP, suing SAP's US subsidiary and Andersen Consulting in the process, is illustrative of what can happen to the largest organizations when implementations go wrong (Kalakota and Robinson, 1999).

These difficulties have lead to some researchers taking a negative view on ERP systems, with Wood and Caldas (2000) using the Orwellian analogy to characterize the goals of ERP systems and questioning whether the current interest in ERP in the business community is justified more by political reasons than by sound managerial reasoning. Indeed, these authors found low levels of satisfaction in their survey of firms having implemented ERP systems with 45% of firms perceiving no improvements whatever from implementation and 43% claiming that no cycle reduction had been obtained.

The difficulty inherent in ERP implementations is largely due to the fact that organizations implementing them should typically only hold on to 20% of their previous applications (Rowe, 1999). But this extensive replacement of previous systems may be a requirement if the major benefits of ERP implementation – greater integration of functional areas and, in the case of multinational firms, greater co-ordination between entities and between sites – are to be obtained. The consequence of this 'clean slate' approach is that organizations find it virtually impossible to revert to their pre-ERP situation and, in any case, their investment either cannot be recouped or generates very low returns. Unisource Worlwide Ltd. and Dell Computer Corp. have both written off very significant amounts invested in aborted SAP implementations and associated consultant fees (Bingi, Sharma, and Godla, 1999).

Finally, there is anecdotal evidence that many companies have been pushed into ERP projects by the much-publicised fears of what may happen to legacy systems during the year 2000 change. This type of argument hardly indicates that ERP systems are necessarily a source of competitive advantage needed by all companies.

These arguments paint a very mixed picture of the potential of ERP packages; portrayed as silver bullets, and other times as villains, managers are rightly worried that there is no sound basis for decision making in relation to ERP systems. In order to disentangle the conflicting reports, we studied the reality of the ERP phenomenon in Ireland focusing on smaller companies. We were particularly interested in the reasons why they decided to go down the ERP road, the difficulties they encountered, and the benefits they reaped.

11.3 The Study – Mixing Case Study and Survey Research

In this study, we used a two-stage research protocol as a means of triangulation (Bonoma, 1985; Jick, 1979; Kaplan and Duchon, 1988). We initially carried out a case study of an Irish SME having just implemented an ERP package (and whose name has been disguised). Based on the results of this case and the discovery that the full implementation of the package had taken only five months in this site, we proceeded to confirm that ERP implementations, at least in SME-type organizations, are not always the very large and dangerous projects described in the current literature. In order to achieve this second step, we carried out a survey of 14 customers of Ireland's sole distributor for one of the leading ERP package.

The organizations studied covered a broad range of activities (from telecom system suppliers to distributors of medical supplies for animals) and degree of expertise with the deployment of IT (from 'green field sites' to organizations that had a complete IS department). They also covered a broad range of sizes (from ten to 500 employees), but 64% of them were SMEs according to European standards (SMEs are defined as having less than 250 employees and a turnover inferior to 40m ECUs[1]). Thus, they constituted a good sample of the variety of organizations implementing ERP solutions in Ireland at this point in time. The selection mechanism employed, whereby the clients of one ERP supplier were studied, was opportunistic but it enabled us to access a significant number of organizations without having to deal with common problems in breaking barriers and identifying knowledgeable respondents. Also, the fact that all organizations had been dealing with the same supplier of software and the same software package guaranteed the comparability of their experience with ERP implementations and helped us concentrate on the key factors that remain true across organizational settings and those that only appear in specific circumstances.

In all 15 organizations, data were collected through semi-structured interviews with key actors who had been involved with the selection and implementation of the software in the client organizations. A questionnaire was

[1] Sixth Periodic Report on the Social and Economic Situation and Development of SMEs, http://www.inforegio.cec.eu.int/wbdoc/docoffic/official/radi/page31_en.htm (viewed June, 2000).

used to ensure that all relevant issues were dealt with, but interviewees were free to volunteer whatever information they felt was important given the goals of the study. The empirical data collected about each site and about each implementation were then reviewed in collaboration with the staff members in the software house who had delivered the software to this site, providing the opportunity for corroboration/clarification of the problems encountered and the time spent on the different phases of implementation. The analysis of the findings and the conclusions of the research were undertaken in collaboration between the academic researcher and the most senior project leader in the software distributor (the co-authors of this paper) in order to give equal importance to the opinions of all parties involved and reduce the possibility of bias.

11.4 A Case Study of ERP Implementation

History of ABC Communication

ABC communication Ltd. is a supplier of communication software. At the time this study took place, it had a £30 million turnover and 200 employees. The company has established a strong global position in Europe and Australia in integrated network products. As the company was expanding rapidly and had acquired another company supplying complementary products, the Directors decided that an ERP package should be purchased to rationalize internal operations and help the increasingly complex management of inventory. The company initiated the selection process in November 1998 on the basis of a number of specific management targets which are listed below:

- Support of the launch of new products on the European and UK markets
- Implementation of 'Business Excellence' strategy
- Integration of operations with DEF Ltd; the newly acquired company
- Full Y2K compliance well in advance of the new Millennium

The goals of the ERP implementation also included better data access and better visibility on order completion and delivery, reduced operating costs, ability to cope with the increased volume of data and further extension towards the company's future e-business strategy.

Before the project started, the directors also spelt out that the ERP package had to go live before the end of May 1999 in the headquarters of the company (which meant the package had to be selected and implemented within seven months) and before the end of September 1999 in all the other sites.

The implementation was actually complex, because a number of essential (and non-replaceable) applications had to be integrated with the ERP package. In order to make the tight deadline achievable, it was also decided that no customization of the package selected would be allowed. As a result, the selection process was primarily aimed at finding a 'best fit' solution – that is one that would immediately offer functionalities able to support the current business processes or that would suggest business processes that could immediately be implemented in ABC. Whenever possible, the project was hoped to provide the opportunity to implement re-designed processes, but the maintenance of the ISO 9000 accreditation was also an overriding objective.

After one month of negotiations and a thorough demonstration of how this ERP package worked, one of the leading ERP packages was selected and the local supplier of the package was assigned the task of implementing it. The implementation took four months from the time hardware was purchased to the time the 32 users were able to use the system. One more month was required before all the functionalities of the systems were in full use, mainly because of training issues. The targets initially set by top management were therefore easily achieved and the schedule initially suggested was strictly followed.

The evaluation interviews carried out with management and staff at the end of the project revealed the high level of satisfaction of both users and managers with the system. In particular, a number of tangible benefits were identified including:

- greater flexibility in work practices;
- better business approach *vis-à-vis* customers;
- increase in staff efficiency, meaning that similar numbers are able to cope with a much increased volume of business;
- better linkages with vendors and suppliers;
- reduction in the number of errors;
- improved inventory control;
- better access to information;
- better sales support.

The interviews with managers provided other insights into the kind of benefits that can accrue to organizations that implement ERP systems. The ability of the software package to automatically generate high-level reporting on the key performance indicators of the company was particularly noted and resulted in much higher-quality reporting across the board. The improved response time to customer inquiries, while not quantifiable in financial terms, was also put forward as a major source of satisfaction for top management.

Another interesting facet of this project was that the accountant of the company had been assigned to the leadership of the project. This manager had no specific responsibilities in terms of information systems at ABC, but he had been the champion for the ERP projects from the start. This indicates that the push for ERP projects does not always come from the IS area and that sound business reasons may be used to justify ERP projects. As a result of the project, the accountant actually gained much status and was largely credited with the success of the project.

Commentary

This case study is interesting in many respects and deserves some commentary. First of all, the small size of the organization involved must be noted. With a turnover of IR£30 million and a workforce of 200, ABC is a medium-sized organization and not the typical target of ERP vendors. Cases of ERP implementations previously reported routinely involve companies worth several billion dollars and multiple sites. In this respect, this project was very relevant to the evolution of the ERP market, which is being redirected towards SMEs (Jeanne, 1999).

The short duration of the implementation is another key aspect of this case. Five months (including training) makes this ERP project short in comparison to other cases reported in the media. Our observations revealed that this ERP project was a full-scale project in that it involved major modifications in the business processes of the firm and yielded a broad range of benefits quite in keeping with the theory behind the ERP concept. It therefore cannot be said that this project was shorter because of the incomplete nature of the implementation or the restriction to a small number of targets. Rather, it seems that the smaller size of the company and the no-customization stance of top managers are likely explanations for the duration of the project. Smaller companies deal with smaller number of customers and suppliers, and their inventories are less complex. They also have fewer staff and require less training. As a result of the smaller number of people involved, the projects are easier to co-ordinate and manage, which enables projects managers to adopt a more aggressive schedule. Deciding to adopt a 'zero modification' approach was evidently a critical aspect of the rapid implementation.

However interesting these observations, we cannot state with certainty whether the case at ABC is representative of ERP projects taking place in Ireland at this point. The low generalization value of case studies has been noted by previous authors (Lee, 1989). In order to strengthen our beliefs in the findings of this case, we carried out a survey of 14 other cases of ERP implementations. This survey is described in the next section.

11.5 Survey Research Findings

Analysing the Profile of the Organizations Studied

The first step in this research project was to analyse the profile of the organizations studied. In total, there were 14 companies at various stages of maturity with the ERP software, though all sites had at least started to implement some of the modules of the software. As in Wood and Caldas's (2000) survey, some of the companies were subsidiaries of foreign multinational firms and could benefit from specific implementation guidelines and from the past experience of staff at their headquarters. This enabled them to implement the software more quickly than other companies. Table 11.1 presents the key characteristics of all the companies studied.

The figures in Table 11.1 indicate that the average size of the companies we studied is 150 employees and a turnover of £115m., which is far smaller than most of the instances reported in the current literature (nine out of 14 were SMEs). For instance, the respondents in Wood and Caldas (2000) were described as large or medium-size organizations and case studies commonly cited in textbooks involve implementation projects that have a specific budget greater than the turnover of the organizations we studied. These figures reflect the small average size of organizations in Ireland (roughly two-thirds of Irish organizations have less than 250 employees), but they come as a surprise because the current wisdom in relation to ERP systems (as abundantly reported in the media – for example, Jeanne, 1999) is that ERP systems are only for large organizations because of the costs and implementation times involved. Our data indicate that SMEs are already involved in the ERP phenomenon to a large extent, as most of the organizations we studied undertook their implementation projects before 1999.

Typical Implementation Project – Impact on the Organization

The second step in our investigation was to study the key aspects of the projects in which these firms had been involved. Table 11.2 presents the key parameters measuring the size, duration, and impact on the IT capability of the organizations of the ERP projects we studied. The impact of the ERP implementation on the company is a key aspect of ERP projects as reported in previous research (Besson, 1999; Kalakota and Robinson, 1999; Rowe, 1999)

Table 11.1 Profile of organizations studied

Company	Industry	Size	Turnover £ per annum	Status of project
Company A	Medical packaging supplier	50	2m	Post-implementation
Company B	Animal medical products	100	25m	Post-implementation
Company C	Telecom network products	200	30m	Post-implementation
Company D	Mining and industrial products	600	250m	Post-implementation
Company E	Sheet plastic producer	30	2m	On-going (phase 1 completed)
Company F	Procurement specialist	100	20m	On-going (phase 1 completed)
Company G	Plant protection	40	15m	On-going (phase 1 completed + phase 2 started)
Company H	Fitters for oil/gas installations	50	4m	Close to completion
Company I	Medical supplies	10	n/a	Post-implementation
Company J	Tobacco products	n/a	160m	Close to completion
Company K	Food products	50	n/a	On-going (phase 1 completed + phase 2 started)
Company L	Lens colouring	500	n/a	Post-implementation
Company M	Electric connectors	180	500m	Post-implementation
Company N	Entertainment products	25	250m	Post-implementation

Notes: 'phase 1 completed' means that the organization is already using the software but is expected to further develop its usage of ERP functionality (e.g. by acquiring more modules).

and we wanted to establish to what extent the implementation of ERP software had been a source of significant change for the structure and staffing of the organizations we studied.

Table 11.2 indicates that the projects undertaken by the organizations in the sample were of short duration in comparison with durations cited in current literature. The 14 projects we looked at ranged from one month to 24 months and averaged eight and a half months in duration with just over a third of the companies taking nine or ten months. Even though these projects are

Table 11.2 Key characteristics of the ERP projects studied

Company	Start date	End date	Number of users (start/end)	In-house IS capability (start/end)
Company A	Nov 97	Aug 98	5/10	None/limited
Company B	Nov 97	Jul 98	16/32	None/2 staff
Company C	Dec 98	Mar 99	32/32	2 staff/3 staff
Company D	Mar 99	Nov 99	100/150	8 staff/8 staff
Company E	Nov 98	Jul 99	5/5	None/none
Company F			50/50	None/contractors
phase 1	Mar 2000	Apr 2000		
phase 2	Apr 2000	Aug 2000		
Company G			8/8	1 staff/2 staff
phase 1	Nov 97	Jan 98		
phase 2	Apr 2000	May 2000		
Company H	Early 98	End 99	10/10	None/none
Company I	Mar 98	April 98	5/5	None/none
Company J	Mar 2000	May 2000	10/10	None in Ireland
Company K			8/8	None/none
phase 1	Early 97	Mid 97		
phase 2	Apr 2000	June 2000		
Company L	Jan 98	Nov 98	32/32	2/2
Company M	May 97	Oct 97	16/30	None/1 staff
Company N	Early 97	Mid 98*	5/5	None/none

Notes: *The exceptional duration of the project in company N was due to slow decision making at the choice stage and does not mean that the software implementation took longer than other cases in the sample.

shorter than those previously reported, these figures confirm that ERP implementation projects are large IS projects. Fitzgerald's (1998) survey of IS development in Ireland found that the average duration of in-house development, outsourced and purchased software projects was only 5.7 months, which means that 65% of the projects we studied are above the average reported by Fitzgerald and 14% are four times as long. This confirms that ERP projects are above average in terms of duration and complexity.

It is also interesting to analyse the number of users of the software in each of the sites. The cases studied had on average between 21 and 22 users initially and between 27 and 28 at the time this study took place (that is post-implementation). Significantly, only four companies have increased their number of users after the software went live (less than 30%) even though many sites implemented their software as far back as 1998 or 1999. These figures suggest small user populations quite in line with the small size of the organizations studied.

We also tried to evaluate the organizational impact of ERP projects in other ways. In particular, we focused on the effects of these projects on the in-house IS staff organizations possess. This is a matter of worry for SMEs because managers rightly regard the 'hidden cost' side of ERP projects as a major threat. As can be seen in Table 11.2, ten companies (70% of the cases studied) had no resident IS expertise prior to their implementing ERP and five companies (35%) maintained this situation at the end of the project. In total, only five companies hired IS staff as a result of the ERP implementation, three of these being 'green field sites' where there had been no IS department before. When IS staff were hired, the increase was measured as no company hired more than two staff as a result of the project. These figures are significant in that they indicate that ERP software may not have a great impact on the IT staff requirements of organizations.

Table 11.2 also indicates that many companies are still able to rely totally on IS service providers for all their IS needs even after implementing ERP software. These figures are somewhat surprising in that they suggest that a large proportion of Irish SMEs do not have full-time resident IS expertise. Unfortunately, it is not possible to compare this result with Fitzgerald's (1998) study (although its results include average size of IS departments) because the sample used in his study was biased towards firms that had an identifiable IS manager. More research would be required to ascertain whether the sample used in this study is biased towards companies that are weak in their deployment of IT and rely on outsourced IT services. However, it must be noted that the presence of four subsidiaries of international groups (with IT support from headquarters) influences this result. Alternatively, it could be suggested that companies that have not developed any IT capability see in ERP projects the opportunity to catch up in a rapid and robust manner by acquiring a total system that can easily be installed and maintained by an experienced outside implementing partner. Managers may think that such acquisitions can put them in touch with the best practice in their industry without having to experience the usual growing pains of incremental IS development.

Goals of the Projects and Project Leadership

One of the key issues with ERP is whether the benefits commonly associated with ERP implementations actually materialize or whether they are exaggerated by media hype (Caldas and Wood, 2000). As a first step we tried to establish the goals pursued by managers when they purchased ERP software. The data pertaining to these issues are presented in Table 11.3.

Table 11.3 Goals pursued and project leadership

Company	Goals pursued	Project champion	ERP manager
Company A	Initially: 'a' system to be able to show the big boys Now: e-business stock reduction	Finance director	Finance director
Company B	Initially: replace old systems (Y2K) common ERP benefits Now: e-business Euro transactions	Operations manager	Operations manager
Company C	e-business replace old systems (Y2K) data access/visibility reduce costs cope with volume of data	Accountant	Accountant (gained much status after project)
Company D	Y2K Integration of systems visibility supply chain improvements centre of excellence e-business	Finance controller	Financial controller
Company E	Initially: boost company value stock right-sizing Now: visibility accountability process improvements	Production manager	Accountant (after change of owner)
Company F	Phase 1: replace old systems (Y2K)	Operations director (hands-off) + Contract IT	Operations director

Table 11.3 *(cont.)*

Company	Goals pursued	Project champion	ERP manager
	Phase 2: e-business CRM expansion Executive information system MRP and inventory control	Lack of true champion as project leader for the implementer feels must act as champion.	
Company G	Phase 1: world class system Phase 2: e-business Euro transactions manufacturing processes	IT manager	IT manager
Company H	group decision 'flair'	Finance manager	Finance manager
Company I	bought by large group using ERP	Operations manager	Operations manager
Company J	instruction from group	IS manager for Europe	IS manager for Europe
Company K	Phase 1: world class system ERP functionality Phase 2: manufacturing MRP costs tracking	Accountant	Accountant
Company L	world-wide implementation	Financial controller	Financial controller
Company M	a system required common ERP functionalities	Financial controller	Financial controller
Company N	replace existing systems MRP cope with growth in volumes	Financial controller	Financial controller

The first observation is that organizations in the sample were pursuing specific goals in their projects. The most common goals pursued were the implementation of a robust transaction processing system (nine companies) – a 'world class system' for two companies. For nine companies, the traditional benefits of ERP systems, in particular cost reduction benefits, were also sought, while for 28% of our respondents, Y2K had been a major incentive to buy an ERP system, although in no case had it been the only rationale. Three companies were also seeking improvements in the visibility of their transactions while four companies bought ERP software under instruction from their foreign headquarters and put forward no specific goals for the implementation. Finally, e-business was a long-term objective for six companies.

In three cases, actually implementing the ERP software developed the perception of the managers and led to additional goals being pursued. In two of these cases, the availability of a solid platform for order acquisition and processing gave managers enough confidence to experiment with e-business. In the third case, the ERP system had been intended to boost the value of the company prior to its sale, but the availability of the ERP software led to managers attempting to capitalize on the good work done to reap additional benefits. There were also encouraging signs in three of the four subsidiaries, as local managers had taken ownership of the project and were eager to see it succeed. This meant that, even in the absence of clear 'local' managerial rationale initially, the ERP implementation was being taken seriously and the potential of the ERP system was understood by organizational actors.

These findings are divergent from those of Caldas and Wood (2000) who found that 'following the trend' was the motivation of 77% of their respondents. They also found that 41% of companies had implemented ERP systems under pressure from their IT departments and this contrasts strikingly with our finding that, in only two companies (14%) had the IT department been the champion or even the manager of the project. Indeed, only 35% of the companies we investigated had an IS department prior to starting their ERP project, which indicates that the motivation behind the ERP movement often comes from other areas. Thus, the accounting/finance area had championed the ERP project in 57% of the companies, while 21% of projects had been led by operations managers on the client's side.

It is difficult to assess why these results diverge so markedly from these reported in Caldas and Wood's study, but the fact that 85% of their companies were subsidiaries of multinational groups (and therefore were asked to implement ERP systems by their headquarters) may have had a negative influence on the perception of managers regarding their ERP projects. Also, the fact that most companies in our study were SMEs means that coordination and consensus about such projects were easier to achieve for these firms.

Outcomes of the ERP Projects

In order to assess the degree of success of the projects studied, we then looked at whether the ERP system had been widely adopted by staff in the organization and whether staff were convinced by the benefits brought by the software. The degree to which managers have taken ownership of the ERP system was another key aspect of the study. Project leaders in the client organizations were asked whether they were satisfied with the implementation of the software and also to what extent the goals pursued (as outlined in Table 11.3) had been achieved. All sites were satisfied with their ERP software and its implementation, apart from a certain amount of anxiety which staff experienced as they began to realize the extent of change that the ERP system was going to require. This included concerns about trying to replicate existing functionality rather than meeting business requirements with the new system, migrating data and designing workarounds. Although migrating data have been greatly simplified by emerging common standards, the lack of data integrity in legacy systems had proven a big problem. In addition, in some cases, no workaround was available for a specific requirement, which gave rise to tough negotiations between the implementer seeking 'zero modification' and the client seeking '100% functionality'. By and large, all these problems had solutions and the implementer and the client collaborated in designing them.

However, it was not straightforward to assess whether goals had been achieved. Some companies were more able than others to tell us about the benefits gained because not all had attempted to specifically measure the outcome of the project in managerial terms. Company C was an interesting site because managers had carried out a specific analysis of the consequences of the ERP implementation. They reported a broad range of improvements in business efficiency, including: ability to handle increased volume of transactions with the same staff, easier linkage with customers and suppliers, less errors and re-runs, better access to information and easier after-sales support. They also reported additional benefits such as improved quality of business reporting and easier access to key performance indicator analyses for top management. Managers were particularly impressed with the improvements relating to the speed and accuracy of enquiry responses in general, which they perceived as a key factor in their business. The manager interviewed concluded that the ERP system had enabled top management to implement major growth plans for the company.

Other companies reported problems that occurred after the implementation was completed and the ERP team was disbanded. One of the main concerns of managers was that, in the short term, their organization was more vulnerable to any staff member involved in the ERP project leaving the company too soon. If transfers of key knowledge about the ERP system have not had time to occur, then the company stands to waste much time in achieving the objectives pursued. However, introduction of proper incentive schemes was perceived as a solution to this problem. Indeed, the shortage of experienced staff in the area of ERP in Ireland is such at this point that staff who have taken part in a successful ERP implementation are tempted to capitalize on their newly gained experience and seek higher salaries. Management must acknowledge this and update the status and reward level of staff involved to reflect the additional skills gained in the ERP project and the resulting modifications in their job content.

Managers also referred to the *sigh effect* which affects staff immediately after the implementation of the software has been completed. In this crucial phase, staff must return to their normal duty and have less time to tackle operational problems linked to the ERP systems. This results in a drop in motivation that has been experienced by most of the staff we interviewed, as discussed in detail in the next section.

Key Aspects of the Relationship with the Implementer

Some key points about the relationship between the implementer and its clients were identified by staff in the software implementation firm. These helped the researchers make sense of some of the opinions expressed by the interviewees. It emerged from our investigations that, as time goes by, organizations' perceptions and expectations about the software they have acquired fluctuate. Typically, the mood of the staff in the client organization follows a curve that is represented in Figure 11.1.

In the initial section of the curve, expectations rise steadily and the perception of the software product that is being purchased come in sharper focus as a state-of-the-art solution to old problems is in sight. Managers are excited about getting new tools to achieve their goals and their belief in the ERP system rises. Thus, organizational actors become increasingly committed to the project. Then, as the two organizations (client and implementer) begin to negotiate the finer detail of the implementation plan, staff realise the extent of work and change to current practices involved in the project and their fears

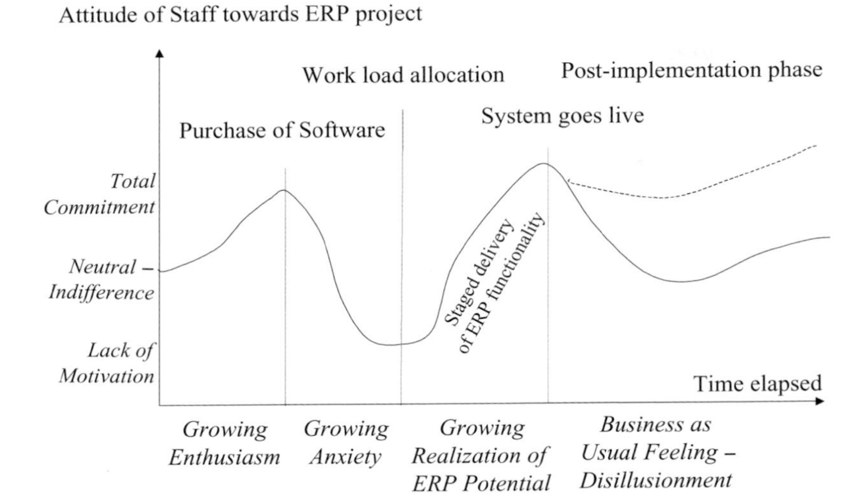

Figure 11.1 Evolution of staff and management motivation for the ERP project

grow steadily. They culminate when the workload allocation is completed and they must face new responsibilities, while strangers invade the organization asking all sorts of detailed questions about the way staff handle their duties.

However, as the implementation gets under way, early benefits are being delivered and, when obstacles are encountered, staff in both organizations (who know each other better by then) actively try to solve them. In this phase, staff in the client organization report that they get a real feeling of being at the leading edge of IT and collaboration develops between the two sides. Spirits are high and the commitment of staff to the project reaches its highest. Unfortunately, this dynamic period does not last as, in the post-implementation phase, the ERP team is disbanded and staff must return to their usual tasks of dealing with customer queries or processing sales orders.

The turning point in the relationship between the two firms is the management of this post-implementation phase when the relation moves from an active hands-on role to a support role through the intermediary of a service desk or call centre. At that stage, the implementer loses track of the general perception of the product they have installed because the communication that develops as part of the support relation is too rigidly structured around the capture and processing of the problems with the software and cannot capture the business dimensions of the project – that is whether the organization is achieving its targets or whether the site is ready to move to the next stage of development of its IT capability.

Clients all noted how significantly the mood of users and project leaders alike dipped in the post-implementation period. As the burden of their everyday tasks took over and customer inquiries became a priority again, their perceptions of the system tended to crystallize around the problems that emerged and the attention of all actors involved became focused on the problems resulting from the software. The challenge for the implementer, as perceived by the implementing staff, is in maintaining the active relationship as long as possible in order to redress the post-implementation dip as quickly as possible. This is done by maintaining a type of communication that is based on a discussion of what the software has achieved and how to take it further rather than move to a restricted and typically tension-laden software support situation (as is suggested with the dashed curve in Figure 11.1).

The project leader for the implementer firm was of the opinion that the fact that so many of the client organizations had no resident IT staff was another incentive to try to stay in touch with the evolution of the site. In organizations that have full-time IS staff, there is an obvious channel of communication, but in organizations where the ERP leader is essentially a user of the system as opposed to an administrator, the communication can quickly be restricted to matters of proper operation of the system that may be crucial in terms of the operations of the firm, but are not conducive to an analysis of how the usage of the ERP functionalities can evolve or what kind of business development it could allow. Thus, a rich, continued and business-oriented communication between the implementer and the managers in charge was seen by both organizations as required for a more effective deployment of IT in the client organization. This may be the only way that the low morale of the post-implementation phase can be reversed quickly so the company can take full advantage of its newly acquired ERP functionality and develop more effective business processes.

Such communication is also necessary for a proper review of the projects to take place. The review phase of decision-making processes has long been highlighted as critical to proper evaluation of the effectiveness of organizational decision making and to learning about key managerial issues, most notably by Simon (1977) and Mintzberg, Raisinghani, and Theoret (1976). There is empirical evidence that not enough organizations undertake the review of their IS investments (Ward and Griffiths, 1996). Indeed, IS investments are largely regarded as badly analysed and justified from the outset (Farbey, Land, and Targett, 1992). These findings were confirmed in our study where some managers could not indicate to what extent the benefits sought had been achieved. It seems that implementers can play a crucial role in this stage of the project[2] because they can help managers identify the benefits that may materialize and those that are unlikely. Later in the project, they can also help managers identify the next steps in deploying IT capabilities and exploiting the ERP systems acquired to the fullest.

11.6 Conclusions

This study has used a two-phase approach in order to capture some of the reality of ERP implementations in Ireland. It focused on the experiences of 14 small and medium organizations which purchased QAD's MFG/PRO package from Ireland sole distributor of this software package. As a result, the applicability of our findings must be reviewed in light of the likely bias which this sampling method may have brought. All the organizations we studied were successful cases of ERP implementation which had been brought to a happy conclusion from the point of view of both vendor and client. Whether our findings can be extended outside of this context cannot be determined before more research is carried out.

Our data indicate that ERP projects do not have to be as long and complex as previously described. They are comparatively large IS projects, but companies can use a phased approach as a way to minimize the impact of the ERP implementation on the firm's business activities. This approach yields other benefits because it helps managers to get a better understanding of the potential of ERP systems in their case. This study also shows that SMEs are just as likely to be interested in ERP as the larger organizations that have been previously studied. The only reason why SMEs have not been the primary target of software vendors so far seems to have been that vendors and consultants perceived that the arguments in favour of ERP would be stronger in companies such as Microsoft, Dell, or McKesson. However, in Ireland where SMEs are a primary target on the software market (because organizations are much smaller on average), SMEs have already gone down the ERP route in significant numbers and have successfully implemented ERP systems. The duration and complexity of large ERP projects as portrayed in the literature may be due as much to the complexity of the organizations where they are implemented as

[2] The implementers could obviously not be relied upon exclusively in justifying the investment because of the conflict of interest inherent in the investment decision. It is however clear that project leaders in the implementer organizations must, as part of their task, differentiate between sites where implementing ERP is likely to succeed and sites where it should not be undertaken. It is ultimately in no one's interest to get involved in a project that will never succeed and will generate a lot of problems.

to the complexity inherent in ERP systems. This is good news for managers in SMEs who are interested in ERP because it means that the route to ERP may well be shorter and cheaper for them.

The relatively short duration of ERP projects for SMEs is also good news for ERP vendors because it means that their market can be extended to a large group of organizations with far less-developed IT infrastructures in place and where the potential for selling total software solutions is quite high. This development of the ERP market may actually be quite timely for the suppliers of ERP solutions if the predictions reported in the media that large multi-million IT projects will soon become a thing of the past turn out to be accurate (Robineau, 1998).

Our case study at ABC provided additional conclusions regarding the steps that can be taken to reduce the risks inherent in ERP projects. In particular, the notion that top management can impose a no-customization approach deserves more research. ERP packages are based on an underlying business model and, naturally, not all business models can adequately support the needs of a company. However, if the offer on the ERP package market is sufficiently flexible and varied, companies may be able to find a package that exactly fits their needs or at least that can support their business processes without major changes in the services or products offered. In this case, a no-customization approach appears a better option because it enables a quicker and cheaper acquisition and implementation of ERP functionalities. We are led to conclude that the selection phase companies must go through in identifying the ERP package that best suits their needs is the crucial phase. It is our contention that there is a trade-off between the time and effort invested in the selection phase and the time and effort required for the implementation of the ERP software. Companies that do not spend the time analysing their requirements in detail and rely from the outset on the ready-made solutions proposed by consultants and vendors may find themselves regretting their lack of preparation when it comes to sacrificing unique assets of their organization to shoe-horn their ERPs in their operations.

REFERENCES

Adam, F. and Cahen, F., D. (1998) L'achat des systèmes d'information informatiques comme alternative au développement spécific: le cas Socrate. *Systèmes d'Information et Management*, **3**(4), 79–100.

Adam, F. and Doyle, E. (2000) Enterprise Resource Planning at Topps International Ltd. In Johnson and Scholes (eds), *Exploring Corporate Strategy*, 6th edn. Prentice Hall.

Appleton, E. (1997) How to Survive ERP. *Datamation*, **43**(March), 50–53.

Bancroft, N. (1996) Implementing SAP/R3: How to Introduce a Large System into a Large Organisation. London: Manning/Prentice Hall.

Berger, P. (1998) PGI: les services valent cher. *Le Monde Informatique*, 25 September, 779.

Besson, P. (1999) Les ERP à l'épreuve de l'organisation. *Systèmes d'Information et Management*, **4**(4), 21–52.

Bingi, P., Sharma, M., and Godla, J. (1999) Critical Issues Affecting an ERP Implementation. *Information Systems Management*, Summer, 7–14.

Bonoma, T. V. (1985) Case Research in Marketing: Opportunities, Problems and a Process. *Journal of Marketing Research*, **22**(2), 199–208.

Caldas, M. and Wood, T. (2000) How Consultants Can Help Organizations Survive the ERP Frenzy. http://www.gv.br/prof_alunos/thomaz/ingles/paper6.htm.

Eglizeau, C., Frey, O., and Newman, M. (1996) Socrate: An Implementation Debacle. Proceedings of the 4th European Conference on Information Systems, Lisbon, Portugal, 1233–1243.

Farbey, B., Land, F., and Targett, D. (1992) Evaluating Investments in IT. *Journal of Information Technology*, **7**(2), 109–122.

Fitzgerald, B. (1998) An empirical investigation into the adoption of systems development methodologies. *Information and Management*, **34**, 317–328.

Forrest, P. (1999) Genealogie des ERP et gestion des flux physiques. *Systèmes d'Information et Management*, **4**(4), 71–90.

Jeanne, F. (1999) Progiciels: Pas d'ASP sans SAP. *Le Monde Informatique*, 24 September, 822.

Jick, T. (1979) Mixing Qualitative and Quantitative Methods: Triangulation In action. *Administrative Science Quarterly*, **24**, 602–611.

Kalatoka, R. and Robinson, M. (1999) *E-Business – Roadmap to Success*. Reading, MA: Addison-Wesley.

Kaplan, B. and Duchon, D. (1988) Combining Qualitative and Quantitative Methods in Information Systems Research: A Case Study. *Management Information Systems Quarterly*, **12**(4), 571–586.

Kelly, S., Holland, P., and Light, B. (2000) A Departure from Traditional Systems Development Methodologies: Enterprise Resource Planning, ERP, Systems and the use of Process Modelling Tools. Proceedings of the 9th Annual Business Information Technology Conference, 3–4 November, Manchester.

Lampel, J. (1995) Innovation as Spectacle: Dramaturgical Construction of Technological Change. Conference on the Social Construction of Industries and Markets, Chicago, USA.

Lee, A. (1989) A Scientific Methodology for MIS Case Studies. *MIS Quarterly*, **13**(1), 32–50.

Mintzberg, H., Raisinghani, D. and Theoret, A. (1976) The Structure of 'Unstructured' Decision Processes. *Administrative Science Quarterly*, **21**, 246–275.

Miller, J. (2000) Grainger Says It Will Miss Estimates After Installing Complex ERP Software. *Wall Street Journal*, 10 January, Eastern Edition.

Robineau, O. (1998) Les nouvelles Facettes de l'infogerance. *Le Monde Informatique*, 16 October, 782.

Rowe, F. (1999) Cohérence, Intégration informationnelle et changement: esquisse d'un programme de recherche à partir des Progiciels Intégrés de Gestion. *Systèmes d'Information et Management*, **4**(4), 3–20.

Simon H. (1977) *The New Science of Management Decisions.* Englewood Cliff, NJ: Prentice Hall.

Shakir, M. (2000) Decision Making in the Evaluation, Selection and Implementation of ERP Systems. Proceedings of the 6th Americas Conference on Information Systems, 10–13, August Long Beach California, 1033–1038.

Shanks, G., Parr, A., Hu, B., Corbitt, B., Thanasankit, T., and Seddon, P. (2000) Differences in Critical Success Factors in ERP Systems Implementation in Australia and China: A Cultural Analysis. Proceedings of the 8th European Conference on Information Systems, 3–5 July, Vienna, Austria, 537–544.

Stefanou, C. (2000) The Selection Process of Enterprise Resource Planning, ERP, Systems. Proceedings of the 6th Americas Conference on Information Systems, 10–13 August, Long Beach California, 988–991.

Ward, J. and Griffiths, P. (1996) *Strategic Planning for Information Systems.* Chichester: Wiley & Sons.

Wood, T. and Caldas, M. (2000) Stripping the 'Big Brother': Unveiling the Backstage of the ERP Fad. http://www.gv.br/prof_alunos/thomaz/ingles/paper5.htm.

White, B., Clark, D., and Ascarely, S. (1997) Program of Pain. *Wall Street Journal*, 14 March, 6.

12 The Role of the CIO and IT Functions in ERP

Leslie P. Willcocks and Richard Sykes

12.1 Introduction

By 1999 the ERP 'revolution'[1] was generating over $US20 billion revenues annually for suppliers and a further $20 billion to consulting firms. If 2000 and 2001 saw a fall-off in business globally, then some believed that once the distractions of Y2K deadlines were over, new ERP business plus the need to support and capitalise on the sunk investment in IT infrastructure ERP already represented would guarantee further take-off.[1] Once again, at first without many noticing, information technology (IT) not only raised itself high above the cost parapet, but also set off traditional alarm bells about questionable business value. For many firms ERP represents the return of the old IT catch 22 with a vengeance – competitively and technically it is a 'must-do', but economically there is conflicting evidence, suggesting it is difficult to cost justify, and difficult to derive benefits from.[2]

The problem has been further complicated by the coming of web-based technologies. Thus Sauer and Willcocks (2001) found many organizations struggling to integrate their legacy and relatively new ERP systems with e-business initiatives and technologies. The goal of a relatively seamless e-business infrastructure seemed particularly difficult to achieve in 'bricks and mortar' companies trying to move to the web (Sauer and Willcocks, 2000), a finding explained by Kanter (2001) and Willcocks and Sauer (2000) in terms of cultural, organizational, and political issues, together with less than good organization and project management for e-business.

Critical success factors, and reasons for failure in ERP implementations, have now been quite rigorously and widely researched, and are also discussed

[1] In our view an accurate term, which we borrow from Ross (1998).

[2] The existing dilemmas with, and future possibilities for ERP are neatly caught in the title and content of a 1999 report on 62 Fortune 500 companies: Deloitte Consulting (1999).

in detail by other contributors to this book.[3] Authors such as Markus and Tannis (1999), and Ross (1998), point to some distinctive characteristics of ERP that require different treatments from many other IT implementations: packages requiring a mix of old and new skills, often a 'whole organization' suite of applications; software embodying generic best practices that imply large-scale business process reengineering; integrated software requiring further assembly of the *technology* platform; the constant need to keep up with evolving functionality, software, and technology related to ERP; recognition of the degree of customization that is possible or prudent.

However, what is more noticeable is how the difficulties experienced in ERP implementations and with their business value are not untypical of most IT projects, especially where they are large and complex, expensive, take over a year or more to install, use new technology, and impact significantly on the organizational culture and existing business processes. Consider a typical finding represented by a 1995 Standish Group study of 175 000 projects in the USA. It concluded that 30% of projects, representing $US81 billion expenditure, delivered no net benefits. Our own 1998 UK study found 26% of projects producing very disappointing results, while 5% were complete failures. Regularly, such studies show that 70% plus of IT-based (and 90% of ERP) projects are 'challenged', meaning over budget and late. Introducing new IT, even just to support the existing business model, emerges as a high risk, hidden cost process. The risks are even greater when, as for example with many ERP and internet applications, real *business* innovations are also being looked for. The top reasons for disappointment cited in IT and ERP studies also remain stubbornly familiar: incomplete definition of business requirements and of the business case, insufficiently detailed technical specifications, changing business requirements, and lack of business user input into the development process.[4]

Our own work on ERP success and failure factors complements the findings already cited, and documented in other chapters in this book, but differ in one essential respect. We have identified serious neglect in ERP implementations in identifying the most effective roles for the CIO and IT function.[5] Moreover,

[3] Some of the most convincing, rigorous, independent research includes work by the following researchers: Markus and Tanis (1999), Ross (1998, 1999), Holland, Light, and Gibson (1999), Parr, Shanks, and Darke (1999), Hirt, and Swanson (1999), Francalanci (2001).

[4] See Markus, and Tanis (1999), Parr, Shanks, and Darke (1999), Willcocks, and Griffiths (1996), Avital, and Vandenbosch (2000).

[5] Our own ERP research, on which we draw here, are based on firstly acting as IT vice-president throughout the 1990s in a major multinational with multiple ERP implementations, and secondly on interviewing through 1999–2001 senior business and IT and supplier executives participating jointly in 27 ERP projects in organizations throughout Europe, USA and Australia.

in case studies we have found failures in this area to be correlated strongly to subsequent difficulties in achieving delivery and business value. This chapter will spell out the several ways in which we have found the CIO and IT function, and relatedly it must be said, often senior business executives, 'asleep at the wheel' on ERP, and the more effective capabilities and practices some organizations have been drawing upon. Our work also enables us to describe a consistent picture of how business innovations based around ERP, and IT generally, occur, how the technology-related, organizational and project management risks can be mitigated and managed, and how IT-based innovation can be translated into real business advantage. In practice, our ERP and broader research into a range of IT-based business innovations and transformation projects identifies enduring organizational and cultural factors that explain success. Irrespective of the differing technologies being implemented, we have indeed found, with Markus and Tannis (1999) and Ross (1998), that there are still consistent principles to be applied to ERP, and many lessons to be had from history.

12.2 ERP: Efficiency or Transformation?

The roots of the ERP 'revolution' technologically were two-fold: first, prewritten software available off the shelf with sufficient flexibility to match most needs; and, second, an underpinning of data structures shared across many applications (so that data on a given invoice were not passed step to step but accessed from a common data structure). The first step promised a great improvement in software implementation productivity – no time wasted reinventing the wheel and writing endless new software. Seen in that particular IT context, the promise has been delivered, but the context has been one of IT resource productivity alone. The second step promised both IT and operational productivity improvement (including quality improvement and less room for error) through more simple implementation of seamless processes.

The challenge of the ERP revolution has been how best to use the new capabilities. This is where the failure to deliver begins to be real in many organizations we and others have studied. The opportunity was to automate and simplify complex paper-based transactional systems. In reality the *status quo* was more than a paper-based system – it was a whole generation of business processes and how business was done. The real value-adding opportunity was therefore to radically reshape how business was done (business process reengineering) and exploit the new automated, seamless ERP capabilities in the process. 'How a business is done' is, however, not simply the transactional processes that they supposedly serve. The human processes reflect an accumulation of human skills, human processes and preferences (human ways of going about things), and so on. So the real route to value in ERP includes major human and organizational change as a necessary part of the process. Simply put, it is about a transformation involving identifying business strategy and objectives, and designing integrated processes, technologies, information systems and skills to deliver on these.

A Case In Point: Guinness

Consider a successful implementation we studied. Guinness is the brewing division of Diageo, formed in 1997 through the merger of Guinness plc and Grand Metropolitan plc. Guinness operates in 150 markets, and its turnover exceeded £2.5 billion in 2000. Needing to enhance its logistics and systems to increase sales growth, make better use of its assets, sell new products globally, and reduce stock levels and back-office costs, senior management resolved to transform and establish new common business processes and IT systems throughout its five trading divisions.

Both cultural and organizational changes were needed as the company had very separate businesses each operating globally, with their own IT departments, systems and processes. IT architecture was regional in nature in both governance and organization. While the infrastructure in terms of email, UNIX, and desk top was reasonably consistent across regions, the applications base was very mixed, with, for example, Ireland having a lot of legacy systems, and Europe a lot of locally purchased packages. A strong point was the selection of SAP R/3 as the new core ERP platform, since this had already been implemented in the UK between 1997 and 1999.

From February 1998 Guinness ran an Integrated Business Programme and a year of implementation work saw SAP R/3 going live at Guinness UK, extended to Ireland in July with Customer Order Fulfilment going live in both places by November 1999. Throughout 2000 both were further rolled out in the USA and elsewhere. Guinness used R/3 implementor Druid to help put in finance, sales and operations planning, procurement, customer order fulfilment, and product supply suites. But Guinness also had to implement workflow, data warehousing, advanced supply chain solutions, and integrate some legacy systems as well. It also resolved to continue to use Manugistics for production planning. Guinness also changed its organization architecture to a shared service model, with, for example, finance being one dedicated department servicing all business divisions. Centralisation of most back-office processes has been implemented and has been found to work satisfactorily, and is reflected in having only one corporate IT function, with only some systems infrastructure and software development decentralized.

Guinness has tended to achieve its goals with the ERP implementation by making some radical organizational, skill and business process changes, at the same time as restructuring the IT base and the way IT was organized and managed. The strong factors enabling success mentioned by management were: creating a project team out of staff drawn from all parts of the business globally; detailed communication to and buy-in from all relevant stakeholders from the beginning and throughout; strong leadership from the top; a focus on changes to the business as a whole rather than just on the ERP implementation, for example breaking down local cultures, building secure and stable networks, and a single IT department; and implementing in a fast, focused manner. A further aspect was being flexible about some aspects of implementation. Thus IBP left out operations in the Far East and Africa that remain fundamentally stand-alone. IBP also recognized that sometimes localization should take precedence. For example, in Ireland most pubs were still owner-managed rather than owned by big chains as in England. This meant their order-processing systems had to be different, with Ireland not using SAP R/3 exclusively, but having a Siebel front-end.

In these respects Guinness can be located as following the orientation and practices typical of the second change equation illustrated in Figure 12.1. The dangers seem to come when an organization allows itself to follow only the first change equation, or deludes itself into believing that it is delivering on the second, without adopting the necessary transformation practices this requires.

12.3 The CIO and IT Function: Less Effective Practices

In this context, on what occasions, and in what ways can it justifiably be said that the CIO and IT functions have been, on ERP, 'asleep at the wheel'? Our research has identified three 'asleep' modes and one main 'wide-awake' set of capabilities and practices in ERP implementations. The three 'asleep' modes are: technological determinism, supplier/consultant driven, and absent relationships and capabilities. These are now described.

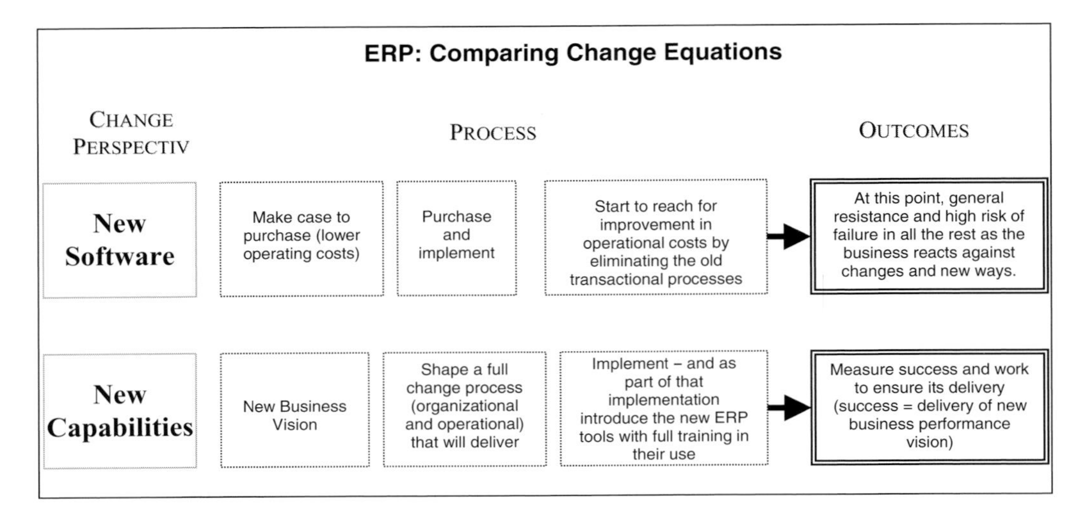

Figure 12.1 Enterprise systems: two change equations

Scenario 1: 'Technological Determinism'

ERP software has particularly lent itself to the notion of being a packaged total solution to a whole range of technical and business problems. This itself has become a problem when the CIO is technologically focussed, the IT function's skill base is technical and the organization is functionally driven, with the IT function seen by itself and the rest of the business to be the prime 'owner' of IT issues. A relatively mediocre previous IT track record on business systems delivery and value becomes the basis on which ERP is handled and implemented. Invariably ERP is regarded as a 'new software system' and the change equation that results, typically and disappointingly, is that shown in Figure 12.1. In our study we found examples of such outcomes, especially in the early implementations in the 1995–1997 period in two banks, an aerospace company, and several manufacturing companies and public sector organizations. As one example a senior business executive in a UK-based multinational manufacturing company commented on a two-year ERP project begun in 1996:

> When I think back, the whole thing was seen as a technology infrastructure development. We were also promised a solution to Y2K problems. I recall the business case premised on infrastructure renewal, there was little there on business benefits . . . it was assumed they would come . . . (but) a lot on streamlining of technical functionality, indeed on a global basis.

In such examples typically the ERP project is abandoned into the often eager hands of the IT function, whose view is to deliver technical capability

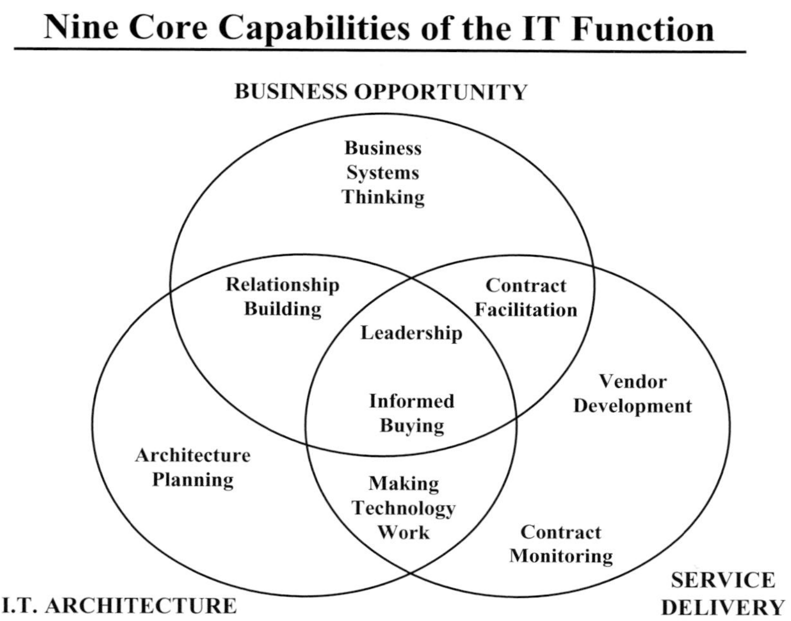

Nine Core Capabilities of the IT Function

BUSINESS OPPORTUNITY
Business Systems Thinking
Relationship Building
Contract Facilitation
Leadership
Informed Buying
Vendor Development
Architecture Planning
Making Technology Work
Contract Monitoring
I.T. ARCHITECTURE
SERVICE DELIVERY

Figure 12.2 Emerging core capabilities of the IT function

on time and within budget. The change equation posited is the first shown in Figure 12.1. In itself this often proved more difficult than at first appeared, because of lack of relevant in-house IT skills, attempts to customize the ERP, and the complexity of linking with legacy data, systems, and technologies.

Scenario 2: 'Supplier/Consultant Driven'

A second scenario has been where senior business executives have taken the significant decisions without meaningful CIO and IT input, either because they see ERP as 'too important and enterprise wide' for the IT function to be left with the responsibility, or because the IT function has a poor track record and ERP is seen, somewhat like large-scale IT outsourcing, as both a way of replacing the IT headache with a packaged solution, and also a way of giving substantial IT work to external suppliers and consultants whose core capabilities are seen to lie in this area. This tendency has often been fuelled by suppliers and consultants increasingly selling ERP not into the IT function, but into the boardroom itself. One noticeable feature of this scenario is how far the CIO is regarded neither as a peer within the top team, nor as an ally with loyalty to the business rather than to the IT function and to the technology.

Too often an outcome here is considerable cost overruns: one respondent mentioned the cost being ten times that posited in the first feasibility study with suppliers. A second is lack of buy-in by the organization. The ERP gets implemented, but the reengineering and training necessary becomes long and painful, subsequently; while the continued use of external contractor staff cause a sharp rise in the maintenance and development budgets for ERP. While change equation 2 is posited (see Figure 12.1), the means to deliver it are not really present. Respondents in two-fifths of our case studies recognized this scenario to various degrees. As one example, in one US-based manufacturing company operating globally, the ERP expenditure between 1997 and 2000 was estimated to be $500 million. The Board made the decision to launch a so-called 'global business integration project' largely in negotiations with suppliers and consultants. Subsequently the IT function was supplying some 300 staff to the project, believed little in its value, but was receiving a high level of complaints from business units for their lack of ability to deliver on other pressing IT-related business requirements.

Scenario 3: 'Absent Relationships and Capabilities'

All too many IT functions and CIOs have not been sufficiently prepared to live up to the admittedly difficult business challenges represented by new technologies, increased competitiveness, and the changing, ever-pressing requirements from the business. Quite often, of course, the business itself has not set a suitable context in which CIOs and IT functions can flourish. Where IT is seen as a cost to be minimized, and as a cost efficiency tool and not as a potential strategic resource, and where the CEO and/or senior business executives are permanently disengaged from a business transformation agenda through IT, there is in fact little chance for the IT function to contribute significantly to business imperatives. Inflexible human resource policies on pay and career and contracts will also exacerbate the skills shortage problems many even large corporations face in today's volatile IT labour markets.

Translated into ERP development and implementation, it can be seen how a third scenario has developed. Either change equation may be posited. In either case, the IT function is made largely responsible for ERP, but is in no real state to be successful. It lacks the technical skills, but also lacks the capabilities needed to manage the external suppliers hired to fill the planning and operational skills gaps. It also has not managed to build up relationships with the business side at CIO and operational levels, nor helped to reorientate business thinking and motivation sufficiently to cause the business to live up to their key responsibilities in the reengineering, change and transformation aspects of ERP implementations.

In practice this scenario was the most common one we found, even to certain degrees in the ERP implementations that were proving successful. As one CIO in an insurance company commented:

> In the last three years I have not had time to rethink the skills base . . . it's been one thing after another . . . in some ways ERP was a blessing, there were so many problems I could shovel into it, but I can't say we were ready for it, the complexity, the management of the suppliers, the problems with the business.

12.4 On Being Wide-Awake: (1) Key it Capabilities

In practice, however, the need to identify and build key in-house IT capabilities before entering into ERP projects emerges as one of the critical – and neglected – success factors we have identifed throughout our practical and research experiences. Earlier research has already been strongly indicative on this issue, and the relevance of the capabilities displayed in Figure 12.1 for ERP implementation has been confirmed both in the multinational, of which one of the authors of the present Chapter was CIO for six years, and with respondents in the organizations whose ERP projects we have been studying. Thus Feeny and Willcocks (1998) have identified the following nine core IT capabilities:

- *IT leadership:* to devise the strategy, structures, processes, and staffing to ensure that the IT function delivers value for money. Our own work on ERP supports that by Ross and Feeny (1999). In successful ERP the CIO must at least have built strong relationships with her/his business executive peers, must behave as a knowledgeable strategic partner with the business, and be able to align investments in IT with strategic business priorities. The IT function will have built, and will continue to strive to retain credibility for its ability to deliver on promises.
- *Business systems thinking:* to ensure that IT capabilities are envisioned in every business process. In the ERP context the failure to have integrated business, process, technology, and skills thinking and planning at what Markus has called the chartering phase has been an all too frequent reason for sub-optimal ERP business performance.

- *Relationship building:* to establish understanding, trust, and cooperation among the business users and IT personnel. ERP potential cannot be leveraged without strong coordination of effort and goals across busines and IT personnel.
- *Architecture planning:* to create the technical platform to meet current and future business needs. Again this cannot be abandoned to ERP suppliers, despite the seeming attractions of a total packaged solution. The client organization needs to maintain control over its IT destiny, its ability to plan for the future technology requirements of the business, while retaining the ability to have informed discussions with ERP suppliers.
- *Technology fixer:* to rapidly troubleshoot problems and to create workable IT solutions to business needs. This capability to handle non-routine technical problems as they are disowned in the technical supply chain must be retained in-house. Ross reports that this happened at Dow Corning, and we saw it also in several of our examples of success. With ERP, suppliers and consultants cannot be expected to have the commitment or the specific business and technical knowledge to deal with the unusual company-specific problems that arise. Retaining such talent in-house may involve paying them more to stay than they would receive as contractors, but several respondents reported that this was precisely the trade-off between cost, stability, and effectiveness they were willing to make.
- *Informed buyer:* to develop a sourcing strategy and to evaluate and negotiate sound contracts with suppliers. This and the remaining three core capabilities are vital in ERP implementations, where so much use of external contractors, supplier staff, and consultants is made.
- *Contract facilitation:* to ensure success of existing contracts through user-supplier management and multiple supplier coordination.
- *Contract monitoring:* to hold suppliers accountable to existing service contracts and to the developing performance standards of the market through benchmarking and other means.
- *Supplier development:* to identify and seek added value from supplier relationships by looking beyond contractual arrangements to explore long-term mutual benefits.

These nine core IT capabilities must be retained in-house. Where, as frequently is the case in ERP projects, these key skills may be lacking one solution is to build them over time by 'insourcing' – buying human resources off the market to work closely with the in-house team and ensuring that a transfer of learning takes place. These capabilities represent the minimal IT function needed to plan for and implement ERP for business advantage.

12.5 On Being Wide Awake: (2) Managing ERP Implementations

Once these capabilities are in place, the overriding factor then becomes the implementation process adopted. The main features of difficult projects, ERP or otherwise, continue to be large size, long time scales, complex, new or untried technology, and lack of clear, detailed project staffing and management structure. Moreover, traditional 'waterfall' methods of systems development seem particularly inappropriate for implementing ERP-type projects if they are to be genuine business innovations.

For these reasons, in highly competitive sectors, many organizations – such as Ford Europe, Daiwa, British Airways, Woolwich Bank – both with ERP or other IT-based projects have moved to a different 'time box' approach with very positive results. The significant differentiator is not new tools for faster systems development, such as CASE, OOPs, and RAD tools, but innovations in staffing, culture, project disciplines, and learning. Let us distill out the rationale, and major features, of an 'implementation as innovation' approach.

User vs. specialist focus. Projects like ERP that hope to embody IT-based business innovation are, first, business projects and, second, are inherently unstable. Detailed business requirements, as opposed to the overall objective, are unclear and subject to rapid change. Flexibility for further learning and innovation is required. Additionally, the technology itself (less so the ERP software) may be under-developed, lacking stability and detailed technical specification. Technology maturity refers to where a radically new technology is being utilized, or where a radically new business application of an existing technology/software is being made, or where relevant in-house technical expertise is lacking. Increasingly in fast-moving business environments, as IT becomes increasingly organizationally pervasive, development can no longer be left primarily to IT specialists or external IT suppliers. If second wave ERP is to happen then ERP-based business innovation requires a user-focused approach involving multi-functional teamwork, personal relationships, and business goals. Only when technology maturity is high, and a detailed contract and delivery can be pre-specified does it become low risk to hand over those aspects of development and operations to in-house IT specialists or outsource to external IT suppliers.

Governance and staffing. Our present ERP study, and earlier IT-related studies, consistently show that effective IT-based business innovations require a high-level sponsor, and a project champion, both taken usually from the business, not the IT side. The former will provide only up to 5% of his/her time, but be involved in initiating the idea, underwriting the resources required, and protecting the project into business adoption and use. The project champion will provide between 20 and 60% of his/her time. The role involves communicating the vision, maintaining motivation in the project team and the business, fighting political battles, and remaining influential with all stakeholders, including senior management.

A small 'dolphin' team is needed to implement the time box philosophy described below. Effective project managers have three distinguishing characteristics: credibility with the salient project stakeholders, a track record of success with *this* size and type of project, and skills in controlling the detailed actions needed to keep a project on its critical path. Additionally potential users of the IT, drawn from the best people available, will be assigned to work full time on the project, along with in-house IT specialists. External IT resources may be needed to fill skills gaps. Certain users and managers may need to be brought in to provide additional knowledge and opinion on an occasional basis. The multi-functional team will also contain people with 'bridge-building' interpersonal skills. Co-location of team members also helps the key processes of team building, knowledge sharing, and mutual learning.

Time box philosophy. The primary discipline here is that the ERP-based business innovation must be delivered within a six-to nine-month period. Moreover the IT must be developed within the overall IT architecture of the organization, and not as a separate 'portakabin'-type system. If after a hard look this cannot be achieved the project does not start. However it can be decomposed into smaller projects, each of which will deliver tangible business benefits. Time discipline reduces the risks of the project not meeting business requirements, ensures that big projects or 'whales' are reduced to a series of more manageable 'dolphins', means that business benefits flow regularly rather than being delayed, and ensures the team has to remain focused and fully staffed over a more realistic, limited period. Within the project further time boxing will place time limits on each part of the development, to reduce drift from the overall business delivery target.

Development proceeds on the basis of the 80/20 rule, with the business accepting in advance that the first systems release may well only provide 80% of the functionality originally demanded. As the marketing director for Ford Europe puts it: '"80% systems are OK" is now part of the culture here.' As the head of distribution development at the Woolwich comments: 'We decided to build it as quickly as possible and accept that we would make mistakes along the way.' However in the new culture mistakes and failure are treated as positive contributions to learning. Iterative development by prototyping ensures close interaction between developers and users with a workable system built quickly and constantly refined. The knowledge exchange and incessant testing and learning leads to a culture of rapid improvement, with sometimes radically new discoveries on business utilization of the systems. Prototyping and user involvement also sees the developing system becoming business-owned, thus easing acceptance problems traditionally experienced when IT specialists hand over IT to users.

The Supplier/Consultant Role in ERP

In ERP, external perspectives and knowledge can contribute much to the process of technical and business innovation. Furthermore, with the need for rapid delivery of systems to the business and all too typical in-house shortages of both routine and key skills, suppliers can perform an important 'fill-in' role. More routine, easily defined tasks within the overall project can be outsourced. Our own research suggests that, with IT-based innovation, suppliers are most effectively utilized as resources brought in to work under in-house direction and control. The responsibility and detailed direction for innovation must stay with the business, but 'insourcing' external skills can, if properly managed, release valuable transfer-of-learning affects. These suggestions are mapped in Figure 12.3, which is adapted from Feeny, Earl and Edwards (1997), which also summarizes the argument made above on the roles of users versus suppliers/specialists.

The alternative – of outsourcing to a third party the management and resourcing of IT-based innovation – places the external supplier in an invidious position. Consider one European insurance company, intending in the mid 1990s to achieve competitive advantage through transforming its policy and administration systems. A major supplier was given aggressive time scales to deliver detailed business requirements for the systems. However, the supplier had no great in-depth knowledge of insurance and greatly underestimated the complexity of insurance work and information. The method used for driving out requirements was new and relatively untried. Project management was also made the supplier's responsibility. In practice a business innovation was being abandoned to the supplier. After one year the insurance company invoked penalty clauses for non-delivery to cancel the project, with the supplier incurring significant costs.

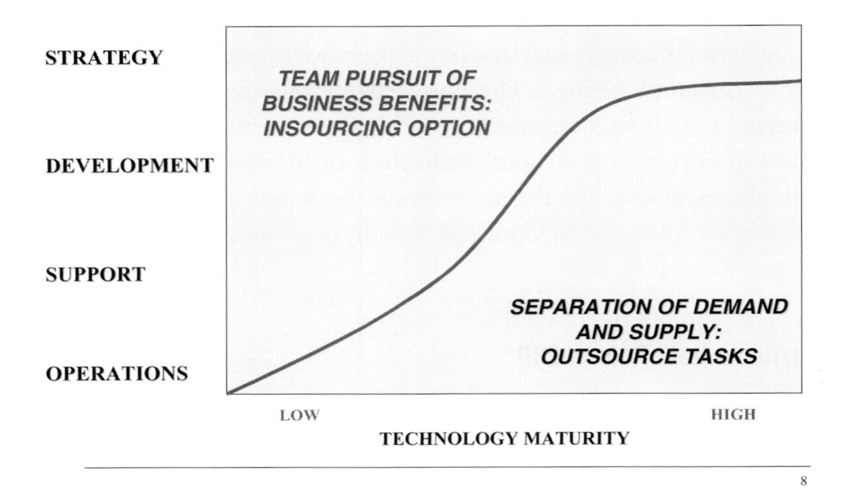

Figure 12.3 Sourcing and the roles of IT suppliers/specialists

A useful comparison can be made with the successful ERP implementation at ICI Polyurethanes between 1993 and 1998. The company started with a complex series of operations across Europe and a complex customer base covering for example auto manufacturers, shoe manufacturers, and white goods makers. A clear change programme was led by a very systems literate CEO and backed by a very IT literate finance director. The internal IT function was very heavily involved in what was conceived as a business reengineering project from the start. The main steps were

1 Revisioning a detailed planning exercise to position the business as a single pan-European operation
2 Negotiating with many national tax authorities to allow the business to operate as a single legal entity across Western Europe
3 Developing a change plan to revise the 'footprint' of manufacturing plants, systems, processes. The human dimension tackled as an integral component of the plan
4 Redesigning Business processes to align with the new streamlined business 'footprint'.
5 Implementing SAP R/3 to support and underwrite the new processes. Heavy emphasis was placed on training and implementation exercises. Very strong focus throughout was placed on delivery of business performance metrics.

In practice the business has grown strongly in a very competitive marketplace. ERP achieved considerable business advantages including allowing vital tax negotiations to take place and enabling the eradication of 125 000 internal transactions involving the company doing business with itself. The IT function's role was integral to this success, and it followed many of the effective practices on in-house capabilities, 'user' focus, governance and staffing, time-box philosophy, and use of external supply emerging from our study.

12.6 Conclusion

Much of this, of course, assumes that the business itself has matured in its ability to manage ERP and similar IT-based business innovations, thus setting the context in which the CIO and IT function can thrive. In these circumstances there emerges from our practical and research experiences eight critical enabling factors if ERP-supported business innovations are to stand a chance of succeeding:

- Senior level sponsorship, championship, support and participation
- Business themes, new business model and reengineering drives technology choice
- 'Dolphin' multi-functional teams, time-box philosophy, regular business benefits
- CIO as strategic business partner
- Nine core IT capabilities retained/being developed in-house
- In-house and 'insourcing' of technical expertise preferred
- Supplier partnering – strong relationships and part of team
- ERP perceived as business investment in R&D and business innovation rather than primarily as a cost efficiency issue.

Does the arrival of web-based technologies change these imperatives in any way? Actually, from the additional research we have carried out, so far we have found the opposite. The difficulties of integrating ERP with legacy and web-based technologies makes it even more critical to make sure these these enabling factors are applied. This theme is pursued in more detail in the final chapter of this book.

REFERENCES

Avital, M. and Vandenbosch, B. (2000) SAP Implementation at Metalica: An Organizational Drama in Two Acts. *Journal of Information Technology*, **15**(3), 183–194.

Deloitte Consulting (1999) ERP's Second Wave: Maximizing the Value of ERP-enabled Processes, Atlanta.

Feeny, D., Earl, M., and Edwards, B. (1997) Information Systems Organization: The role of Users and Specialists. In *Managing IT: A Strategic Resource*, Willcocks, L., Feeny, D. and Islei, G. (eds), Maidenhead: McGraw Hill.

Feeny, D. and Willcocks, L. Core IS Capabilities for Exploiting IT. *Sloan Management Review*, **39**(3), 9–21.

Francalanci, C. (2001) Predicting the Implementation Efforts of ERP Projects: Empirical Evidence on SAP/R3. *Journal of Information Technology*, **16**(1), 33–46.

Hirt, S. and Swanson, E. B. (1999) Adopting SAP at Simens Power Corporation. *Journal of Information Technology*, **15**(3).

Holland, C., Light, B., and Gibson, N. (1999) A Critical Success Factors Model for Enterprise Resource Planning Implementation. Proceedings of the European Conference on Information Systems, Copenhagen, June.

Kanter, R. (2001) E-volve: Succeeding in the Digital Culture of Tomorrow. Boston, MA: Harvard Business Press.

Markus, L. and Tanis, C. (1999) The Enterprise Systems Experience – From Adoption to Success. Working paper presented at the 'Enterprise Systems' seminar, AGSM, Sydney, August.

Parr, A., Shanks, G., and Darke, P. (1999) The Identification of Necessary Factors for Successful Implementation of ERP Systems. In *New Information Technologies in Organizational Processes*, Ojelanki, N. et al. (eds), Boston, MA: Kluwer Academic.

Ross, J. (1998) The ERP Revolution: Surviving versus Thriving. Centre for Information Systems Research Paper, MIT, Cambridge, MA.

Ross, J. (1999) Dow Corning Corporation: Business Processes and Information Technology. *Journal of Information Technology*, **15**(3).

Ross, J. and Feeny, D. (1999) The Evolving Role of the CIO. OXIIM Working paper 99/3, Templeton College, Oxford.

Sauer, C. and Willcocks, L. (2000) Building the E-business Infrastructure. Business Intelligence, London.

Willcocks, L., Feeny, D., and Islei, G. (1997) *Managing IT: A Strategic Resource*. Maidenhead: McGraw Hill, particularly Chapters 8, 9, and 10.

Willcocks, L. and Griffiths, C. (1996) Predicting Risk of Failure in Large-scale Information Technology Projects. *Technological Forecasting and Social Change*, **47**, 205–228.

Willcocks, L. and Plant, R. (2001) Pathways to E-business Leadership: Getting from Bricks to Clicks. *Sloan Management Review*, April.

Willcocks, L. and Sauer, C. (eds) (2000) *Moving to E-business*. London: Random House.

13 Enterprise System Management with Reference Process Models

Michael Rosemann

13.1 Introduction

Enterprise Systems (ES) offer configurable business solutions for typical functional areas, such as procurement, materials management, production, sales and distribution, financial accounting, and human resource management (Rosemann, 1999). These functions are typically individualized for industries, such as automobile, retailing, high-tech, etc. Consequently, ES tend to be very comprehensive and complex. This is mirrored in the software documentation, which was often measured in metres before online documentation was developed. To improve the understandability and to stress the process-oriented nature of their solutions, ES vendors have developed reference models which describe the functionality and structure of their systems. Enterprise system reference models exist in the form of function, data, system organization, object and business process models, although the latter is by far the most popular type. The dominance of reference process models results from the increasing popularity of process-oriented management concepts such as business process engineering (Hammer and Champy, 1993) or process innovation (Davenport, 1993), which led to the development of several new process modelling approaches (for example, Kim, 1995).

Enterprise system reference process models describe on different levels of abstraction selected ES processes, that is, sequences of functions supported by the system. Depending on the underlying methodology, these models include details about the control flow (including AND/OR splits or joins), the involved system organizational units, input and output data and business objects (Curran and Keller, 1998). If the models are embedded in the enterprise system, it is usually possible to refer to the relevant part of the online documentation, the required configuration tasks and on the lowest level of abstraction even to the corresponding ES transaction. Figure 13.1 shows an extract from a simple Enterprise System reference process model – a part of

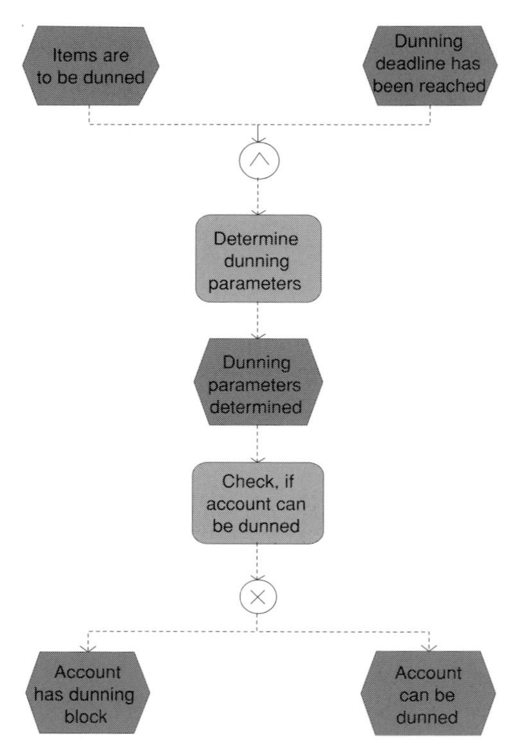

Figure 13.1 Extract from SAP's reference process model 'Automatic Dunning'

the dunning process within SAP R/3 accounts receivable. The model describes the procedure for generating and sending out reminder notices to customers with outstanding payments. This includes the specification of the parameters, the selection of accounts and open items, grouping open items and the creation of reminder notices with an update of the customer master data. The modelling technique in this case is the event-driven process chain (Curran and Keller, 1998; Scheer, 1998). It consists of events (hexagons) and functions (soft rectangles) as well as control flow connectors (AND, OR), which describe the joins and splits of a process. Figure 13.1 shows an AND-join (first connector) and an exclusive OR-split (second connector).

It has to be stressed that these models are designed for the end-users of the enterprise system, and not only for the implementation team. End-users benefit from these models as they comprehensively and quickly inform about the functionality of the software. These reference models are part of the ES solution and do not have to be purchased separately. Surprisingly, however, few companies are using these models. In interviews, users of enterprise systems indicated that they are not aware of these models, that they do not have the required access rights to study these models, that they have problems with the

ES-specific terminology (for example, 'dunning'), or that they believe these models were only of relevance for the implementation team. One motivation of this chapter is to increase the awareness of the models and the benefits of their application.

In the next section, current shortcomings of enterprise System reference models are discussed. The following chapters are structured along the enterprise system lifecycle. After an overview about proposed lifecycle models, separate sections elaborate on the use of reference process models in the lifecycle phases: business engineering, system selection, system implementation, and system use and change. This chapter ends with a conclusion and outlook.

Shortcomings of Enterprise System Reference Models

The existence of reference models highlights a difference from the traditional software development process. Instead of starting from scratch and continuously adding functionality, ES solutions require a narrowing of the scope of the system. It is necessary to select the necessary functions and to decide during the configuration process between alternatives (for example, reporting in financial accounting or controlling). In comparison with the development of software the ES implementation process has a different starting point. It starts with the 'big picture', which is then reduced to the relevant parts. Reference models can be used as a description of this big picture.

Although enterprise system reference models have contributed significantly to the understandability of the software functionality, they still have significant shortcomings.

- As the models are focused on the description of the process execution and the data structure, it is not obvious what *configuration alternatives* exist. The analysis of a reference model shows what is possible in general, but not what might be a recommended alternative. They represent the entire functionality from the viewpoint that the complete system is used. The models are not designed for configuration. Reference modelling techniques do not support constructs that cover possible decisions during the implementation phase, that is, decisions at buildtime. In summary, enterprise systems typically do not differentiate between decisions on instance level and type level.
- Besides the missing transparency regarding possible choices during the configuration process, it is also not clear what *consequences* a configuration of one process or data structure has *on other processes or data structures.*
- Reference models concentrate on the elements that are of importance for a specific enterprise system. *Enterprise-individual aspects* of an organization, business objectives, or manual tasks cannot be seen in these models.

- Enterprise System reference models *do not include any reference to the knowledge* that is required for the evaluation, implementation, execution or change of the described processes. In the context of enterprise systems, we can differentiate between business, technical, company-specific, product, project, and communication/collaboration knowledge (Rosemann and Chan, 2000). Without such a reference it is, for example, unclear how complex the process configuration is and to what extent external knowledge would be required.
- The models do not have any *link to the actual process execution*. Thus, it is not possible (for example, in the form of model attributes) to see the process performance expressed in key performance indicators such as processing time or resource utilization.

Though most of the Fortune 1000 companies currently use enterprise systems, the IS literature has ignored the conceptual problems related to the model-based configuration of enterprise systems. This area of research can be divided into requirements engineering for the *development* of enterprise systems (Brinkkemper, 1999; Daneva, 2000) and requirements engineering for the *configuration of enterprise systems*. The latter one is the focus of this chapter. Theoretical contributions in this field are still the exception. As an example, Rolland and Prakesh (2000) suggest a map including ERP goals and objectives for the identification and evaluation of user needs. Gulla and Brasethvik (2000) introduce three process modelling tiers to manage the complexity of process modelling in comprehensive enterprise resource planning (ERP) system projects. Their functional tier dimension deals with the functionality of the enterprise system. However, they do not discuss how to differentiate reference models in this tier. Most current literature on systems engineering and conceptual modelling still focuses on the classical system analysis and design process. The ES lifecycle is significantly different from the software development process.

The Lifecycle of Enterprise Systems

In comparison with research conducted on software development and related systems analysis and design activities, the management of enterprise systems has received little academic attention (Gable *et al.*, 1997, Klaus, Rosemann, and Gable, 2000). One result is the absence of a generally accepted enterprise system lifecycle model. Several models have been developed for the traditional software engineering process (for example, waterfall model, spiral model) but corresponding discussions regarding standardized ES lifecycle models are non-existent.

Further, most of the work until now has concentrated on implementation issues. An overview of ES-related research in June 2000 (Klaus, Rosemann, and Gable, 2000) showed that about 30% of publications deal with implementation issues. This corresponds with the focus on enterprise systems by the trade press, which also deals mainly with implementation and associated issues. Several publications (Bingi, Sharma, and Godla, 1999; Holland *et al.* 1999a; Stefanou 1999; Sumner 1999; Nah, Lau, and Kuang, 2001) attempt to identify critical success factors of enterprise system implementations. Shanks *et al.* (2000) strongly recommend taking national cultural issues into account, since critical success factors may vary depending on the country in which an implementation is carried out.

Implementations have also been investigated through case studies with varying objectives: to explore strategic options open to firms beyond the implementation of common business systems (Holland et al., 1999b); to avoid ES project failures (Scott, 1999); to identify issues of alignment (Smethurst and Kawalek, 1999; Volkoff, 1999), business process reengineering (BPR) (Slooten and Yap, 1999), and change management (Pérez et al., 1999). Implementing ES with or without BPR has been surveyed and analysed (Bernroider and Koch, 1999) and an enhancement of process engineering related to Enterprise Systems has been proposed (Sato, 2000). The analysis of ES-related publications shows also that case studies are the dominating research methodology.

Only few publications discuss enterprise systems beyond the cost-intensive system implementation phase and try to develop a comprehensive lifecycle model. The following list gives an overview of some proposed lifecycle models.

- Bancroft (1997) proposes an ES lifecycle with a concentration on the early stages that includes focus, as-is, to-be, constructing and testing, and actual implementation.
- Gable, Scott, and Davenport (1998) suggest a lifecycle that consists of the consulting process, selecting the ES software, implementing the software, and learning and knowledge transfer.
- Markus and Tanis (2000) differentiate along the ES lifecycle between chartering, project, shakedown, and onward-and-upward.
- Ross (2000) discusses in an analysis of the perceived organizational performance into design, implementation, stabilization, continuous improvement, and transformation of ERP.
- Sandoe, Corbitt, and Boykin (2001) propose the phases initiation, planning, analysis and process design, realization, transition, and operation.
- As one suggestion for a consolidation of some of these models, Shanks et al. (2000) suggest to distinguish between planning, implementing, stabilization, and improvement.

- An example of a software-specific approach is ValueSAP (SAP, 2000), a framework of methodologies, tools, content, and programs. ValueSAP aims to increase the benefits derived from SAP's ES solution during the entire lifecycle and consists of the three phases discovery and evaluation, implementation, and operations and continuous improvement. The embedded AcceleratedSAP (ASAP) focuses on the actual system implementation. ASAP's roadmap includes the five sequential tasks of project preparation, business blueprint, realization, final preparation, and going live.

These approaches have in common more or less detailed treatments of the pre-implementation and post-implementation stages. However, most of them lack an explicit stage for the use of the system. This is surprising, as this is the longest phase of the ES lifecycle and the one in which an organization is supposed to benefit most from the ES system.

Process Models in the Enterprise System Lifecycle

The enterprise system lifecycles described above can be consolidated and simplified into four phases: business engineering, system selection, system implementation, and system use and change. The following discussion focuses on how process models and enterprise system reference models can be used in these phases.

Business engineering

In addition to tasks specific for project management (such as forming a project team and scheduling the project tasks) the business engineering phase includes the creation of an awareness of the required IT and organizational change, the documentation and analysis of the current situation, and the development and selection of possible process improvements. Although this phase is independent of specific ES solutions, it usually includes a general comparison of integrated ES solutions with best-of-breed-approaches (Dewan, Seidmann, and Sundaresan, 1995; Light et al., 2000). This decision is a strategic one and can be made before the detailed selection of systems.

Process modelling has a critical role in the business engineering phase. Beyond the central aim of documenting the current business, including weaknesses, it helps to develop a common understanding of the domain. The appropriateness and acceptance of process modelling can usually be tested in a pilot study. In order to 'unfreeze' the organization, it is recommended that a process be depicted with many organizational and IT interfaces, such as invoice verification or customer complaints handling. These first process

Table 13.1 Advantages and disadvantages of as-is modelling

Advantages	Disadvantages
Same problem understanding	Results are obsolete as soon as new processes are designed/implemented
Same terminology	Danger of narrow-focused process design (thinking in constraints)
Supports acceptance for the project (unfreezing)	Time and cost consuming
Base for a migration strategy towards the redesigned processes	
Completeness of to-be processes can be evaluated	
Results of as-is analysis can be used as to-be, if no or only minor changes	
Shows weaknesses and restrictions	

models have to demonstrate that they offer a new way of understanding the business and, particularly, of identifying weaknesses.

After the formal project kick-off, but before the more comprehensive documentation of the current processes (as-is modelling) takes place, a careful consideration of the advantages and disadvantages of as-is modelling is required. These are listed in Table 13.1.

The core benefit of as-is modelling is that all project members develop the same problem understanding and terminology. During the discussion of possible improvements, as-is models serve as a kind of a benchmark and completeness check. Parts of or complete as-is models can often be declared as to-be models, if no major process changes are required or possible. Finally, descriptions of existing processes highlight weaknesses and the potential for improvements, but also existing constraints. Models that depict many weaknesses may convince the project team to follow a process-oriented approach. On the other hand, intensive as-is modelling carries the danger that the project team will lose the capability of 'thinking out-of-the-box'. Further, as soon as new models are valid, as-is models only document history. As-is modelling can also become time and cost consuming as it requires general agreement among participants. This is not necessarily the case for to-be models.

As-is and to-be process modelling can become a complex task, as the number of designed models usually grows quickly. For example, in a former project with a facility management company belonging to the German Telecom, we designed more than 600 process models (Becker, Kugeler, and Rosemann, 2001). In order to manage the comprehensiveness and complexity of this task, it is required to have *precise modelling conventions* that standardize the relevant modelling techniques, object types, relationship types, attribute types, level of abstraction, naming and layout conventions (Becker, et al., 2000). Figure 13.2

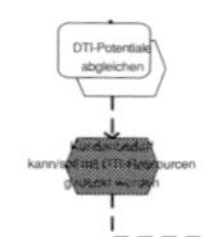

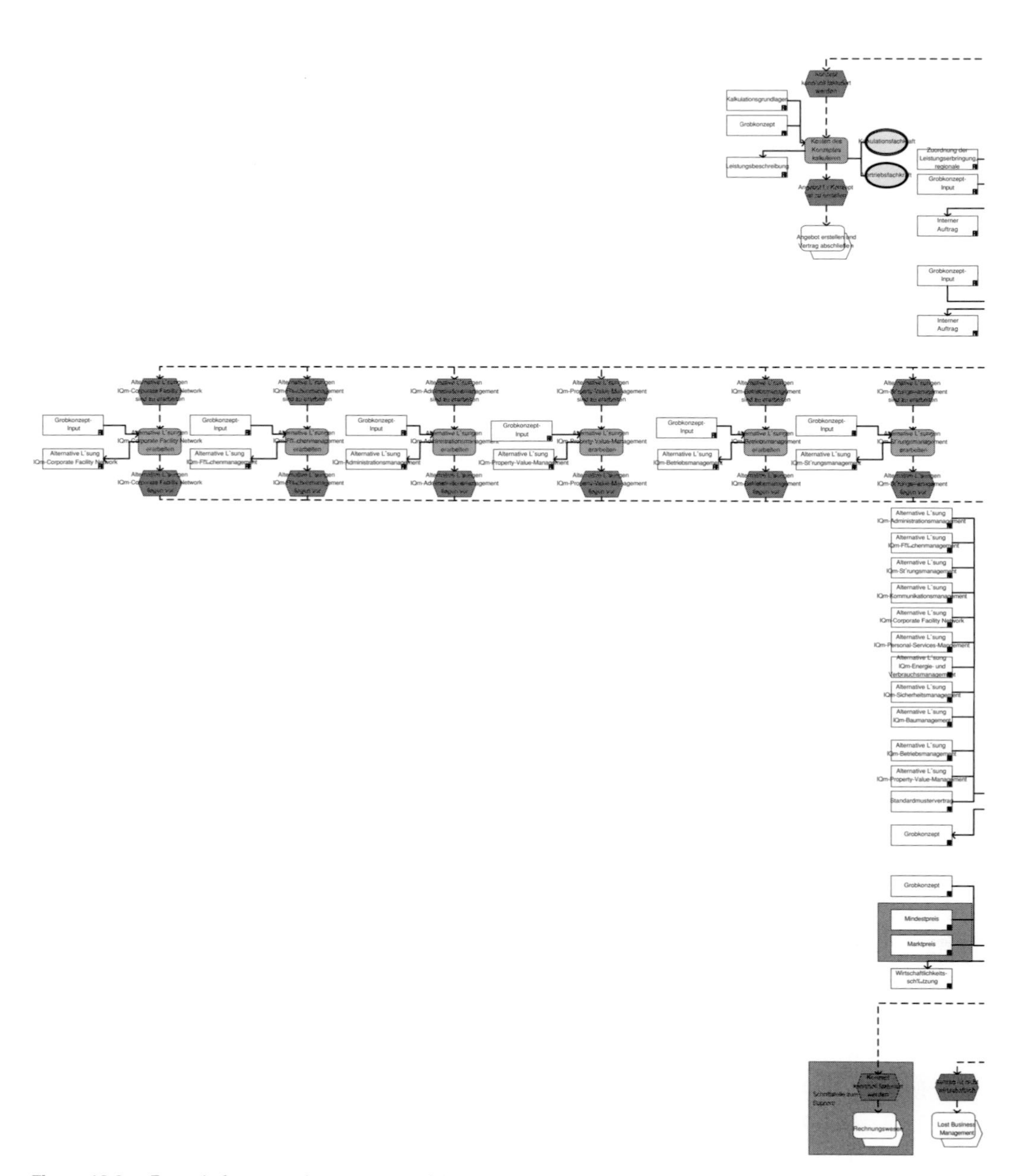

Figure 13.2 Example for a complex process models

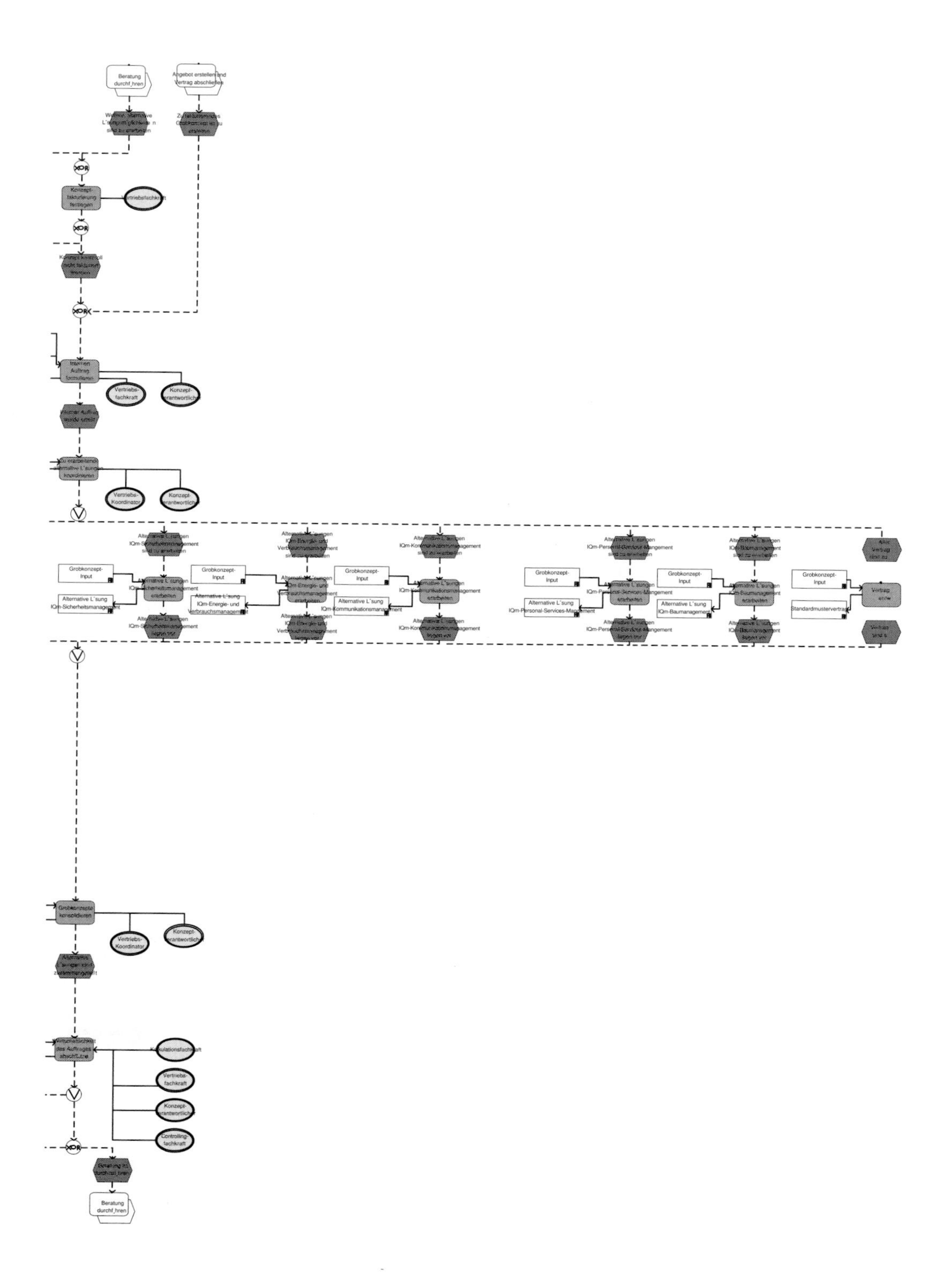

Beratung durchf_hren
Grobkonzept-Input
Alternative L`sung IQm-Sicherheitsmanagement
Grobkonzept-Input
Alternative L`sung IQm-Energie- und Verbrauchsmanagement
Grobkonzept-Input
Alternative L`sung IQm-Kommunikationsmanagement
Grobkonzept-Input
Grobkonzept-Input
Grobkonzept-Input
Standardmustervertrag
Vertriebs-fachkraft
Vertriebs-Koordinator
Vertriebs-Koordinator
Vertriebs-fachkraft
Controlling-fachkraft
Beratung durchf_hren

gives an indication of the complexity, but also the well-defined structure of such process models.

Based on the designed process models of the existing situation, a team of business analysts, IT staff, involved business representatives and, usually, external consultants analyse the processes in detail. They identify the objectives of the process and list current shortcomings. At this stage, the project team is dealing with the following questions:

- Which aim does the process have?
- Why is the process executed as it is?
- Which organizational units and application systems are involved – and where are the interfaces between them?
- What are the current problems?
- Which changes independent from the process engineering project will take place soon (internal/environment)?
- What technology is (will be) available?
- What are the current benchmarks?

System selection

Process models that describe the core requirements of a company are input for the system selection phase. Regarding enterprise systems, it is more important to have a precise description of the *critical* and *unique* requirements than a complete description of all processes. In general, it can be expected that ES software supports typical business processes, such as basic payment procedures or maintaining material master records.

The selection of enterprise systems follows existing procedures for software selection. Process models can help when taking the special requirements of the desired sequence of activities into account. For example, an enterprise system might offer support to purchase orders that refer to a cost center. However, this might be mandatory in one enterprise system, while a company may be interested in refering to a cost center at a later stage in the process. In such a case, the constraint is not the functionality itself, but the sequence of functions. Process models can help to add this requirement to the selection process. A major benefit is that the existence of a certain functionality is checked, *and* that the ES software is confirmed as supporting the desired sequence of functions (Kirchmer, 2000).

A strong motivation for enterprise system vendors in developing reference models for their solutions was to support the model-based selection of their systems. In such an approach the enterprise-individual models are compared with the ES reference models. Such a model comparison has to deal with:

- different levels of abstraction in the models,
- different scope (length and width) of the processes,
- differences in additional information (organizational units, input data, documents, related transactions), and
- different naming.

These differences make an (semi-) automated comparison of models very difficult. Thus, enterprise system reference models are in most cases only used to provide insight into the software functionality. As ES reference models are typically on a more detailed level of abstraction that the models designed during the business engineering stage, they can be used to further detail the models. This approach takes into account that ES software will offer new, unknown, but often better solutions than those designed during the business engineering stage. Consequently, the most appropriate ES solution is not the one that supports the defined requirements in the best way, but the one that goes beyond satisfying the requirements and adds to the business. The results of these process comparisons are, in weighted form, input for the entire software selection process.

Independent of how the individual and the ES-specific reference models are compared, the different situations depicted in Figure 13.3 can be distinguished. Point A represents the ideal, in which the requirements of the company are fulfilled through the complete use of software functionality. Point B offers additional opportunities. As the requirements are fulfilled, but the ES software functionality includes potential beyond the requirements, the initial requirements might be redefined. Point B describes the situation, in which ES software and its promise of at least 'better practice' contributes to business

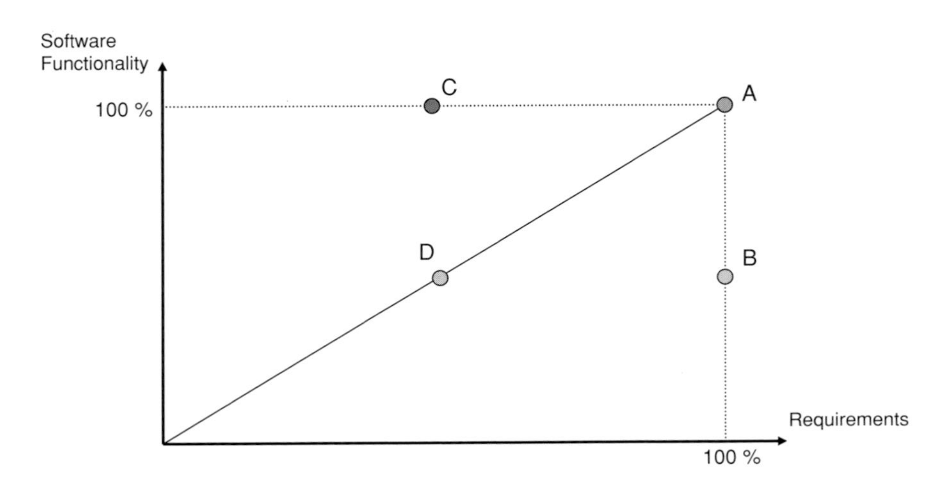

Figure 13.3

improvement. In contrast to this, point C describes the opposite situation. Although the software is used up to its functional limits, the requirements are not fulfilled. Usually, these are individual business requirements. A decision is then required on whether the processes should follow the restrictive software or whether the software should be modified. This decision can only be based on knowledge about the strategic importance of the process. Many companies in this situation decide to obey the principle that 'processes follow software'. Finally, point D is the case in which the requirements are not fulfilled, but the software is not yet completely utilized. This is a temporary situation. An example of this would be a procurement process in a company that currently runs only the financial accounting (accounts payable) module of an enterprise system. As soon as the procurement module is implemented as a part of materials management, D would change to one of the three points A, B, or C.

System implementation

Once the system is selected and necessary project management tasks are completed, system implementation and the required organizational changes can take place. The process of individualizing enterprise systems is called *configuration management*. This phase requires process models which are more detailed than the ones designed during the system selection.

As a system has been selected, it is possible to use existing Enterprise System reference models. Various implementation tools (for example, ARIS Toolset) offer a comfortable navigation through enterprise system reference models and enable changes to be made to these models. It is possible to trigger ES transactions from the modelling tool. *Vice versa*, relevant process models can be opened from the enterprise system.

Available reference models focus on a description of the *execution* of processes. This is important, for example, for the documentation of the new processes for system users. The project team, however, is interested in a process-oriented description of the possible *configurations* of the ES processes. This would support the discussion of alternative process scenarios and highlight required decision points in the reference models.

In describing the inherent configuration opportunities and constraints, two extensions of existing reference modelling techniques are suggested. First, reference process models could be enriched with further symbols so that the configuration potential becomes transparent. Second, reference process models can be linked to highlight dependencies between processes. Both cases will be discussed in the following, using event-driven process chains as an example.

The objective of this extended modelling technique is to describe alternatives in one reference model. As an example, the configuration of the optional system

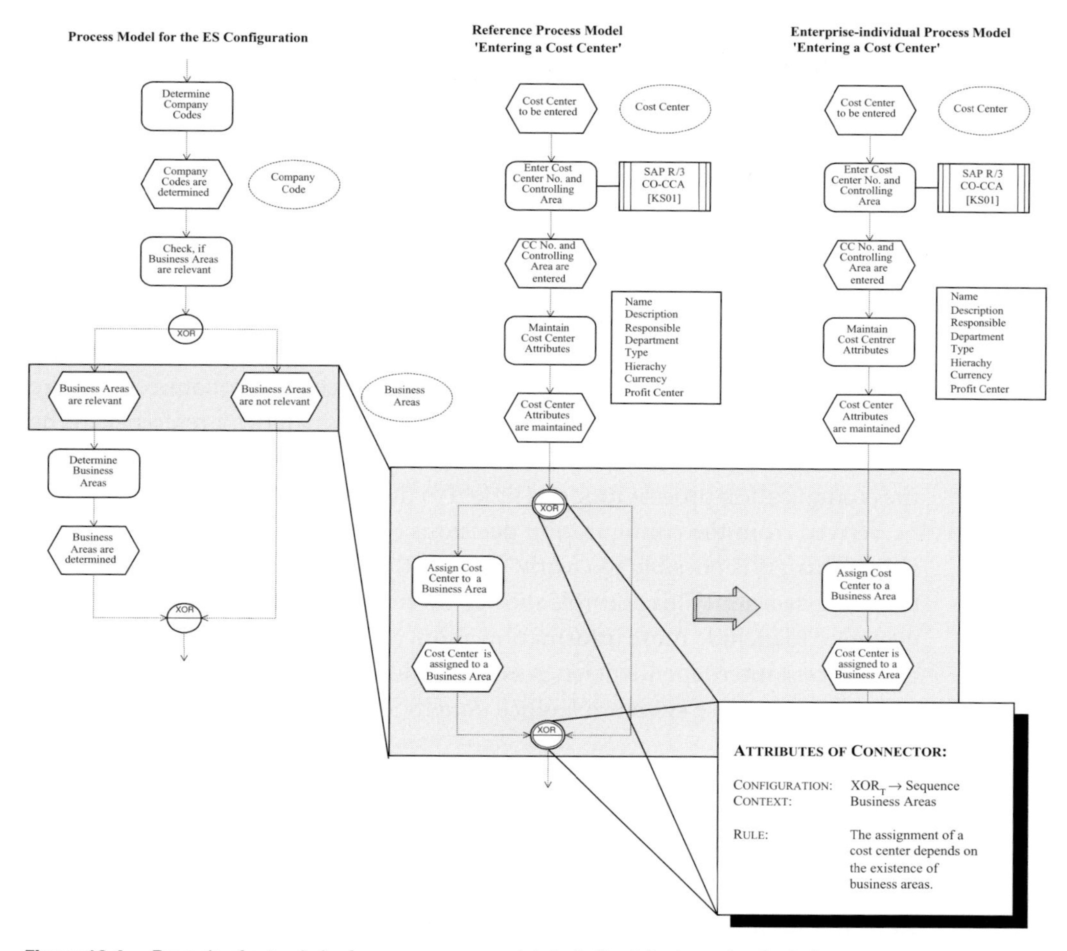

Figure 13.4 Example of extended reference process models including interdependencies between processes

organizational unit 'business area' within SAP's financial accounting solution (SAP-FI) is used. Business areas in SAP R/3 are defined as 'the organizational unit in external accounting that corresponds to a selected area of activity or responsibility within an organization to which the value movements entered in financial accounting can be assigned' (SAP online documentation).

Although business areas are defined in SAP-FI, they are a part of most SAP modules. Consequently, the decision about the business area influences many processes in several areas of SAP R/3. This influence, however, is not shown in the SAP reference process models.

Figure 13.4 (left side) includes a (new) model for the relevant configuration process, in which the decision about the use of business areas is made. This process is strictly sequential as long as mandatory organizational units are configured (for example, for the names of legal entities, the SAP term is 'company code'). Decisions about optional organizational units are depicted as 'check functions'. After the decision has been made (for example, 'Business areas are not relevant'), the configuration process for the organizational units takes place automatically.

Entering a new cost center is a process which can depend on the decision about the business area. If the business area is active, every cost centre may have to refer to a business area. Therefore, the configured model for entering new cost centers either includes the assignment of a cost center to a business area, or it does not. Consequently, the reference process model 'entering a cost center' has to depict both possibilities. A special new connector – the 'XOR' connector in two circles – is required, which includes a reference to the configuration process in which the decision has to be made. On the other side, the configuration process model is linked to the operational processes that can be derived from the configuration decisions (Figure 13.4, model on the right side). Thus, it is possible to clearly identify the influence of a particular customizing decision. This example shows how reference process models could be extended to include more information about actual configuration possibilities and process interdependencies. A corresponding discussion for the configuration of enterprise system reference *data* models can be found in Rosemann (2001b).

System use and change

At this stage process models are helpful for end-user documentation. Standardized reference process models have to be extended with individual organizational units, objectives, customer interfaces, required knowledge, and further forms. These models can be offered on the Intranet with links to the relevant ES transaction or the ES online help.

If no comprehensive process modelling took place in the system selection or implementation phase, process models can be derived from the flow of document types in the enterprise system (IDS, 2001). Elaborated monitoring and controlling tools identify the flow of objects, such as sales orders or purchase requests through the system, and derive graphical models based on the sequence of transactions that are typically performed on these objects.

Further, process models can capture relevant process attributes. The standardized process execution within enterprise systems enables storage of process-related logfiles, in order to get useful process performance data out of the system. These data can be clustered in three areas, which are all maintained over time:

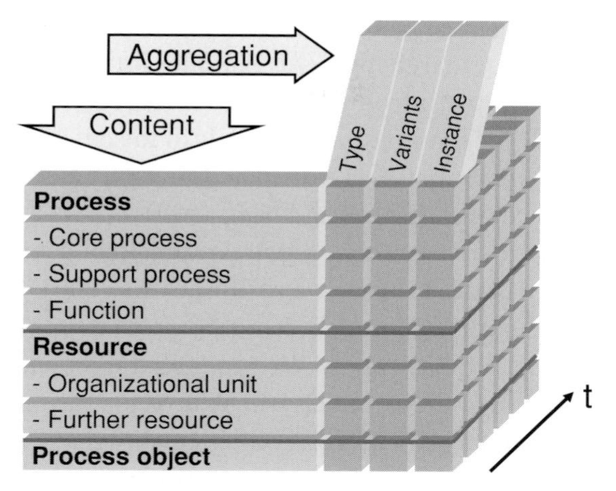

Figure 13.5 Dimensions of a process performance database

- *Process-related data* are directly linked to the business processes and include information about the time, the costs, or the quality of a process. A possible indicator for the process quality might be the number of customer complaints. This data can be further differentiated regarding the process type (core or support process, entire process, or single function).
- Data related to the *resources* include information about the utilization of the involved organizational units, roles, and further resources (for example, printer) and their appropriateness.
- Finally, the *objects* that are processed, such as orders, invoices, or payments, can be analysed and potential complexity drivers can be identified. These can be, for example, the characteristics of incoming invoices that are continuously incorrect.

Such data can populate parts of a data warehouse dealing with process performance indicators. Figure 13.5 shows some dimensions of such a process performance database. Besides the content (process, resource, or object) and the time, performance can also be differentiated on a third dimension, whether the data are on the level of a process model (type), for a variant of this model (for example, domestic vendors) or for one specific data instance in the model.

In this case, an enterprise system has the role of a knowledge repository from which relevant data can be easily extracted. These process-related performance indicators can be embedded in comprehensive EIS approaches like the Balanced Scorecard (Rosemann, 2001a).

The *system change or system evolution phase* includes minor system modifications, such as a stabilization (Ross, 2000) that directly follows the going-live date, or major changes triggered by organizational changes (for example, acquisition of a company), external changes (for example, introduction of GST), technical changes (for example, introduction of a web-based user interface) or product-driven changes (for example, new upgrade of the enterprise system). Change can be further differentiated as anticipated change, emergent change, and opportunity-based change, (Sieber and Nah, 1999).

Process (model) change management requires a precisely defined responsibility for the models (Becker, Kugeler, and Rosemann, 2001). Only with someone in charge of the process models can it be guaranteed that they are continuously maintained. A typical approach is to offer the process models on an Intranet and to expect that these models will be used for various purposes, such as human resource management, controlling, quality management, etc., at any time. If the process models are available on the Intranet, it is possible to receive feedback about necessary or possible changes or to discuss potential changes in a related newsgroup. Further, related documents (texts, presentations, calculations, etc.) can be attached. Process models are therefore not only of temporary use during the definition of the requirements, but can serve as a continuous process-oriented knowledge repository that consolidates relevant information around a process description.

13.2 Conclusion and Outlook

Research on enterprise systems has been focused on implementation issues and still lacks agreement on a comprehensive lifecycle model. Enormous resources are invested in enterprise system implementation, but these efforts drop rapidly after the going-live date. This chapter proposed the increased use of extended reference process models in all main phases of the management of an enterprise system.

Currently, a second wave of process modelling efforts can be observed, based on the fact that many companies struggle to demonstrate that they are getting value out of their ES implementation. Often they are aware that the ES software is not used in an optimal way, for example, that they are not using all relevant ES functionality. Process modelling activities based on ES-specific reference process models might be an efficient approach to re-document, analyse and improve existing processes.

Two further developments demonstrate the increased importance of process modelling in the context of enterprise sytems.

After most of the *Fortune 1000* companies implemented ES (GartnerGroup, 1999), the relevant market moved towards small and medium-sized enterprises (SMEs). SMEs, however, typically have less methodological know-how about process modelling and a lower budget for related consultancy. Reference process models can help to bring these companies up to speed with software functionality. They also serve as a starting point for more individual process documentation of the company.

Another important new market for ES vendors will be the support of processes that go beyond the scope of one enterprise. Concepts including customer relationship management and supply chain management require new ways of designing models for collaborative business processes. First approaches exist already, which describe the interrelation between the business partners in terms of process flow and also the exchange of services and products.

ACKNOWLEDGEMENT

An earlier version of this chapter appeared in the Australian Accounting Review, Vol. 10, No. 3, November 2000, pp. 19–30, titled 'Using Reference Models within the Enterprise Resource Planning Lifecycle'.

REFERENCES

Bancroft, N. (1997) *Implementing SAP R/3: How to Introduce a Large System into a Large Organization.* 2nd edn., Prentice Hall.

Basu, C. and P. C. Palvia (1999) Towards Developing a Model for Global Business Process Reengineering. In Proceedings of the 5th Americas Conference on Information Systems, Haseman, W. D. and Nazareth, D. L. (eds), 13–15 August, Milwaukee.

Becker, J., Kugeler, M., and Rosemann, M. (2001) *Process Management.* Berlin: Springer-Verlag.

Becker, J., Rosemann, M., and von Uthmann, Chr. (2000) Guidelines of Business Process Modeling. In *Business Process Management: Models, Techniques and Empirical Studies*, van der Aalst, W., Desel, J., and Oberweis, A. (eds), Berlin: Springer-Verlag, pp. 30–49.

Bernroider, E. and Koch, S. (1999) Decision Making for ERP Investments from the Perspective of Organizational Impact – Preliminary Results from an Empirical Study. In Proceedings of the 5th Americas Conference on Information Systems, Haseman, W. D. and Nazareth, D. L. (eds), 13–15 August, Milwaukee.

Bingi, P., Sharma, M. K., and Godla, J. K. (1999) Critical Issues Affecting an ERP Implementation. *Information Systems Management*, Summer, 7–14.

Brinkkemper, S. (1999) Requirements Engineering for ERP: Requirements Management for the Development of Packaged Software. Proceedings of the 4th International Symposium on Requirements Engineering, 7–11 June, Limerick, Ireland.

Curran, T. and Keller, G. (1998) *SAP R/3 Business Blueprint.* Upper Saddle River: Prentice Hall.

Daneva, M. (2000) Practical Reuse Measurement in ERP Requirements Engineering. Proceedings CAISE, pp. 309–324.

Davenport, T. D. (1993) Process Innovation: Reengineering Work through Information Technology. Boston.

Dewan, R., Seidmann, A., and Sundaresan, S. (1995) Strategic Choices in IS Infrastructure: Corporate Standards Versus 'Best of Breed' Systems. Proceedings of the 16th International Conference on Information Systems, 10–13 December, Amsterdam.

Gable, G. G., Heever, R., Erlank, S. and Scott, J. (1997) Large Packaged Software: The Need for Research. Proceedings of the 3rd Pacific Asia Conference on Information Systems, 1–5 April, Brisbane.

Gable, G. G., Scott, J., and Davenport, T. (1998) Cooperative EWS Life-cycle Knowledge Management. In Proceedings of the 9th Australasian Conference on Information Systems, Edmundson, B. and Wilson, D. (eds), 29 September–2 October, Sydney.

Gartner Group (1999) Enterprise Resource Planning Vendors. In Proceedings of Symposium/itExpo, 19–22 October, Brisbane.

Gulla, J. A. and Brasethvik, T. (2000) On the Challenges of Business Modeling in Large Scale Reengineering Projects. Proceedings of the 4th International Conference on Requirements Engineering, Schaumburg, IL., 19–23 June, 17–26.

Hammer, M. and Champy, J. (1993) *Reengineering the Corporation. A Manifesto for Business Revolution.* New York: Harperbusiness.

Holland, C. P. *et al.* (1999a) A Critical Success Factors Model for Enterprise Resource Planning Implementation. In Proceedings of the 5th Americas Conference on Information Systems, Haseman, W. D. and Nazareth, D. L. (eds), 13–15 August, Milwaukee.

Holland, C. P. *et al.* (1999b) Beyond Enterprise Resource Planning Projects: Innovative Strategies for Competitive Advantage. In Proceedings of the 7th European Conference on Information Systems, Pries-Heje, J. et al. (eds), 23–25 June, Copenhagen.

IDS (2001) Process Performance Manager. White Paper, Saarbrücken.

Kim, Y. G. (1995) Process Modeling for BPR: Event-Process Chain Approach. Proceedings of the 16th International Conference on Information Systems, 10–13 December, Amsterdam.

Kirchmer, M. (2000) *Business Process Oriented Implementation of Standard Software.* 2nd edn, Berlin: Springer-Verlag.

Klaus, H., Rosemann, M. and Gable, G. G. (2000) What is ERP? *Information System Frontiers*, **2**(2), 141–162.

Light, B. *et al.* (2000) Best of Breed IT Strategy: An Alternative to Enterprise Resource Planning Systems. In Proceedings of the 8th European Conference on Information Systems, Hansen, H. R. (ed.), 3–5 July, Vienna.

Markus, L., and Tanis C. (2000) The Enterprise Systems Experience-From Adoption to Success. In *Framing the Domains of IT Research: Glimpsing the Future Through the Past*, Zmud, R. W. (ed.), Cincinnati: Pinnaflex Educational Resources Inc.

Nah, F., Lau, J., and Kuang, J. (2001) Critical Factors for Successful Implementation of Enterprise Systems. *Business Process Management Journal*, **7**(3).

Pérez, M. et al. (1999) SAP, Change Management and Process Development Effectiveness (II): Case Study. In Proceedings of the 5th Americas Conference on Information Systems, Haseman, W. D., Nazareth, D. L. (eds), 13–15 August, Milwaukee.

Rolland, C. and Prakash, N. (2000) Bridging the Gap between Organisational Needs and ERP Functionality. *Requirements Engineering*, **5**(3), 180–193.

Rosemann, M. (1999) ERP Software: Characteristics and Consequences. In Proceedings of the 7th European Conference on Information Systems, Pries-Heje, J. et al. (eds), 23–25 June, Copenhagen.

Rosemann, M. (2001a) Evaluating the Management of Enterprise Systems with the Balanced Scorecard. In *Information Technology Evaluation Methods and Management*, v. Grembergen, W. (ed.), Hershey: IDEA Group Publishing, pp. 171–184.

Rosemann, M. (2001b) Requirements Engineering for Enterprise Systems. In Proceedings of the 7th Americas Conference on Information Systems, Strong, D. and Straub, D. (eds), 3–5 August, Boston.

Rosemann, M. and Chan, R. (2000) A Framework to Structure Knowledge for Enterprise Systems. Proceedings of 6th Americas Conference Information Systems, 10–13 August, Long Beach, CA, pp. 1336–1342.

Ross, J. W. (2000) The ERP Revolution: Surviving versus Thriving. Working paper, Center for Information Systems Research, Sloan School of Management, MIT.

Sandoe, K., Corbitt, G., and Boykin, R. (2001) *Enterprise Integration*. New York: John Wiley & Sons.

SAP (2000) ValueSAP. White Paper, Walldorf.

Sato, R. (2000) Quick Iterative Process Prototyping: A Bridge over the Gap between ERP and Business Process Reengineering. In Proceedings of the 4th Pacific Asia Conference on Information Systems, Thong, J., Chau, P., and Tam, K. Y. (eds), 1–3 June, Hong Kong.

Scheer, A. W. (1998) *Business Process Engineering*. Berlin: Springer-Verlag.

Scott, J. E. (1999) The FoxMeyer Drugs Bankruptcy: Was It a Failure of ERP? In Proceedings of the 5th Americas Conference on Information Systems, Haseman, W. D. and Nazareth, D. L. (eds), 13–15 August, Milwaukee.

Shanks, G. et al. (2000) Differences in Critical Success Factors in ERP Systems Implementation in Australia and China: A Cultural Analysis. In Proceedings of the 8th European Conference on Information Systems, Hansen, W. (ed.), 3–5 July, Vienna.

Sieber, M. M. and Nah, F. H. (1999) A Recurring Improvisational Methodology for Change Management in ERP Implementation. In Proceedings of the 5th Americas Conference on Information Systems, Haseman, W. D. and Nazareth, D. L. (eds), 13–15 August, Milwaukee.

Slooten, K. and Yap, L. (1999) Implementing ERP Information Systems using SAP. In Proceedings of the 5th Americas Conference on Information Systems, Haseman, W. D. and Nazareth, D. L. (eds), 13–15 August, Milwaukee.

Smethurst, J. and Kawalek, P. (1999) Structured Methodology Usage in ERP Implementation Projects: An Empirical Investigation. In Proceedings of the 5th Americas Conference on Information Systems, Haseman, W. D. and Nazareth, D. L. (eds), 13–15 August, Milwaukee.

Stefanou, C. J. (1999) Supply Chain Management (SCM) and Organizational Key Factors for Successful Implementation of Enterprise Resource Planning (ERP) Systems. In Proceedings of the 5th Americas Conference on Information Systems, Haseman, W. D. and Nazareth, D. L. (eds), 13–15 August, Milwaukee.

Sumner, M. (1999) Critical Success Factors in Enterprise Wide Information Management Systems Projects. In Proceedings of the 5th Americas Conference on Information Systems, Haseman, W. D. and Nazareth, D. L. (eds), 13–15 August, Milwaukee.

Volkoff, O. (1999) Using the Structural Model of Technology to Analyze an ERP Implementation. In Proceedings of the 5th Americas Conference on Information Systems, Haseman, W. D. and Nazareth, D. L. (eds), 13–15 August, Milwaukee.

14 An ERP Implementation Case Study from a Knowledge Transfer Perspective

Zoonky Lee and Jinyoul Lee

14.1 Introduction

Knowledge transfer is an important topic in management literature. Some studies have focused on the transfer of best business practices in multinational corporations as they transfer accumulated knowledge from their headquarters to new foreign affiliates (Ghoshal and Nohria, 1989; Gupta and Govindarajan, 1991; Kogut and Zander, 1993). Other studies have focused on the transfer of technology, business processes, and best practices inside a firm (for example, Argote, 1999). However, current studies have not focused on the unique knowledge transfer practices involved in an Enterprise Resource Planning (ERP) implementation.

ERP is an enterprise-wide application software package. In ERP, all necessary business functions, such as financial, manufacturing, human resources, distribution, and order management, are tightly integrated into a single system with a shared database. While customization is not impossible, the broad scope and close connectivity of all related functions make customization very costly (Davenport, 1998; Davis, 1998). The high cost and long implementation process of customization result in most organizations needing to align their business processes with the functionality provided by the ERP program rather than customizing the ERP package to match their current processes. According to Forrester Research, only 5% of organizations, among Fortune 1000 companies that had purchased an ERP application, customized it to match their business processes (Davis, 1998). Therefore, the implementation of ERP entails using the business models included in the package (Slater, 1998). In other words, business knowledge incorporated in the basic architecture of the software is transferred into the adopting organization.

Most ERP systems contain business process reference models. These are often are called 'best practices' because they have been implemented and proven in many world-class companies. For instance, SAP R/3, one of the most popular ERP packages on the market, contains as many as 1000 business reference models from different industries. When these reference models are mapped to an organization's business, process, knowledge is transferred.

This process of knowledge transfer has two unique characteristics. First, it is a computer-package-based knowledge transfer in which only coded business processes are transferred. In previous studies of multinational corporations and internal transfer of best practices, business practices are transferred either through relocation of human resources, training, and documentation, or by a combination of these vehicles. Studies have indicated the organizational transfer of knowledge is much more complex than the mere imitation of visible processes (Lam, 1997; Szulanski, 1996). Second, the scope of knowledge transfer is company-wide as the term enterprise system suggests. In ERP systems, all organizational business functions are tightly connected with each other. Knowledge transfer is not confined to a specific business function and the degree of adoption in one functional area will greatly influence other functional areas.

This chapter is a part of longitudinal study based on the ERP implementation at the University of Nebraska. Based on an in-depth analysis of the early stages of this implementation, this chapter identifies the types of knowledge transferred during an ERP implementation and the factors affecting this transfer. It also investigates how the business rules, transferred from an ERP implementation, interacted with existing environments. The findings in this study propose a new approach to analysing ERP implementations from a knowledge transfer perspective. They also contribute to a better understanding of how to gain a competitive advantage based on process knowledge when standardized business processes are implemented by an organization.

14.2 Knowledge Transfer of Business Processes

Previous Studies on Knowledge Transfer

Previous studies (Brown and Duguid, 1991; Nonaka and Konno, 1998; Polanyi, 1962, 1966) indicated that there are two types of knowledge in the context of knowledge sharing and transfer. Based on its visibility and expressiveness, knowledge can be said to be either explicit or tacit. If knowledge is visible and expressible, it is called explicit knowledge. Explicit knowledge is transmittable in a formal, systematic way. Explicit knowledge can be 'processed' by a computer, transmitted electronically, and stored in a database (Nonaka and Takeuchi, 1995). Tacit knowledge, on the other hand, is associated with individual experiences and cannot be codified. Because it is subjective and intuitive, it is not easily processed or transmitted in any systematic or logical manner.

Can knowledge be transferred when business processes are mapped and imitated? If so, what knowledge is transferred and what is not? One important step in answering these questions is to understand how business processes differ. Brown and Duguid (1991) presented two different classifications of business processes: canonical and non-canonical. Canonical processes, based on an abstract representation of the organization, originated with Taylor's scientific management in which complex tasks are mapped to a set of simple canonical steps. Non-canonical processes, on the other hand, are what actually happen during work. They are the informal processes defined by the relationship, communication, and coordination of on-the-job practices.

A similar classification of processes can be found in Sachs' (1995): 'organizational, explicit' and 'activity-oriented, tacit' processes. The organization-based process, represented by sets of defined tasks and operations, differs from the activity-based process that determines how employees actually make the business function more effectively. Sachs observed that it is through workers' relationships in 'communities' and within their human systems that problems are discovered and resolved, and work is effectively accomplished.

Process-based Knowledge Transfer in ERP Implementation

The implication of this conceptual clarification is that ERP-based knowledge transfer has two different aspects. On the one hand, explicit knowledge transfer is based on canonical processes. The reference models in ERP systems are formal chains of activities represented by process modeling tools. These processes are visible and explicitly coded in the software. Studies have shown that a high degree of codification can increase the speed of transfer (Zander and Kogut, 1995).

On the other hand, ERP-based knowledge transfer entails non-canonical processes associated with tacit knowledge. Although an ERP system is a software package, it embodies established ways of doing business. Studies have begun to observe that an ERP system is not just a pure software package to be tailored to an organization but an organizational infrastructure that affects how people work (Hanseth and Braa, 2000) and it 'imposes its own logic on a company's strategy, organization, and culture' (Davenport, 1998). Non-canonical processes are related to a difficult-to-migrate portion of organizational knowledge, which is deeply embedded in complex social interactive relationships within organizations (Badaracco, 1991), and have a property of stickiness (Szulanski, 1996). ERP implementation, therefore, transmits business knowledge related to the business processes, and this type of knowledge is not necessarily easily transferred to the adopting organization. In sum, implementation of ERP entails tacit knowledge transfer as well as explicit knowledge transfer.

Studies have also reported that knowledge transfer is best achieved through intimate communication or relationships such as franchise (Darr, Argote, and Epple, 1995), research alliance (Powell, Koput, and Smith-Doerr, 1996) and strategy alliance (Uzzi, 1996) between a source and a recipient. There is no collaborative social arrangement such as narration, communication, or collaboration between a source and a recipient in the ERP knowledge transfer process. This means that the ability of a recipient to adjust their existing business fundamentals (for example, culture, values, and norms) to the invisible part of principles embedded in the business models will play a crucial role in this type of knowledge transfer.

The clear distinction between the two different types of knowledge transfer related to an ERP implementation leads us to identify two different ERP implementation phases: implementation and internalization. Implementation is defined as the 'degree to which the recipient unit follows the formal rules implied by the practice' Internalization refers to the 'state in which the employees at the recipient unit attach symbolic meaning to the practice' (Kostova, 1999).

When an ERP package is implemented by an organization, the canonical processes can be relatively easily mapped and transferred: the process of implementation. The non-canonical processes transmitted from reference models should be merged with the existing values of the organization to complete the knowledge transfer: the process of internalization. In this internalization process, the current values of the organization may conflict with the new values. The conflict results in organizational resistance, creating a gap between the designed model and the actual process (Berger, Sikora, and Berger, 1994). Only when the non-canonical processes have been internalized with existing organizational practices, will organizational members begin to appreciate the value of the system.

Following the conceptual development discussed earlier, we suggest analysing an ERP implementation from a knowledge transfer perspective as a research model (see Figure 14.1 below). First, canonical knowledge is transferred based on reference models accompanied with business rules. Second, canonical knowledge is accepted in the organization but the business rules associated with the reference models begin to conflict with existing business

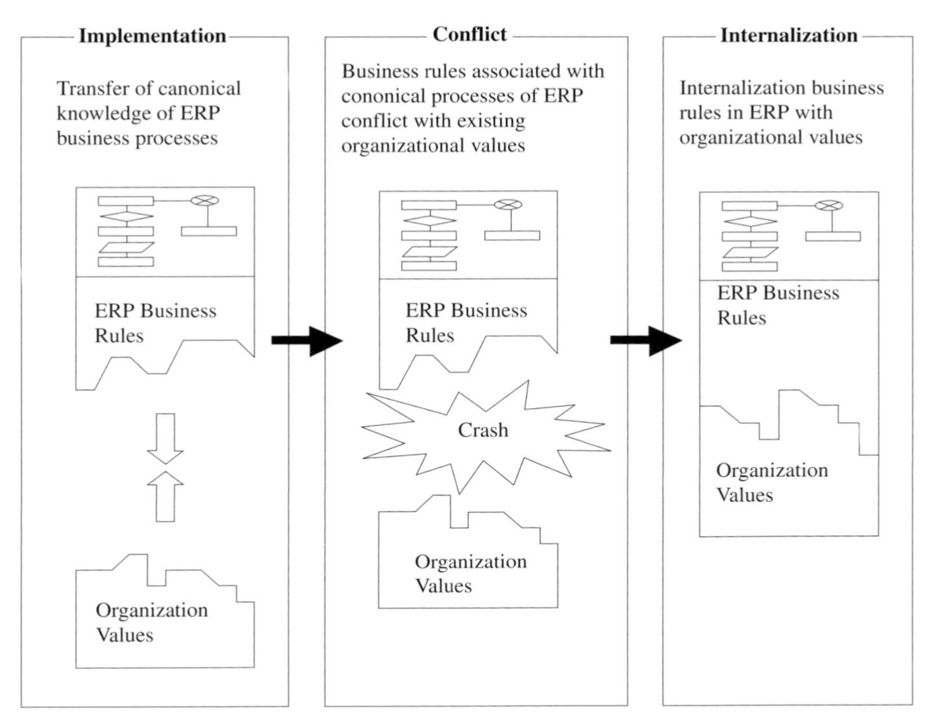

Figure 14.1 Research model

fundamentals of the organization. Third, the organization tries to internalize the business values from the ERP package with its existing business practices. Such a process was observed at the University of Nebraska when they implemented SAP R/3.

14.3 Research Method

On 5 September 1997, the University of Nebraska decided to purchase the R/3 client–server software system from SAP in order to resolve their legacy system issues, such as limited functionality and Y2K problems. On 1 July 1999, after two years of extensive consulting and system configuration, the University of Nebraska's four-campus system began using SAP R/3 for its external and internal accounting functions as well as its purchasing and inventory functions. A more detailed case study is available from Sieber et al. (2000). The study reported here is based a three-month, in-depth analysis conducted during the early implementation phase at the University of Nebraska, Lincoln campus.

Among the different business processes implemented, this study focused on the purchasing process. Since this process involved limited system modification, it provided us with an ideal case to observe direct knowledge transfer from ERP reference models. Additionally, the process was selected because it is relatively easy to understand and is cross-functional, impacting end-users as well as the accounts payable and purchasing departments.

This study is based on in-depth interviews, process analysis, and documentation analysis. First, for the purpose of understanding and analysing the business processes, relevant documents such as requirements analysis, training materials, fit and gap analysis, future business process impacts, material management configuration, and material management procedures were obtained. Both old and new processes were mapped and analysed based on the materials. Second, both existing documents and interviews were used to identify types of knowledge transferred and new organizational requirements under the new model. A total of ten interviews were conducted: five with departmental users (mostly department secretaries), three with purchasing employees, and two with accounts payable employees. Each interview lasted 60 to 90 minutes. All interviews were recorded with permission of the interviewees and transcribed at a later date. The format of the interviews was semi-structured; prepared questions were asked and answered in an open-ended manner.

14.4 Analysis

Analysis of Old Process vs. New Process

For the purchasing function, the University adopted standard ERP processes that were different in many aspects from the old processes. One primary difference between the two processes is that the new process is computer-based while the old process was paper-based. In the old process, for instance, the purchasing department (PU) hand entered a purchase order based on a requisition submitted manually by departmental users (USER). In the new process, a requisition is entered and electronically stored by the USER. The PU creates a purchase order instantly by pulling forward the information from the requisition to a purchasing order.

The difference in the two processes is not limited to automation. One of the most distinctive improvements is the elimination of the invoice matching process by the accounts payable department (AP). This was probably the most

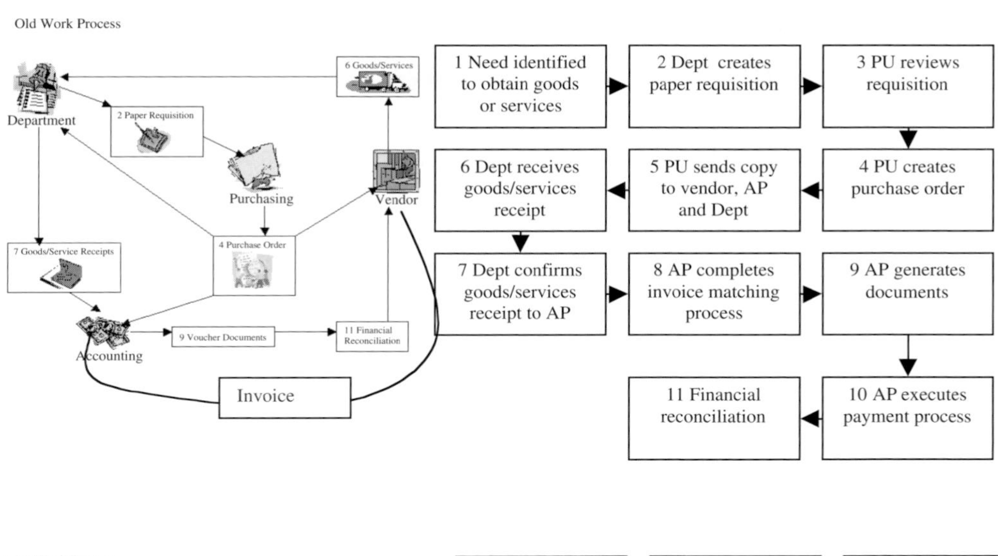

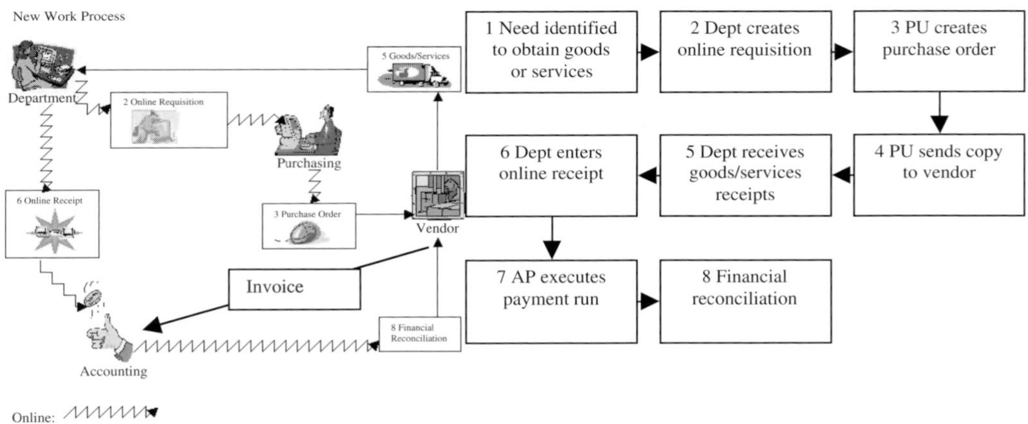

Figure 14.2 Purchasing process (Old vs. New process)
Notes: AP – Accounts payable
Dept – User department
PU – Purchasing department

time-consuming process in the entire purchasing process. In the old process when the USER received goods, the USER received a copy of the receipt and sent it to the AP. The AP then matched it to a copy of the original purchase order received from the PU and an invoice from the vendor. The payment process was then executed. In the new process, the USER reconciles the goods they receive with an electronic invoice, and AP executes the payment process. Figure 14.2 shows both processes, old and new.

Analysis of Interviews

We began our research shortly after the University had begun implementing the system. Initially, a parallel conversion strategy was required where users operated both the new and the old system. Adaptation to the new system was not easy for most users. They not only had to deal with the new system but also with the new business process that often required new business rules. We analysed the implementation process based on the three-staged research model in Figure 14.1.

Acceptance of the explicit business processes from the system (implementation process)

When tasks were paper based, people had much more flexibility. When a USER sent a requisition, the PU would sometimes contact vendors before processing a purchase order and then ask the USER to modify the requisition based on that contact. This procedure is different from their process of sending a requisition to PU and sending a purchase order to vendors (see the old process in Figure 14.2). Now once a requisition is entered into the computer system, the PU generates a purchasing order in the system, and then the PU faxes the purchase order to the vendor directly from the computer screen. One purchasing manager says:

> We don't have paper floating around. Because it's all in the system. When we create a purchase order, what is going to happen, is we create a purchase order to, say, WALMART, and the purchase order will be faxed from the computer to WALMART. We will not even get a hard copy of the P.O. It'll just be faxed.

Departmental users (USER) also experience the same inflexibility due to the automation. Traditionally, users identified and created vendor files for themselves. The new process, however, requires them to use vendors from an existing file. If they want, they can ask the PU to create a new vendor file but that takes time. The inherent business rule behind the process gives them little choice but to follow a strict new business process, unlike the old system, which allowed for many different process variations.

The most challenging dilemma the University of Nebraska SAP implementation team faced was how to resolve gaps that were identified between SAP's functionality and the University's desired business processes (Sieber et al., 2000). While the project implementation team tried to apply the functional models in the ERP package directly to the desired processes, not all organizational requirements were compatible with the business processes provided from the package. However, due to the limited technical flexibility in the package, the University has to reconcile their process to the ERP package. One manager in PU described it in this way:

> there are a few areas we had to go and customize. Sometimes, users told us that they don't like the way we do it, and they want to do it in totally a different way. But, we can't change code. . . . SAP gives you these options to pick a choice. So, what configuration is that SAP gives you, say, 10 ways to do something. You go through what they call repository and you choose.

Under the old system, all departments were very loosely coupled. Changes in one data set did not affect the processes of other activities. Under the new system, however, all the inputs and outputs are tightly connected. Data are shared from the same integrated data dictionary and they are all interconnected. Furthermore, the system often restricts changes to ensure integrity. A manager in AP said:

> It is more difficult to decide information in online. SAP just wants you to do cancellation or reverse and show every single cancellation or reverse. Because it is online, it is just automatically done. Then you have to clear some items out of financial systems because they are in purchasing modules. . . . You have to figure out that did it really get paid or canceled or what do you need to do with that.

New business knowledge from the ERP package (conflict process)

Our observations indicated that not only are explicit business processes of ERP imposed on an organization, but also the implicit non-canonical processes inherent in the ERP model begin to transmit into the organization. Often, this enforcement leads to different types of organizational requirements: role and responsibility redistribution, new knowledge requirements for people involved in the process, and a new knowledge structure in the organization.

After the ERP implementation, people in PU and AP generally felt that their jobs were deprived of meaning and responsibilities. In particular, the people in AP began to feel helpless as their important matching processes were discontinued. This feeling of helplessness may explain a higher-than-expected turnover rate following the implementation.

As some of the PU functions move to the USER, role redistribution is occurring. Within a specified budget limit, a USER can now create a purchase order directly. Previously, this was a PU function. The implementation of purchasing cards, a pre-approved card with which USER or individual users purchase goods also gives more control to the USER, leading one PU manager to believe that the PU department will eventually disappear.

The cross-functional interconnectivity requires people involved in a process to have more cross-functional knowledge of the process. A department staff member put it this way:

> you definitely have to have the bigger picture view now. Accounting and Purchasing have been real helpful on trying to get the information to you and everything, but um, and I think maybe that's one of the drawbacks of it too, is that people are so niched into their little place and function that, that was my problem anyways, I didn't ever see the big picture, I didn't really understand how everything was working together.

But the very fact that they should now be equipped with more cross-functional knowledge makes some users frustrated:

> Learning SAP is big. My focus is on purchasing. I don't know a lot about the accounting side or the project side. So, in first starting, just a core group of our people is getting involved with SAP and it is very overwhelming because, the software does everything. How do we integrate it so it still works between all the modules we need it to. It is difficult.

Organizational efforts to resolve conflicts (internalization process)

While the new business processes that accompany ERP involve new organizational requirements, organizations can make some changes to resolve those conflicts. Faced with the requirements for role and responsibility redistribution, for instance, the University began to make some adjustments. After the ERP implementation, people in PU and AP generally believed their jobs were deprived of meaning and responsibilities. In recognition of the problem, the University recently assigned the purchasing audit function to the AP. This function used to be accomplished by the state government.

Sometimes the adjustment may require changes to what people value in the organization. Since many employees are now required to have a broader scope of knowledge, they tend to actively seek knowledge sources for their new inquiries. One manager in AP, for instance, asserts that his knowledge of purchasing, vendors, and accounting resulted in more calls from departmental users than ever before. In the past, the inquiry calls made by departmental users were regarding vendor invoices they had received. Now, under the new system, users can verify the invoice from the system, making the inquiry process unnecessary. But some experienced managers, who are knowledgeable about the entire purchasing process, are actually receiving more inquiry calls since all processes are now more interconnected.

The University also made a systematic effort to help users share knowledge and learn about other business processes and technical knowledge related to the system. User group meetings in which users from different functional areas get together and share knowledge about other businesses were held regularly. Also the university developed an Internet website (http://slugo.uneb.edu) to assist users in using the system. The features of the site include not only online help, computer-based training and SAP news but also an interactive knowledge base that tracked help-desk submissions and their resolution.

14.5 Implications

There are many different ways to analyse an ERP implementation. One can use a technology diffusion perspective or a business process reengineering point of view. This study investigates the process from a knowledge transfer point of view. This view is consistent with current expert views on ERP implementations in which the implementations are believed to involve a process of mapping existing business processes from the ERP model (Parr and Shanks, 2000). First we conceptually discerned two different types of business process knowledge in relation to an ERP package. We then discussed how the concepts of implementation and internalization could be applied, from a knowledge transfer perspective, to the context of an ERP implementation along with two different knowledge types. Figure 14.3 shows factors related to ERP implementations.

This analysis began by looking at how ERP users accept explicit business processes specified from the package. Overall, users are required to follow the explicit business processes inherent in the ERP package for three reasons: the process being automated, the package has limited flexibility, and the cross-functional nature of the package. Since ERP systems are, by their very nature, integrated, it is not surprising to observe that these three factors play a role in implementing the system. What is especially interesting is how these factors force users to adopt the business processes inherent in the system. This enforced adoption is similar to 'process restriction' observed by Gattiker and Goodhue (2000), and the 'formative context' structure from Ciborra and Lanzara (1994). Initially, users are restricted by the nature of the system and the system becomes an emerging infrastructure that governs what they are doing (Hanseth and Braa, 1998).

In the next part of the analysis, we looked at how the business rules, associated with explicit business processes in the ERP package, conflict with existing organizational structures and values. The results indicated that the processes often require new roles and responsibilities for the organization and that the

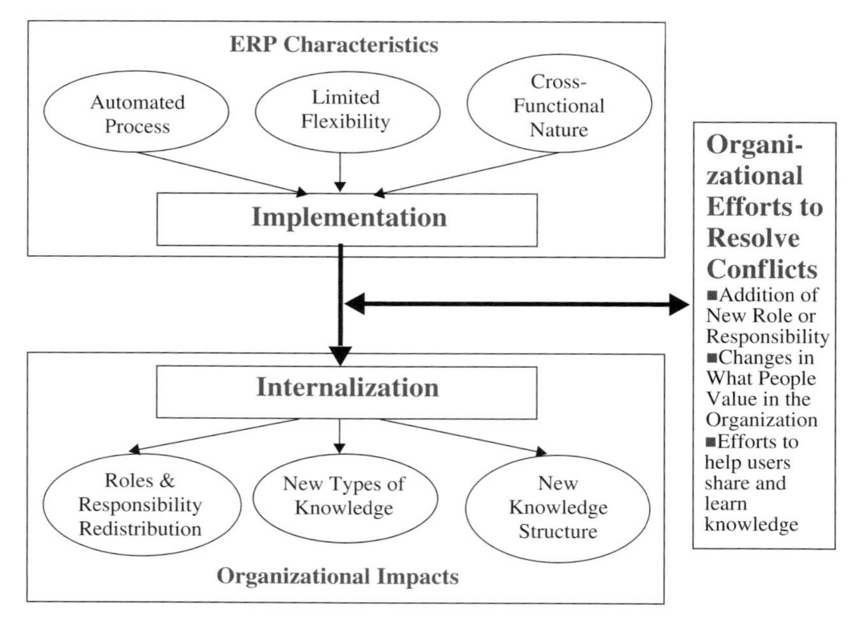

Figure 14.3 Factors related to ERP implementation from a knowledge transfer perspective

integrated nature of the ERP package requires more knowledge and coordination with other functional areas. Our data also illustrated some organizational efforts to resolve those conflicts as a result of new business knowledge imposed on the organization. Although this was an early stage of an implementation, we found both structural and value-change adjustments. The organization allocated new roles to the accounts payable department when its major invoice reconciliation role was eliminated.

Organizational members began to understand the requirements of a broader scope of knowledge, and people actively began seeking knowledge sources for their inquiries. This seeking process generates an organizational atmosphere that values organizational meta-knowledge, or knowledge about who knows what knowledge. This process is very similar to Baskerville's (1999) observation of ERP's impact on knowledge management. Users become more knowledgeable on what others are doing and the organizational knowledge structure changes to be more 'convergent' as they share the same types of knowledge. Each individual's knowledge becomes more 'divergent' as each subject expert needs to learn and understand other business areas: an IT expert needs more knowledge about business processes, and an expert in customer billing needs to understand other related functional areas as well as technical knowledge about the IT system.

14.6 Conclusion

Although this was an exploratory study based on one case, insights from this study provide three important contributions to our understanding of ERP implementations. First, we applied the models developed from knowledge literature (Badaracco, 1991; Brown and Duguid, 1991; Szulanski, 1996) and provide a new approach for observing ERP implementations. This view helps us observe differences in ERP implementations from other technology implementations. The question of technology diffusion has been of major interest in many organizational studies for quite some time. A number of studies have proposed implementation stages similar to implementation/internalization (for example, Cooper and Zmud, 1990). ERP implementations, however, differ from other technologies with respect to the extent to which an ERP influences organizational practices. Barley's observation of the adoption of CT scanners in the hospital (Barley, 1990) or Orlikowski's investigation of computer CASE tool's in system development (Orlikowski, 1993) process, for instance, showed how the technologies interact with existing organizational values, practices, and agents, gradually changing the relationships among the agents and structures of the organization. In both cases, the technologies did not specifically govern the organizational processes, and, in fact, these studies mainly showed how interactions between organizational factors and technologies could produce different results. From a knowledge transfer perspective, on the other hand, ERP implementation is seen as a process of bringing explicit business processes to the organization along with other business rules. Through this process we could conceptually distinguish two different types of knowledge transfer and knowledge transfer processes in an ERP implementation.

Second, this study provides an initial explanation of how an organization can gain a competitive advantage based on a standardized business process solution. Current practices of ERP implementation raise the question of how one organization can design more competitive processes if most organizations accept standardized processes from reference models. The results from this study suggest that the ERP implementation process should be understood by distinguishing the implementation process from the internalization process. In the first implementation process, organizations adopt the 'best processes' by configuring to their environment, and their explicit processes are easily transferred to the organization. But when it comes to internalizing the process, the adopted processes conflict with existing business values and rules, and it is the organizational capability to adjust to the conflicts that provides a process-based competitive advantage. Each organization has a variety of ranges of 'capability' and options in integrating systems, which determine its process-based competitiveness.

Finally, this study adds to our understating about the organizational capability to internalize business processes in light of an ERP implementation. The results suggest that, while new requirements based on the ERP adoption create a challenge, the organizational capability to adapt to the new rules is described as a process of internalization. For instance, in dealing with the issue of redefining roles and responsibilities, organizations might adopt a broader set of options from 'leave it as it is' to 'actively restructure the organization to redistribute roles and responsibilities'. The same is true for new knowledge and knowledge structure requirements. Organizations can develop new training programs to use more cross-functional training as well as new knowledge management systems that indicate who knows what information.

REFERENCES

Argote, L. (1999) Organizational Learning Crating. *Retaining and Transferring Knowledge.* Kluwer Academic Publishers.

Badaracco, Jr. J. L. (1991) *The knowledge Link: How Firms Compete through Strategic Alliance.* Boston, MA: Harvard Business School Press.

Barley, S. (1990) The Alignment of Technology and Structure through Roles and Networks. *Administrative Science Quarterly*, **35**(1), 61–103.

Baskerville, R. (1999) Enterprise Resource Planning and Knowledge Management: Convergence or Divergence? Information Systems Working Paper 99–6, Georgia State University Department of Computer.

Berger, L., Sikora, M., and Berger, D. (1994) The Change Management Handbook: A Road Map to Corporate. Burr Ridge.

Brown, J. S. and Duguid, P. (1991) Organizational Learning and Communities of Practice: Toward a Unified View of Working, Learning, and Innovation. *Organization Science*, **2**(1), 40–57.

Camp, R. (1995) *Business Process Benchmarking: Finding and Implementing Best Practices.* Milwaukee, WI.

Ciborra, C. U. and Lanzara, G. F. (1994) Formative Contexts and Information Technology. *Journal of Accounting, Management and Information Technology*, **4**, 61–86.

Cooper, R. and Zmud, R. (1990) Information Technology Implementation Research: A Technological Diffusion Approach. *Management Science*, **36**(2), 123–139.

Davenport, T. (1998) Putting the Enterprise into the Enterprise System. *Harvard Business Review*, **76**(4), 121–131.

Davis, Jessica. (1998) Scooping up Vanilla ERP: Off-the-shelf versus Customized Software. *InfoWorld*, **20**(47), 1–4.

Darr, E., Argote, L., and Epple, D. (1995) The Acquisition, Transfer and Depreciation of Knowledge in Service Organizations: Productivity in Franchises. *Management Science*, **41**(11), 1750–1762.

Gattiker, T. and Goodhue, D. (2000) Understanding the Plant Level Costs and Benefits of ERP: Will the Ugly Ducking Always Turn into a Swan? Proceedings of the 33rd Hawaii International Conference on System Sciences, Hawaii.

Ghoshal, S. and Nohria, N. (1989) Internal Differentiation Within Multinational Corporations. *Strategic Management Journal*, **10**(4), 323–337.

Gupta, A. K. and Govindarajan, V. (1991) Knowledge Flows and the Structure of Control Within Multinational Corporations. *Academy of Management Review*, **16**(4), 768–792.

Hanseth, O. and Braa and K. (1998) Technology as Traitor: Emergent SAP Infrastructure in a Global Organization. *International Conference on Information Systems.*

Hanseth, O. and Braa, K. (2000) Who's in Control: Designers, Managers – or Technology? Infrastructures at Norsk Hydro. In *Control to Drift: The Dynamics of Corporate Information Infrastructures*, Ciborra, C. et al. (eds), Oxford: Oxford University Press, pp. 125–147.

Kogut, B. and Zander, U. (1993) Knowledge of the Firm and the Evolutionary Theory of the Multinational Corporation. *Journal of International Business*, **24**(4), 625–645.

Kostova, T. (1999) Transnational Transfer of Strategic Organizational Practices: A Contextual Perspective. *Academy of Management Review*, **24**(2), 308–324.

Lam, Alice. (1997) Embedded Firms, Embedded Knowledge: Problems of Collaboration and Knowledge Transfer in Global Cooperative Ventures. *Organization Studies*, **18**(6), 973–996.

Nonaka, I. and Konno, N. (1998) The Concept of 'Ba': Building a Foundation for Knowledge Creation. *California Management Review*, **40**(3), 1–15.

Nonaka, I. and Takeuchi, H. (1995) *The Knowledge-creating Company: How Japanese Companies Create the Dynamics of Innovation.* New York: Oxford University Press.

Orlikowski, W. (1993) CASE Tools as Organizational Change: Investigating Incremental and Radical Changes in Systems Development. *MIS Quarterly*, **17**(3), 309–339.

Parr, A. and Shanks, G. (2000) A Taxonomy of ERP Implementation Approaches. Proceedings of the 33rd Hawaii International Conference on System Sciences. Hawaii.

Polanyi, Michael. (1962) *Personal Knowledge: Towards a Post Critical Philosophy.* New York: Harper Torchbooks.

Polanyi, Michael. (1966) *The Tacit Dimension.* New York: Anchor Day Books.

Powell, W., Koput, K., and Smith-Doerr, L. (1996) Interorganizational Collaboration and the Locus of Innovation: Networks of Learning in Biotechnology. *Administrative Science Quarterly*, **41**(1), 116–145.

Sachs, P. (1995) Transforming Work: Collaboration, Learning and Design. *Communications of the ACM*, **38**(9), 36–45.

Sieber, T., Siau, K., Nah, F., and Sieber, M. (2000) SAP Implementation at the University of Nebraska: Teaching Case Article. *Journal of Information Technology Cases and Applications*, **2**(1), 41–66.

Slater, D. (1998) A CIS and a Prayer. *CIO*, **11**(9), 24.

Szulanski, G. (1996) Exploring Internal Stickiness: Impediments to the Transfer of Best Practice within the Firm. *Strategic Management Journal*, **17** (Winter Special Issue), 27–43.

Uzzi, B. (1996) Sources and Consequences of Embeddedness for the Economic Performance of Organizations. *American Sociological Review*, **61**, 674–698.

Zander, U. and Kogut, B. (1995) Knowledge and Speed of the Transfer and Imitation of Organizational Capabilities: An Empirical Test. *Organization Science*, **6**(1), 76–92.

15 Knowledge Integration Processes Within the Context of Enterprise Resource Planning System Implementation

Jimmy Huang, Sue Newell, and Robert Galliers

15.1 Introduction

An increasing number of multinational enterprises (MNE) have adopted ERP systems in the hope of increasing productivity and efficiency as a means of leveraging organizational competitiveness (Davenport, 1998; Wagle, 1998). While some are starting to harvest the benefits from their initial investments, others are still struggling to release the promised potential of their ERP systems. This can be seen as an illustration of the 'productivity paradox' (Fitzgerald, 1998), that is that firms face a significant problem in measuring the return on their IT investments. While there have been several accounts that have examined the adoption of technology, including ERP, few of these accounts have considered this from the perspective of cross-functional knowledge integration (Grant, 1996). Against this backdrop, a case study was conducted as a means of exploring and theorizing the dynamics of knowledge integration underlying the process of ERP implementation. In this paper we focus not only on presenting the theoretical framework but also on describing the stages that were undergone to derive this framework.

15.2 Current Debates and Perspectives

There is a great deal of literature conceptualizing the phenomenon of how firms implement new IT systems. This study draws upon research in four distinctive areas, including (1) the development of technology, (2) the process of technology implementation, (3) enabling and inhibiting factors, and (4) management of process innovation.

Firstly, addressing the process of technological development in an organizational context, one of the major contributions made by the social construction of technology (SCOT) perspective is to discard technological determinism's linear model in explaining the technology development and diffusion processes (McLoughlin, 1999). In particular, emphasis is placed upon the social dynamics that influence the development of technology, and upon unveiling the complexity that can be regarded as a 'seamless web' (Woolgar, 1991). In addition, Latour (1987) actor-network theory perceives the process of introducing new technology as a 'black box'. Opening up this 'black box' involves exploring the changes in social and technical networks which follow from this introduction. These research traditions provide a persuasive rationale for exploring the dynamics of technological development which take into account the importance of the social context.

The second stream of studies focuses primarily on the process by which new technology is implemented. Much of this literature highlights the critical influence of managerial actions on end users acceptance of new systems, as well as the need for understanding the dynamics underlying the social interaction between managers and end users (Kwon and Zmud, 1987). Additionally, Attewell (1996) argues that the process of new IT implementation can be considered as a phenomenon of communication and uncertainty reduction, and understood by investigating the information flow between participating parties. Major contributions made by this stream of research are twofold. First, they provide an alternative conceptualization to conventional project management studies that perceive the implementation process as a sequential series of steps. Second, they address the on-goingness and interconnection of implementation processes that cannot be studied in isolation.

In terms of enabling and inhibiting factors related to the implementation process, there is a wide diversity of findings. This diversity demonstrates the difficulty of providing a holistic account and also emphasizes the depth and width of issues related to the success of technology implementation. It is possible to cluster the identified factors into two distinctive but interrelated categories, namely internal factors and external factors. In terms of the internal factors, research has tended to focus upon issues such as organizational culture (Holfman and Klepper, 2000), user-friendliness (Markus and Keil, 1994), and managerial influence on end users (Leonard-Barton, 1987). Also, organizational innovation characteristics are explored by Rogers (1962), and further extended by Tornatzky and Klein (1982) who emphasize the 'compatibility' between technology and the nature of the task, as well as by Davis (1989) who proposes the importance of 'perceived usefulness' by end users and stakeholders. In terms of external factors, Swan, Newell, and Robertson (1999) suggest that understanding technology diffusion relies on investigating the roles of central agencies, namely professional associations and technology suppliers, and their influence on firm's technological decisions.

Finally, this paper draws upon the stream of research that emphasizes the need for a more radical management approach to the implementation of new IT systems. For example, the need for redesigning business processes to enable system implementation can be traced back to work, such as Lenoard-Barton (1987) and more recently Markus and Keil, (1994). One of the underlying assumptions shared by these authors is the need to integrate the design of new technology with business processes and radically renew an organization as a basis for leveraging the strategic value of the technology. However, while the need for a radical approach may be clear, there are relatively few accounts which explore the processes and dynamics underlying such an approach, in particular from the focus of cross-functional knowledge integration. This is the focus of the study examined in this current paper.

Both Lawrence and Lorsch (1967) and Grant (1996) argue that what firms do is integrate their functionally specific knowledge. This knowledge is often dispersed, differentiated and embedded. However, while studies have examined the impact of knowledge integration (for example, Pisano, 1994) and its implications (for example, Boland and Tenkasi, 1995), there are few studies which have explored the processes of knowledge integration (Hauptman and Hirji, 1999). The need for understanding knowledge integration processes can be seen as a theoretical gap in the existing literature on IT implementation. In other words, the investigation of knowledge integration processes in the context of technology implementation, in particular the implementation of ERP systems, is the main objective of this study. However, in addition this paper focuses on illustrating the research method through which these knowledge integration processes were investigated. The following section thus outlines research methodological issues, namely research design, data collection, and data analysis.

15.3 Research Methodology

The research methodology is based on grounded theory (Glaser and Strauss, 1967; Orlikowski, 1993; Strauss and Corbin, 1990; Turner, 1983) with an objective of generating a theory that depicts the process of knowledge integration in the context of ERP implementation. The rationale behind adopting this approach is twofold. Firstly, the inductive and theory discovery characteristics

Table 15.1 Type and number of interview conducted at Company A

Role of the Interviewee	1st Interview	Follow-up Interview	Total
Project Sponsor (Senior Manager)	2	1	3
Steering Group (Head of Division)	3	1	4
Project Team Members	11	6	17
End User	6	3	9
Consultant (Vendor)	3	1	4
Total	25	12	37

of grounded theory can help us to develop a theoretical account in a research area where there has been no or very little previous examination (Eisenhardt, 1989). Thus, as seen, despite the fact that there have been numerous previous accounts which conceptualize IT implementation, less emphasis has been placed upon the dynamics and processes of knowledge integration in the implementation of ERP. Secondly, the ground theory approach takes into account the complexity of organizational context and can, therefore, lead to an in-depth analysis (Orlikowski, 1993). As seen from the earlier discussion, it is clear that the significance and complexity of the organizational context has often been neglected, in particular when considering the issue of knowledge integration.

Orlikowski (1993) notes that, 'the methodology of grounded theory is iterative, requiring a steady movement between concept and data, as well as comparative, requiring a constant comparison across types of evidence to control the conceptual level and scope of the emerging theory' (p. 311). To enable this iterative comparison, various sources of data were collected through semi-structured interviewing, on-site observation and documentation. The rationale of using multiple data collection methods is not only for the purpose of triangulation (Bryman and Burgess, 1999; Denzin, 1988) through cross checking (Eisenhardt, 1989). Also, it ensures the richness of the emerging theory by obtaining various perspectives on an issue through iteration (Orlikowski, 1993; Strauss and Corbin, 1990). As will be seen, the data that were collected were coded and analysed iteratively, following Glaser and Strauss (1967). This allowed the authors the flexibility to adjust the data collection process, in particular in relation to the critical selection of interviewees.

The research started with a three-month on-site observation. In addition, 25 semi-structured and 12 follow-up interviews, as shown in Table 15.1, were conducted each lasting on average 90 minutes. Further on-site observation was spread over an overall period of 15 months so as to acquire first-hand knowledge of how ERP was implemented through social interactions between organizational members, consultants, and the software vendor. Documentation about the company and the ERP project were examined through reviewing letters, written reports, administrative documents, newspapers, archives, as well as the company Intranet.

Prior to the coding, the preparation stage consisted of activities, such as transcribing interview tapes, typing and filing research notes (taken during and after interviews as well as on-site observation) into the database, summarizing documents and categorizing them based on the sources. There were, therefore, four main categories of data grouped during this stage – interview transcripts, research notes, photocopied documents, and information downloaded from the case company's Intranet. Brief notes that described the content of each file and potential linkages with different files were inserted in virtually every file stored in the four categories of data. These data served as the foundation for open, axial, and selective coding. The initial stage of analysis – open coding (Strauss and Corbin, 1990) – aimed to reduce the amount of data. This stage of analysis was open-ended and generative (Orlikowski, 1993). Themes emerging from the open coding were used as a basis to proceed to the axial coding stage. Here the primary aim was to generate interconnections between these themes. Finally in the last stage of selective coding, similarities and differences between categories and concepts were further explored. This coding technique enabled the authors to identify richer concepts and more dynamic relations between these concepts. These coding processes fulfilled the requirement of multi-level analysis in the study of the dynamic relations between processes and contexts. Findings generated from the different stages of the coding processes are outlined in the following section.

15.4 Research Results

Open Coding

The critical categories and concepts that emerged from the initial open coding are outlined in Table 15.2. In particular this first stage of coding focused on identifying the various contexts which can help to offer an explanation in terms of 'an interaction of contextual conditions, actions, and consequences, rather than explain variance using independent and dependent variables' (Orlikowski, 1993, p. 311).

Environmental context

The increasing popularity of using ERP systems in the engineering industry was found to be one of the main drivers influencing the case organization to

Table 15.2 Research outcomes generated from open coding

Categories	Concepts	Data from the research site
Environmental context	Customers	• Custom built engines and components for power generation related purposes and functions
	Competitors	• Some offer standardized engines and solutions at relatively low price
	Technologies	• Increasing number of firms adopt ERP systems
Organizational context	Organizational structure	• Project-based structure with the majority of staff in R&D, engineering, and production working on various projects
	Culture and sub-culture	• Innovative culture enhanced by working closely with clients and suppliers
	Corporate strategies	• Providing world-class solution and products by continuous innovation
IS context	Role of IS in firm	• In addition to the intensive use of ICT for dispersed project team members and divisions, IS has long been aligned with strategic business development
	IS-related staff	• A decentralized and dispersed IS division had over 400 staff in total among six main sites world wide
	IS and user relationships	• An increasing number of members of staff attended courses (over 1900 staff in 1998) and on-line forums (over 4100 staff in 1998) organized by the IS division
Project context	Project scope	• Estimated project length 36 months (excluded evaluation study) • ERP system was implemented globally and covered all major divisions
	Overall cost	• Estimated saving over £116 millions over five years • Software cost: £2.2 millions • System implementation (including evaluation and training): estimated cost £17.8 millions and actual cost over £19 millions

Table 15.2 (*cont.*)

Categories	Concepts	Data from the research site
	Internal project participants	• Internal man hour cost: estimated at £13.5 millions • Six project sponsors consisted of senior managers • 11 steering group members formed of division heads
	External project participants	• 18 project team members (seven from IS division, and one representative from each division) • SAP software, evaluation study, business case, and implementation services were all provided by firm X
Internal conditions for ERP implementation	Awareness of ERP	• A reasonable number of staff in IS, HR, accounting, production, and R&D are aware of ERP system
	Support for adopting ERP	• From various hierarchical levels, in particular engineers, accounts, and procurement managers
Implementing ERP systems	Analysing	• A consultant team from Firm X carried out the evaluation study and proposed business case was accepted
	Planning and design	• Decisions were made to implement the ERP with global coverage
	Implementation	• Co-led by the external consultant team and internal participants

adopt ERP. As noted by the Chief Operating Officer (COO), 'ERP has become more or less an industrial standard in our business. Given the dynamics and complexity of our environment, we need a system that is capable of not only efficiently providing information for decision making, but also effectively integrating all systems in our company.' In addition to this rationality for adopting ERP, the evidence also suggests that the adoption of ERP was seen as inevitable because of the need to manage very large quantities of information and data both intra- and inter-organizationally. As explained by the Procurement and Logistic Director: 'the sheer amount of suppliers we get has driven the department into chaos. Given each project leader the autonomy to select their own suppliers and process their orders, we have wasted enormous amounts of money which can be otherwise saved by having strategic partnership with fewer suppliers.'

Organizational context

Internally, the need for an integrative system, such as ERP, to effectively manage information and support decision making was found to be another source influencing the adoption of ERP. Sustaining technological advancement was reflected in the case organization's culture that strongly emphasized continuous process and product innovation. With a great emphasis on innovation, the case organization had a strong record in keeping up with the latest technology. For instance, the case company had introduced a common IT infrastructure in 1992, had undertaken a major system upgrading in 1993, and implemented company-wide video conferencing in late 1994. To enhance product innovation, project teams were intensively used. Three interviewees argued that the use of project teams was to stimulate creativity through pooling together different expertise dispersed around the globe. In addition to that, they also indicated that on-going process innovation was also critical, in particular in relation to the way in which product-related information and suppliers were managed. The ERP system was adopted to simplify the process of project management by reducing some of the project leaders'responsibilities, such as recruiting new members and dealing with suppliers.

IS context

As illustrated in Table 15.2, the role of IS in the firm was critical for managing client-related information and enabling continuous process innovation. In addition to the management of software and hardware, the IS division was also responsible for proposing and planning various IS investments and developments. The involvement of the IS division was not only reflected in the head of IS participating in all major ERP-related decisions, but also in the fact that other IS staff were involved in collecting information and acquiring knowledge related to ERP. According to an internal document, more than 70% of IS staff had attended IS-related conferences or seminars in the past two years, including more than 20% of staff who had attended ERP-related ones. Positive relationships between IS staff and end users were evident in the company survey that indicated a growing level of satisfaction from end users. This was also seen as a critical element in implementing new technology. In particular, such relationships were seen to be beneficial in changing the way information was disseminated and acquired within the organization. For instance, there was an increase in the number of users who used the company Intranet to obtain information. For example, the ERP site on the Intranet was visited more than 4000 times two weeks after the project was launched. This depicts not only the increasing popularity of the Intranet as an information distribution tool, but also how ERP awareness was generated during the early stage of the project.

ERP project context

In addition to the initial ERP evaluation conducted by external consultants, the availability of capital for investing in new technology was found to largely influence how the project scope was defined. For instance, in the last five years, there had been an average of £12–15 million per annum budgeted for IT/IS investment. This was sufficient to support the on-going adoption of new technology. Also, the availability of capital permitted the implementation of new technology that reached an organization-wide coverage. In terms of internal participants, the project structure was composed of the ERP team and the steering group – which included representatives of different stakeholder groups and end users. The composition of the ERP team and steering group reflected not only the variation between different divisions' involvement, but also the power differences in managing and controlling the project. Even though the IS staff argued that project ownership was equally shared, it was evident that the IS division played the leading role together with the external consultant team.

Internal conditions for ERP implementation

As elaborated in Table 15.2, individuals' awareness of and support for adopting ERP were identified as the two main internal conditions influencing the process of technology implementation. Despite the fact that the acquisition of ERP-related knowledge from the external environment was paramount, the mechanism used to diffuse such knowledge internally was equally critical. In particular, this first level of analysis identified the importance of the diffusion mechanism used to raise ERP awareness across various organizational levels. For instance, IS staff played the pioneer role in introducing the concept of ERP to the organization. ERP-related knowledge was further diffused across the organization through the Intranet, written memos, and newsletters. It was found that staff's awareness not only influenced how the usefulness of the new technology was perceived, but also affected how ERP was implemented. The case analysis suggested that user support was generated depending on their awareness of the new technology, their past experience, as well as their relationships with the IS division, which was generally positive as indicated earlier.

Processes of ERP implementation

The ERP project began with an evaluation study conducted by an external consultant team. Following this, the analysis identified three processes of the ERP project – analysing problems and opportunities, planning and design, and implementation. The evaluation study suggested the need for a global implementation project over three years and proposed a potential cost saving of £116 million (including £56 million saving due to the improvement of management) over five years. The business case was sanctioned while some minor adjustment was made. However, the main discussions at this stage centered on the issues of system integration and defining responsibility. During the planning and design stage, cross-departmental debates and conflicts emerged. These emerged due to the ambiguity of information ownership and the integration of the product data management (PDM) system with ERP. These two interconnected issues were reflected in the difficulty of converting data between the ERP and PDM systems because of the differences in orientation, as well as the difficulties of ensuring the free flow of information across functions. As explained by one engineer, 'both ERP and PDM have different ways of storing, translating and managing data. ERP is more engineering orientated, while PDM is more manufacturing oriented. Even though you can find an overlap between them, like product structure management, it does not necessarily mean that when we release a PDM product definition to manufacturing, they can automatically convert such information into a version for ERP.' Further conflicts between departments, mainly between engineering and production, were also found during the implementation stage. Even though a translator that converted PDM data into ERP was developed to solve the integration problem, the Engineering Department was still dissatisfied with such a one-way translation, because of losing the ownership and control of information generated by this department. As another engineer recalled, 'it [the one-way translation] means that they get what we have got, but we do not have any legitimacy of benefiting from what they have got'.

In the following sections, this first-level open coding is developed using axial (Figure 15.1) and then selective (Figure 15.2) coding to explore more specifically the knowledge integration processes underlying the implementation of ERP.

Axial Coding

Derived from the analytically iterative coding processes, this section highlights various aspects of knowledge integration, namely its nature, processes, and implications.

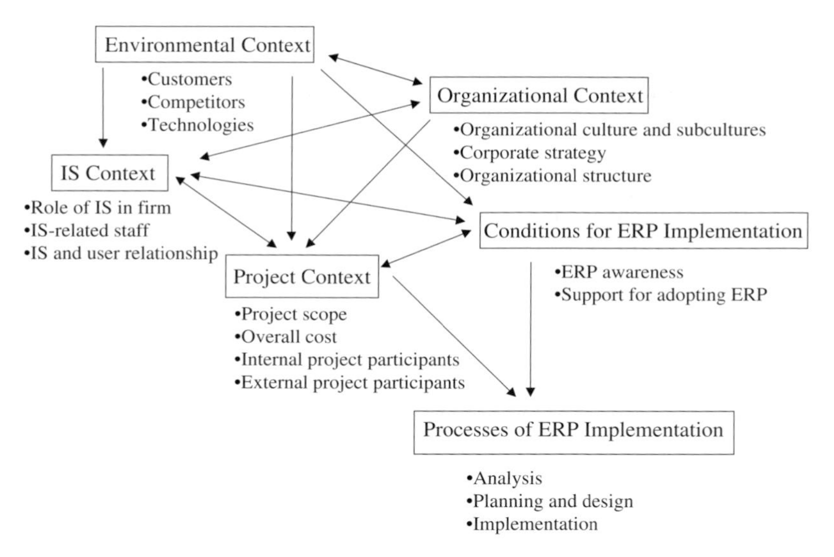

Figure 15.1 Dynamics of ERP implementation

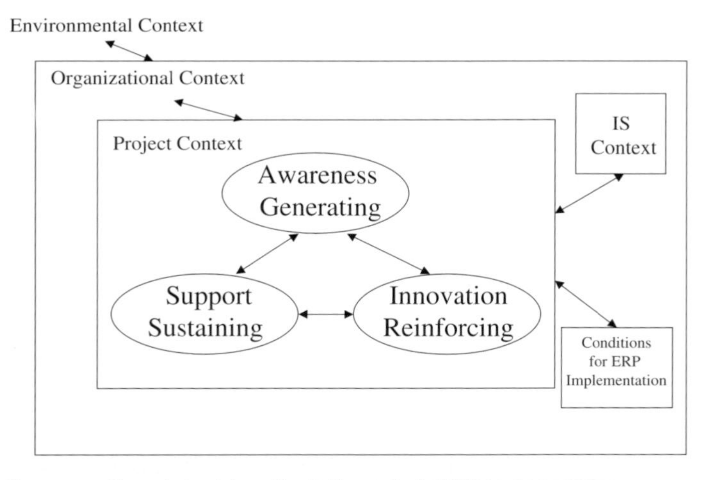

Figure 15.2 Processes of knowledge integration in the context of ERP implementation

The nature of knowledge integration

Evidence abstracted from the analysis suggests that the nature of knowledge integration can be conceptualized based on four distinctive but interrelated dimensions, namely structural, technological, intellectual, and socio-emotional. The nature and philosophy of ERP illustrates the technological dimension of knowledge integration. As elaborated earlier, the adoption of ERP was aimed at installing an integral system that promoted the use of common language in interpreting information generated by different divisions. Also, it was aimed at facilitating and enabling cross-functional knowledge integration through the advancement of technology.

As can be seen in Figure 15.1, the dynamics of knowledge integration in the context of ERP implementation are not limited to within the organizational boundary. Instead, knowledge integration is an ongoing process which takes place intra- and inter-organizationally. In particular, the decision to adopt ERP is triggered through integrating knowledge from the environment. The way in which the awareness of ERP was generated in the case echoes the findings of Swan, Newell, and Robertson (1999) who depict the influence of professional associations on the diffusion of new technology. However, it is evident that the availability of mechanisms, such as an Intranet and newsletters, to disseminate the acquired knowledge, to various organizational levels is equally critical.

The transmission of knowledge further surfaces the intellectual dimension of knowledge integration. It is necessary to create 'common knowledge' (Demsetz, 1991) through synthesizing differentiated expertise (Lawrence and Lorsch, 1967), such as the expertise of ERP implementation possessed by the external consultant team. In addition to the above three dimensions, it is found that knowledge integration is also a socio-emotional process since support from end users and stakeholders had to be generated through developing trust (Nahapiet and Ghoshal, 1998; Newell and Swan, 2000). The socio-emotional dimension addresses the point that socialization is not only a critical ingredient for knowledge creation (Nonaka and Takeuchi, 1995), but also an essential process in achieving 'emotional attachment' (Lembke and Wilson 1997) to overcome departmental boundaries.

The above discussion also brings out the fact that the nature of knowledge integration may have been oversimplified by the current literature (for example Grant, 1996), which has tended to focus on the intellectual rather than socio-emotional aspects. This discussion also surfaces the need to understand the interplay between those dimensions as a means of anticipating the processes of knowledge integration. While the analysis derived from the axial coding helps to explore in more detail the links between the issues derived from the open coding, the final level of analysis is to use selective coding to develop a theoretical framework outlining knowledge integration processes during ERP implementation.

Selective Coding

Three knowledge integration processes generated from the analysis are termed awareness generating, support sustaining, and innovation reinforcing, as shown in Figure 15.2. While these three identified processes are analytically separated, they are highly interrelated.

Awareness generating

The term 'awareness generating' refers to the process by which ERP-related knowledge is acquired and distributed. The reason for using the term is not only to describe the ongoing need for promoting the ERP philosophy across the organization, but also to sketch out how the new technology is perceived through the dissemination of knowledge. Referring to Figure 15.1, it is clear that the generation of ERP awareness portrays the dynamics through which information generated from the external environment is processed and interpreted (Galbraith, 1977). In addition to the influence of professional associations, as studied by Swan, Newell, and Robertson (1999), ERP awareness was also generated through observing competitors and obtaining industry-specific information through various sources, such as seminars and magazines. A long-term strategic partnership between the case company and the service provider was found to be equally vital in generating such awareness.

The above observation also points out the need to address the issue of boundary spanning. In the current literature, despite the fact that the importance of boundary spanning in obtaining strategically critical knowledge is evident (Dollinger, 1984; Jemison, 1984; Lang, 1996), how boundary spanner's knowledge background affects the adoption of new technology and knowledge integration processes is often neglected. The diversity of knowledge background possessed by the organization's 'frontiers' to acquire ERP-related knowledge prior to the implementation was found to be critical. For instance, in addition to the IS Department, some members of staff in R&D, Production and Engineering were aware of the concept of ERP. Such diversity was found to influence how the project scope was defined, and how different aspects of potential usefulness related to ERP were formed. The process of awareness generating not only echoes the importance of perceived usefulness (Davis, 1989) in relation to technology adoption, but also pinpoints how the perceived usefulness can be expanded through the increase of knowledge diversity possessed by an organization's frontiers.

In addition to the importance of IS/user relationships elaborated in Figure 15.1, the approaches used to share and distribute ERP-related knowledge were found to largely influence the process of awareness generating. As indicated by some of the interviewees, they were informed by colleagues and friends about the ERP site on the Intranet, and they further contacted people they knew to inform them about the site and to discuss various issues they had newly discovered. This explains not only the importance of social networks in exchanging knowledge (Burt, 1992; Nahapiet and Ghoshal, 1998), but also the importance of combining formal and informal communication channels in generating awareness required for technology implementation. As shown in Figure 15.2, the generation of ERP awareness was found to further influence how internal and external participants could obtain support from end users and stakeholders, and how process changes required by ERP could be reinforced.

Support sustaining

The term 'support sustaining' is used here to depict the ongoing process by which internal and external project participants obtain resources and support from other organizational members at various hierarchical levels. From the case, it is clear that support needed for ERP implementation is not limited to the management level in the form of obtaining resources. Also, there is a need to obtain support from end users across the organization through collaboration. In contrast with some project management research that regards obtaining resources as a key activity during the initial stages of a project (for example Case and Shane, 1998; Mamaghani, 1999), findings generated in this study suggest that it is critical to ensure support and the availability of resources throughout the project life cycle. This pinpoints the importance of support sustaining not only in energizing the continuity of ERP implementation, but also in enabling the ongoingness of knowledge integration.

While the continuous need for sustaining support from various hierarchical levels has been suggested by some studies (see, for example, Markus and Keil, 1994), the dynamics of this process have been relatively underexplored. As elaborated earlier, the process of support sustaining is an intellectual activity by which the importance and value of new technology is communicated to the potential users. The intellectual dimension highlights the need to increase ERP awareness through communication, as well as the development of common knowledge (Demsetz, 1991) to create shared understanding through 'perspective taking' (Boland and Tenkasi, 1995).

On the other hand, the emotional dimension of knowledge integration, as intertwined with the development of trust and distrust, is found to be equally critical for sustaining support from organizational members at various levels. Resistance and dissatisfaction observed in the case is an example which shows how organizational members' perceptions could form emotional obstacles, even though they were intellectually aware of the strategic importance of ERP. It is clear that disputes between Engineering and Production Departments were not simply an intellectual or technological issue, in terms of how ERP should be implemented. Rather, it was an emotional issue, so that 'psychosocial defenses' (Allcorn, 1995) arose because of the fear of losing project and information ownership. The way in which support is sustained by the participants is also found to influence how new organizational processes and routines are created, as illustrated in the following process.

Innovation reinforcing

The term 'innovation reinforcing' is used here to describe the process by which existing organizational routines are modified through the implementation of new technology. Organizational routines and processes, as the representation of continuous cross-functional knowledge integration and differentiation (Lawrence and Lorsch, 1967), were continuously modified and refined through various initiatives implemented within the organization. This echoes the concept of organizational learning (Huber, 1991) in which organizational memory is continuously reconfigured through incremental or radical change (Miner and Mezias, 1996). Referring to the case, the impact of ERP on the business processes was found to result in the redirection of information flows across the organization. For instance, departmental boundaries, which previously served as 'valves' in controlling the availability and accessibility of information, were gradually minimized through the implementation of ERP. Additionally, the change in knowledge flow (Starbuck, 1992) was also evident. For instance, the procurement procedure was simplified by taking away project leaders' responsibilities in dealing with suppliers. However, such modification also reduced suppliers' involvement in the development of new product. At least two Project Leaders saw this as hampering the quality and efficiency of product innovation.

Even though the reinforcement of innovation facilitates the free flow of information, it is also critical to consider how new business processes can potentially affect the dynamics of knowledge integration and creation and how 'compatibility' can be ensured between the technology and the nature of task (Tornatzky and Klein, 1982). This suggests that an organization may successfully implement ERP and benefit from its ongoing capital investment, and yet improved knowledge integration through the implementation of new technology can remain a challenge.

Referring to Figure 15.1 and the process of awareness generating in Figure 15.2, it is clear that how the organization reinforces its innovation initiative through the implementation of new technology crucially depends on how awareness across the organization can be generated. Also, it is evident that the reinforcement of innovation is largely influenced by how support from various divisions and hierarchical levels is sustained, as elaborated in the second process of knowledge integration. Dynamic interrelationships between these three processes suggests that knowledge integration in the context of ERP implementation is an information processing activity, since awareness is generated through the distribution of information and knowledge across divisions (Campbell, 1985; Combs, 1993; Galbraith, 1977). It also suggests that knowledge integration in the context of ERP implementation is a social construction process (Berger and Luckmann, 1967) in which social interaction between external participants, internal participants, and end users shapes the implementation of new technology. The social construction of technology (McLoughlin, 1999; Pinch and Bijker, 1987) further reflects the importance of a socio-technical perspective. Examining the implementation of new technology must thus focus on understanding the interplay between structural, intellectual, socio-emotional, and technological issues.

15.5 Conclusion and Implications

Based on a ground theory approach, this study has explored the dynamics and processes of knowledge integration underlying the implementation of ERP. Instead of simply presenting the outcomes derived from selective coding, this paper has explored the findings generated through each stage of coding. Table 15.2 illustrates the findings generated from the first stage of open coding, Figure 15.1 illustrates the findings generated from the next stage of axial coding, while Figure 15.2 presents the final analysis derived from the selective coding stage. This paper thus seeks to make both a methodological contribution and a theoretical contribution. In terms of the latter, the paper considers an alternative approach to conceptualize the dynamics of ERP implementation, focusing on processes of knowledge integration. The analysis suggests that the implementation of ERP is a process that involves complex interactions between technological, socio-emotional, intellectual, and structural issues.

By taking into account various contexts, such as environmental, organizational, IS, and project, this explorative account has elaborated on how decisions were made to adopt ERP, and how the implementation of ERP was influenced by the degree of awareness, support, and reinforcement of innovation. More importantly, this study has generated a theory that depicts the nature and processes of knowledge integration within the context of ERP implementation. The framework developed here does not imply that successful

ERP implementation necessarily leads to effective knowledge integration. Instead, the findings suggest that the implementation of ERP helps to trigger the renewal of organizational routine through process innovation, and that the successful implementation of ERP can create yet another challenge for effective knowledge integration.

In addition to the theoretical and methodological contributions, this study also has its practical implications. Lessons learnt from the case are potentially beneficial for adopting and implementing new technologies such as ERP systems. For instance, it is critical for management to consider how awareness can be establishment and support can be obtained throughout the project life cycle. Additionally, it is vital to take into account how the ownership of an ERP project should be located, as well as how participants should be selected, both internally and externally.

REFERENCES

Allcorn, S. (1995) Understanding Organizational Culture as the Quality of Workplace Subjectivity. *Human Relations*, **48**(1), 73–96.

Attewell, P. (1996) Information Technology and the Productivity Challenge. In *Compaliers and Controversy*, Kling, R. (ed.), 2nd edn, Academic Press, pp. 227–238.

Berger, P. and Luckmann, T. (1967) *The Social Construction of Reality: A Treatise in the Sociology of Knowledge.* New York: Anchor Books.

Boland, R. and Tenkasi, R. (1995) Perspective Making and Perspective Taking in Communities of Knowing. *Organization Science*, **6**(4), 350–372.

Bryman, A. and Burgess, R. (1999) Introduction: Qualitative Research Methodology – A Review. In: *Qualitative Research*, Volume I, Bryman, A. and Burgess, R. (eds), London: Sage.

Burt, R. (1992) *Structural Holes: The Social Structure of Competition.* Cambridge: Harvard University Press.

Campbell, N. (1985) Sources of Competitive Rivalry in Japan. *The Journal of Product Innovation Management*, **2**(4), 224–231.

Case, R. and Shane, S. (1998) Fostering Risk Taking in Research and Development: The Importance of a Project's Terminal Value. *Decision Sciences*, **29**(4), 765–783.

Combs, K. (1993) The Role of Information Sharing in Cooperative Research and Development. *International Journal of Industrial Organization*, **11**(4), 535–551.

Davenport, T. (1998) Putting the Enterprise into the Enterprise System. *Harvard Business Review*, **76**(4), 121–131.

Davis, F. (1989) Perceived Usefulness, Perceived Ease of Use, and User Acceptance of Information Technology. *MIS Quarterly*, **13**(3), 319–340.

Demsetz, H. (1991) The Theory of the Firm Revisited. In *The Nature of the Firm*, Williamson, O. and Winter, S. (eds), Oxford: Oxford University Press.

Denzin, N. (1988) *The Research Act.* New York: McGraw-Hill.

Dollinger, M. (1984) Environmental Boundary Spanning and Information Processing Effects on Organizational Performance. *Academy of Management Journal,* **27**(2), 351–368.

Eisenhardt, K. (1989) Building Theories from Case Study Research. *Academy of Management Review,* **14**(4), 532–550.

Fitzgerald, G. (1998) Evaluating Information System Projects: A Multidimensional Approach. *Journal of Information Technology,* **13**(1), 15–27.

Galbraith, J. (1977) *Organizational Design.* Reading, MA: Addison–Wesley.

Glaser, B. and Strauss, A. (1967) *The Discovery of Grounded Theory.* Chicago: Aidine.

Grant, R. (1996) Prospering in Dynamically Competitive Environment: Organizational Capability as Knowledge Integration. *Organization Science,* **7**(4), 375–387.

Hoffman, N. and Klepper, R. (2000) Assimilating New Technologies: The Role of Organizational Culture. *Information System Management,* **17**(3), 36–42.

Huang, J. (2000) Knowledge Integration Processes and Dynamics: An Empirical Study of Two Cross-functional Programme Teams. Ph.D. Thesis, Warwick Business School, The University of Warwick, Coventry.

Huber, G. (1991) Organizational Learning: The Contributing Processes and the Literatures. *Organization Science,* **2**(1), 88–115.

Jemison, D. (1984) The Importance of Boundary Spanning Roles in Strategic Decision-making. *Journal of Management Studies,* **21**(2), 131–152.

Kwon, T. H. and Zmud, R. W. (1987) Unifying the Fragmented Models of Information Systems Implementation. In *Critical Issues in Information Systems Research,* Boland, R. J. and Hirschheim, R. A., Chichester: Wiley, pp. 227–252.

Lang, J. (1996) Strategic Alliances between Large and Small High-tech Firms: The Small Firm Licensing Option. *International Journal of Technology Management,* **12**(7), 796–807.

Latour, B. (1987) *Science in Action: How to Follow Scientists and Engineers through Society.* Open University Press, Milton Keynes.

Lawrence, P. and Lorsch, J. (1967) *Organisation and Environment: Managing Differentiation and Integration.* Boston: Harvard University Press.

Lembke, S. and Wilson, M. (1998) Putting the 'Team' into Teamwork: Alternative Theoretical Contributions for Contemporary Management Practice. *Human Relations,* **51**(7), 927–944.

Leonard-Barton, D. (1987) Implementing Structured Software Methodologies: A Case of Innovation in Process Technology. *Interfaces,* **17**(3), 6–17.

Mamaghani, F. (1999) Information Systems Project Evaluation and Selection: An Application Study. *International Journal of Management,* **16**(1), 130–138.

Markus, L. and Keil, M. (1994) If We Build it, They Will Come: Designing Information Systems that People Want to Use. *Sloan Management Review,* **35**(2), 11–25.

McLoughlin I. (1999) *Creative Technological Change: The Shaping of Technology and Organizations.* London: Routledge.

Miner, A. and Mezias, S. (1996) Ugly Duckling No More: Pasts and Futures of Organizational Learning Research. *Organisation Science,* **7**(1), 88–99.

Nahapiet, J. and Ghoshal, S. (1998) Social Capital, Intellectual Capital, and the Organizational Advantage. *Academy of Management Review,* **23**(2), 242–266.

Newell, S. and Swan, J. (2000) Trust and Inter-organisational Networking. *Human Relations,* **53**(10), 1287–1328.

Nonaka, I. and Takeuchi, H. (1995) *The Knowledge-creating Company.* Oxford: Oxford University Press.

Orlikowski, W. (1993) CASE Tools as Organizational Change: Investigating Incremental and Radical Changes in Systems Development. *MIS Quarterly*, **17**(3), 309–340.

Pinch, T. S. and Biyker, W. E. (1987) The Social Construction of Facts and Artifacts: Or How the Sociology of Science and the Sociology of Technology Might Benefit Each other. In *The Social Construction of Technological Systems*, Biyker, W. E., Hughes, T. P., and Pinch, T. (eds), London: MIT Press, pp. 17–50.

Pisano, G. (1994) Knowledge, Integration, and the Locus of Learning: An Empirical Analysis of Process Development. *Strategic Management Journal*, **15**(Special Issue), 85–100.

Rogers, E. (1962) *The Diffusion of Innovations.* New York: Free Press.

Starbuck, W. (1992) Learning by Knowledge Intensive Firms. *Journal of Management Studies*, **29**(6), 713–740.

Strauss, A. and J. Corbin (1990) *Basics of Qualitative Research: Grounded Theory Procedures and Techniques.* London: Sage.

Swan, J., Newell, S., and Robertson, M. (1999) Central Agencies in the Diffusion and Design of Technology: A Comparison of the UK and Sweden. *Organization Studies*, **20**(6), 905–931.

Tornatzky, L. G. and Klein, K. (1982) Innovation Characteristics and Innovation Adoption Implementation: A Meta-analysis of Findings. IEEE *Transaction on Engineering Management*, **29**(1), 28–45.

Wagle, D. (1998) The Case for ERP Systems. *The McKinsey Quarterly*, Spring Issue, 130–138.

Woolgar, S. (1991) The Turn to Technology in Social Studies of Science. *Science, Technology and Human Values*, **16**(1), 20–50.

Part IV

Cultural Aspects of Enterprise Systems

16 An Exploratory Analysis of the Sources and Nature of Misfits in ERP Implementations

Sia Siew Kien and Christina Soh

16.1 Introduction

ERPs – software packages that manage and integrate business processes across organizational functions and locations – cost millions to buy, several times as much to implement, and necessitate disruptive organizational change. While some companies have enjoyed significant gains, others have had to scale back their projects and accept minimal benefits, or even abandon implementation (Marcus and Tanis, 2000).

Historically, a common problem when adopting package software has been the issue of 'misfits', that is, the gaps between the functionality offered by the package and that required by the adopting organization (Davis, 1988; Lucas, Walton, and Ginzberg, 1988). As a result, organizations have had to choose among adapting to the new functionality, living with the shortfall, instituting workarounds, or customizing the package. ERPs, as a class of package software, also present this problematic choice to organizations.

Even though there are many built-in switches that one can manipulate to customize the software, smooth alignment of the software functionality to business requirements is still far from ideal as many of these issues have to be dealt with at the application architecture level. AeroGroup, for example, has dropped the Apparel Footwear Solution of SAP/R3 because of its inability to model the uniqueness and complexities of the footwear business. Similarly, the failure of Fox Meyer could also be partially attributed to the adoption of an ERP system that was designed more for manufacturers rather than wholesale distributors.

Indeed, the problem of misfits is exacerbated as ERP implementation has added complexity due to cross-module integration, data standardization, compressed implementation schedule, and the involvement of a large number of stakeholders. Furthermore, the 'knowledge gap' among implementation personnel is usually significant. Few organizational users understand the functionality of ERP enough to appreciate the implications of adoption. Similarly, few ERP consultants understand their clients' business processes sufficiently to highlight all critical areas of mismatches.

While the issue of misfit between package software functionality and organizational requirements has been highlighted in the ERP literature, little work has been done to explode the misfit 'black box' to investigate the specific ways in which these gaps surface. Many of the misalignments arise from context specificity. This paper presents a framework for categorizing the misalignments by source or type of specificity, and by the nature of the misalignment. Understanding these systematic sources and types of misalignments will contribute to early identification, which in turn allows for a more complete basis for planning (for example, budget for contingency funds, setting of expectations). Change management and resolution strategies can also be carefully thought through and are likely to be less reactive in nature.

16.2 Methodology

An intensive, multiple case study approach was used to examine the sources and nature of misfits in ERP implementation. Three large, multi-specialty hospitals (with between 800 and 1500 beds) were studied. All three hospitals were implementing the same ERP package. They adopted many of the same modules (financial, materials management, inpatient management), while two of the hospitals also adopted the outpatient management and human resource management modules. All three hospitals also used the same implementation consultants, who prescribed the same sequence of implementation events, following the consulting firm's project management methodology. Finally, the hospitals adopted the ERP package because their legacy systems were not Y2K compliant.

The typology of misfits and examples are drawn from the review of implementation documents and supplemented by interviews with key project team members from these hospitals. Misfits are defined as incompatibilities where ERP system functionalities fall short of an organization's business requirements. We systematically built up the lists of ERP misfits observed by module and tracked them through to resolution by reviewing documents like board papers, contracts, project kick-off slides, project schedules, meeting minutes, issue logs, request for change forms (RFCs). Some of the items on the issue logs and RFCs, for example, technical change requests to interface with existing legacy systems, resolution among users to standardize organizational practices, decisions to create new cost centers, were not considered as misfits and have thus been excluded from our analysis. For consistency of comparison, we have cut off the misfit analysis up to the point of implementation. Total misfits analysed were 136, 115, and 107 respectively.

These misfits were segregated into common (with at least one other hospital surfacing the same issue) and unique (where the issue surfaced in one hospital only) misfits. The comparison across hospitals gave us a sense of the attribution of the sources of misfits, that is, whether they were industry- or organizational-specific factors. Interviews with users, project team members, and consultants were also conducted to provide an understanding of the context in which these misfits were identified and resolved. On average, there were about 20 interviewees for each hospital.

16.3 Misfit Analysis

From the grounded analysis of the misfits across these hospitals, we noted that the misfits in ERP implementations differ along two major dimensions, that is, the sources and nature of the misfits. The sources of misfits are largely due to the specificity of the local context in relation to the original context in which the ERP systems were developed. Moreover, the nature of the misfits from these sources of specificity also appears to be different, ranging from superficial input/output type misfits to the more critical misfits that affect operational functionality.

16.4 Sources of Misfits: Context Specificity

ERP systems are typically developed from some original contexts with industry 'best practices'. However, the underlying business models often have implicit country biases (for example, European or American practices), sector biases (for example, private sector), industry biases, (for example, manufacturing), and even biases in organizational practices (for example, process-oriented workflow). Thus, transplanting ERPs outside their original context can have wide-ranging implications (Soh et al., 2001).

The three Asian hospitals, having evolved in a different cultural, economic, and regulatory context, provided relevant data for analysing these misfits that

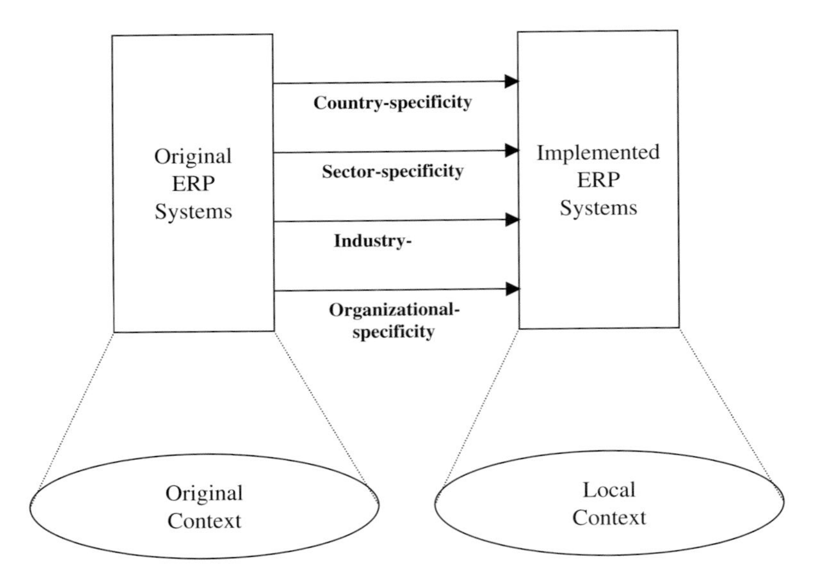

Figure 16.1 Sources of ERP misfits

arise from exogenous (country-specific, sector-specific, industry-specific) and endogenous (organization-specific) sources. Through the intensive implementation process, the respective project teams identified and resolved gaps between the original ERP package and the hospitals' requirements, resulting in implemented versions of the ERP systems that reflected the specifics of the respective hospital context (see Figure 16.1).

Country-Specific Misfits

Country-specific misfits tend to focus on the unique regulatory, social, or cultural practices within a country. One example is the billing and counter collection function. The local healthcare financing philosophy is based on individual responsibility for healthcare costs, while community support and government subsidies help to keep healthcare affordable. This contrasts with the Western healthcare models, where much of healthcare is privately delivered, and government or insurance pays for healthcare services, with the individual bearing little out of pocket costs. In an Asian context, ERP models that are based on Western models give rise to a number of functional deficiencies. For example, the billing and collection of the ERP caters to complex claims submission processes and verification to insurance companies, but not over-the-counter or installment payments by individual patients. The reliance on individuals as payees also creates the need for new ERP functionality to collect and monitor deposits. Similarly, the pricing functionality available in the original ERP systems is relatively rudimentary for an environment where payees of medical bills (that is, individuals) are sensitive to variation in pricing of medical services. Another major source of country-specific misfits is the concept of 'bed-class'. Unlike the West where single-bedded rooms are common, local patients in public hospitals can choose from a range of accommodation types, from one bed to six or more beds per room. The information is needed for claims processing as the government subsidies increase as one moves from a two-bedded to multiple-bedded room. However, the standard ERP patient management reports do not provide patient data by 'bed-class'.

Sector-Specific Misfits

Sector-specific misfits arise from the more stringent accountability requirements and the complex procedures unique to public organizations (for example, specific reporting to regulatory authorities, standard formulas and procedures for government reimbursement to hospitals for services to patients, and civil service human resource practices). Examples of such misfits include requests to expand the functionality of the ERP system to provide for the handling of blood donors scheme (only for the restructured hospitals) and the special hospitalization schemes for civil servants. The hospitals have also inherited highly complex human resource practices in public organizations that cannot be easily handled by the ERP systems.

While some of the sector-specific misfits are clearly rule based and formal, others are driven more by the circumstances or contexts the public organizations are in. For example, the lack of segregation between patient appointment and visit in the data model of the ERP system renders the management of appointment scheduling difficult (for example, comparison between appointments and actual visits). While the private hospitals also face similar problem, this inadequacy is, however, more crucial in the public hospitals, given the large number of patients they have to attend to and the strong external pressure from the government to improve patient services among the public hospitals.

Industry-Specific Misfits

Industry-specific misfits arise from the incompatibilities between the generalized practices embedded in the ERP systems (some of which have originated from the manufacturing environment) and the unique practices specific to some industry. They typically affect the more generic modules of ERP systems, for example, finance, logistic, and human resource. For the hospitals, one unique practice in their operation is the accounting and treatment of fixed assets. The frequency of medical equipment purchase is high. Moreover, the purchases often come from specific approved funds with pre-established conditions. The need for tracking and monitoring this specialized medical equipment demands a much greater depth in terms of fixed asset accounting functionality than the available functionality in the ERP system. Similarly, the generalized human resource functionality fails to take into account the need for managing the complex human resource development requirement in the heavily service-oriented hospital operation. For example, some of the critical attributes of physician competency, for example, seniority rank and medical specialties, have been omitted in the employee master record in the HR module. There is also no functionality for tracking nursing scholarships and bonds. As one user noted, 'the system is fine for handling basic payroll but it is definitely not designed to handle the human resource development for hospitals'.

Organization-Specific Misfits

All the above-mentioned misfits tend to be common across the three hospitals. But despite the fact that the same ERP systems are implemented in the same healthcare industry context by the same consultant, there are other misfits that arise specific to the respective hospitals. These idiosyncratic misfits reflect organizational differences in the medical specialties offered, the pursued hospital strategies, the organizational structure, the management preferences, the user composition, and even the internal power balances that have evolved over time in each hospital. For example, due to the constraint of room shortages, one hospital has to allow several medical specialties to share common rooms. The inability of ERP to link one room to several organizational units has posed a problem for the hospital. Similarly, the different risk tolerances of management has also resulted in varying extents of system modification to ensure validation and default checks. For example, while one hospital insisted on the customization of validation routines for data entry, another hospital chose to 'rely on the empowered role of the users' by providing training and supervision. To a certain extent, the surfacing of these organizational-specific misfits also reflects the power balances among different players in the organization. One ISD project member in one hospital, for example, noted they had to concede to an 'aggressive' user request to change report format. But ever since a new division head took over, the division has reverted to the standard ERP report. The customized report was not used at all.

Overall, we have also noted that the extent of context specificity varies across functional modules. Areas such as accounting and finance, where international accounting standards promulgate some degree of global standardization (across country, sector, industry, and organization), there were fewer misfits. On the other hand, there were many more such misfits for patient care systems, where practices vary more depending on the local implementation contexts.

16.5 Nature of Misfits

While the sources of context specificity provide a framework for thinking through and identifying potential misalignments, they do not provide sufficient guidance about how these gaps will affect system implementation and use. Understanding these implications is important as it provides a sense of the resolution efforts required. In line with a traditional software application perspective (Soh, Sia, and Tay-Yap, 2000), we have thus sub-classified the misfits arising from the different sources of specificity into three broad categories: input, process, and output (see Figure 16.2).

Input misfits are misfits that relate to cumbersome input functionality, inappropriate format for data input, and inappropriate relationships among entities as represented in the underlying data model. The examples noted in Table 16.1 – the need to toggle over multiple screens without input defaults, the inability to expand the first name field beyond 30 characters, and the limitation of assigning an external ID number as the key field for cross-module linkage – are good illustrations of such misfits. **Process** misfits arise from inappropriate or inadequate processing procedures in the ERP systems. Some of required functionalities are control related (for example, missing validation procedures or checking routines) while others are operational in nature. One example is the inadequate pricing functionality in handling the multiple services provided by the hospitals: inpatient pricing (for example, handling the concept of bed-class for room charges), outpatient pricing, surgical pricing (for example, handling split fees for multiple consultants), new-born pricing (for example, higher of accouchement and caesarian charges), package pricing, and even pricing override. **Output** misfits arise from inappropriate presentation format or missing information content in key transaction documents, screen display, or reports. By far, this is the most prevalent form of misfit noted. Table 16.1

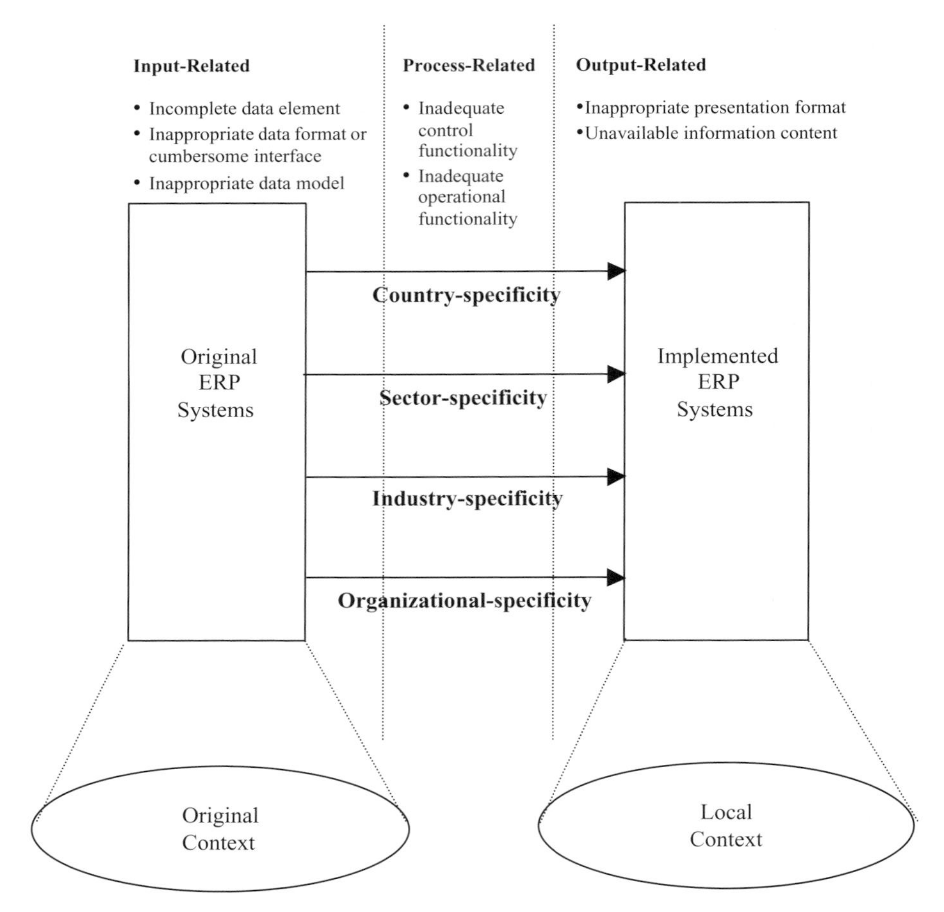

Figure 16.2 Sources and Nature of ERP misfits

provides an illustrative listing of the nature of misfits within each dimension of context specificity.

Although these misfits may appear simple, resolving them in ERP is cumbersome, as it requires changing the structure and relationship of the table objects or even changing the source code at the application architecture level, which is often viewed as prohibitive by ERP consultants. Such changes are generally avoided because of the cost involved and the difficulty of maintaining future upgrades. Even when customizations are needed to provide critical functionality, they are done without changing the source code, through the development of add-on modules that are plugged into the ERP's user exits. Even then, such a strategy may raise some problems during system testing (for example, bugs in the add-on modules). In addition, subsequent versions of the ERP software

Table 16.1 The sources and nature of ERP misfits in the hospitals

	Exogenous sources			Endogenous sources
Misfit type	Country-specificity	Sector-specificity	Industry-specificity	Organizational-specificity
Input (**Cosmetic**) Incomplete data elemen, inappropriate data format, or cumbersome interface	The NAME fields are in the Western format as first, middle, and last name. Inappropriate for the Asian names (Indian, Malay, Chinese).	Capturing of patient addresses to be in MHA (Ministry of Home Affairs) format for ease of integration among public organizations.	Unique to the hospitals, the HR records require a field to indicate whether staff need uniform or if staff are front-line officers.	Setting default for the physician's performing operating units (user preference). Requests to cut down number of screens and to establish automatic screen flow (user preference). Suppressed inactive fields (user preference).
Input (**Substantive**) Inappropriate data model	All citizens of Singapore have unique ID numbers that are used as patient IDs. However, ERP system generates its own internal number for integration across modules, thus making it difficult for enquiry across modules. The ERP system only handles one insurance relationship versus the need to accommodate for multiple insurance relationships locally.	Lack of distinction between patient appointment and actual visits as separate data entities for the management of appointment scheduling, e.g., tracking of walk-in cases.	The hospitals require PO numbers as search keys to relieve asset records. However, PO numbers have not been structured as such in the fixed asset module.	The inability of the system to allow ward sharing across medical specialties (organizational structure).
Process (**Cosmetic**) Inadequate control functionality	Validation checks for external field, i.e., patients' NRIC numbers (unique identification numbers for all citizens).	Inability to default differential pricing for non-resident patients.	Lack of material management functionality to track the receipt and expiry dates of drugs, the classification of standard and non-standard drugs, which is important in the hospitals.	Setting maximum value for service rule (management requirement). Establishing mandatory fields in service masters (management requirement). Validation checks in visit creation (management requirement).

(*cont.*)

Table 16.1 *(cont.)*

	Exogenous sources			Endogenous sources
Misfit type	Country-specificity	Sector-specificity	Industry-specificity	Organizational-specificity
Process (Substantive) Inadequate operational functionality	Deriving from a Western healthcare model, some of key functionalities like service pricing (inpatient pricing based on bed-class, surgical pricing with split fees among multiple consultants, package pricing), billing and collection (estimated provisional bill for CPF claim, counter collection, installment payment), and deposit monitoring are missing.	Handling of blood donor scheme for restructured hospitals. Handling of special hospitalization scheme for civil servants. Inadequate appointment management functionality, e.g., tracking the changes to scheduled appointment.	Lack of fixed asset accounting functionality for monitoring and tracking of the frequent purchases for medical equipment in the hospital industry. HR module also lacks functionality to cater to the complex research and professional developments of nurses, physicians, and visiting consultants.	Functionality for collective processing of the returns of medical records (due to the high volume one hospital has to deal with). Pricing functionality- higher of accouchement or caesarian fee, lodger visits (unique maternity specialty).
Process (Substantive) Inadequate operational functionality	Unlike the West where single-bedded rooms are common, the 'bed-class' concept requires greater functionality in bed assignment, e.g., auto-proposal based on search criteria (sex, specialty, etc.). Unlike the West where accesses to medical specialists are often through referral by general practitioners, individuals can seek direct treatment from specialists locally. The original appointment management module lacks such functionality to search for available slots by medical specialty or doctor's name.			

Output (Cosmetic) Inappropriate Presentation	Report presentation for submission of claims in the Medisave or Medishield schemes.	Inappropriate A&E statistics reporting to the Ministry of Health	Reflecting a business-oriented emphasis, patient cases are referred to as account no. rather than as case no.	Warning rather error messages on screen, missing heading information in reports, e.g., date, time, and page number (user preference). Upper-case for screen presentation (user preference)
Output (Substantive) Unavailable data element	Patients by 'bed-class' report not available for government subsidy reimbursement. Missing information to identify the 'spoken language' of physicians to facilitate communication with patients.	Government tracking of effectiveness of public medical services, e.g., report for stay more than 10 days, report for readmission within 30 days/one year.	Reflecting a bias towards the manufacturing environment, the HR module lacks HR development functionality. For example, the employee master in the HR module has omitted some of the critical attributes of physician competency, e.g., seniority rank and medical specialty.	User request to have the ability to print bills based on date range (customer service) Report on number of times patients have been moved (customer service) Revenue/reconciliation reports (management requirement) Reports for collection agency management (outsourcing of receivable collection)

may not retain the same user exits, and this complicates the upgrade process. Thus, unless it is absolutely critical, package customization is to be avoided. Most resolutions tend to be workarounds. However, while workaround trade-offs may be acceptable for the more **cosmetic** misfits (see Table 16.1, for examlpe, the inconvenience of having to go through more screens, having the users adapt to the standard report format, and having to compensate for inadequate validation checks), the compromise of working-around **substantive** misfits (see Table 16.1, for example, the tedious and error-prone efforts of manually searching for records, performing alternative procedures, gathering unavailable information) is often costly. The sub-classification of misfits by the nature of misalignment thus provides a means to gauge the severity of misfits.

16.6 Sources and Nature of Misfits

We noted a pattern between sources and nature of misfits. As noted in Table 16.1, the misfits arising from exogenous sources (that is, country, sector, industry specificity) tend to more pervasive in nature. The same issue can have simultaneous implications for system input, processing, and output. Many of them are substantive (for example, inappropriate data model, inadequate operational functionality, unavailable data element). Given the high cost of workarounds, there is little the hospitals can do besides opting for customization using available user exits in the ERP systems. The customization of these misfits were handled either by the ERP vendors as contracted localization or by separate consultant engagement to develop them as add-on modules. A few of them were also customized by the internal IS department.

On the other hand, the endogenous organization-specific misfits appear to be more cosmetic in nature. Many of them relate to inappropriate data format, cumbersome data entry procedures, insufficient control checks, and presentation preferences in reporting requirements. The resolutions of these misfits are typically workarounds in terms of users adapting to new ERP functionality, performing alternative procedures either manually or within some other functionality within the systems, or simply accepting the shortfall and compromise on the hospitals requirements.

The implication of the analysis is that misfits arising from exogenous sources are often not within an organization's control. It is thus important for organizations to evaluate the extent of these gaps as it helps us to assess the reasonableness of the implicit '80% adoption–20% customization' assumption in ERP implementation. If the gaps for these non-controllable misfits are significant, it may not make sense to implement ERP system since the cost of customizing these substantive misfits may eventually outweigh its expected benefits. Instead, organizations in highly specific context should perhaps consider the enhancement of an in-house developed system.

On the other hand, the analysis of endogenous organization-specific misfits provides a glimpse of the kinds of trade-offs (for example, more cumbersome work procedures, higher risk of error, less-structured report presentation) an organization needs to make in implementing ERP systems, since these compromises are 'controllable' by the management. Understanding these implications is important as it helps to set the user expectations in planning a smooth transition to a new ERP system.

16.7 Balancing the Perspective: Functional Excess

The focus on misfits as functionality deficiencies (that is, only the scenarios that fall short of the hospitals' requirements) thus far, however, provides only a one-sided view of ERP implementation. Organizations considering the adoption of ERPs should also consider the functionality excesses (that is, scenarios where the available features in ERPs more than satisfied the hospitals' requirements) that the hospitals can leverage upon. For example, the implementation of ERP systems has accelerated the hospitals' transitions to activity based costing, and is providing information on resource usage (for example, through integration) that was previously not available. One manager noted, 'left to ourselves, we would probably have taken longer to move in this direction. But the ERP system forces us to jump in to re-orientate ourselves to the concept of profit centers and learn about the specific cost and revenue of our products.' Another user in the finance module also noted, 'we have not been doing cashflow projection but will definitely do so in the future, given that the forecasting feature is readily available in the system'.

It is understandable that implementation participants focus on misalignments or shortfalls, as their implicit standard of comparison is with the way things have to be done within the organization. Typically, much of the 'excess functionality' will not be appreciated during implementation, but may surface as the organization uses and becomes more familiar with the package after implementation. Ideally, however, as in the case of shortfalls, early identification of functional excesses can contribute to a better initial configuration and design of organizational processes and structure.

Thus, a balanced analysis of the investment in ERP systems should take into account the following:

- analysis of exogenous functional deficiencies arising from country, sector, and industry specificity, typically not within an organization's control
- analysis of endogenous functional deficiencies arising from organizational-specific sources, typically controllable to a certain extent
- analysis of functional excesses available in the ERP system that an organization can potentially leverage on.

16.8 The Lessons Learnt

Our misfit analysis in the ERP implementations of the three hospitals does not dispute the trend in ERP adoption. But as indicated by the misfit examples, there is a need to recognize the various sources of context specificity and to consider the nature of potential misalignments before jumping on the bandwagon to adopt an ERP system.

More generally, we believe ERP vendors need to spend more time explaining the embedded input, process, and output structures to the organization. Organizations need to acquire more skills to ask and probe for such details. We were surprised to find that the reference models that espouse industry best practices are at too high a level for an effective assessment of how the ERP system would actually affect the organizational processes. As one hospital executive observed: 'We needed to go down two or three levels.' Moreover, effective misfit analysis requires both comprehensive understanding of the critical organizational processes (an analysis activity) and detailed knowledge of this very complex software (a design activity). There is thus the need to merge the traditional system development separation of analysis and design phase for ERP implementation.

Fundamentally, the misfit analysis reveals the severity of the knowledge gap in ERP implementation. The three key parties to this process – key users, IS department personnel, and ERP vendor – have different and specific knowledge (organizational requirements, existing IT infrastructure, package functionality, respectively) that is difficult to transfer to one another. While frequent interaction and joint problem solving appear to be the logical way to bring the disparate knowledge together, the varied backgrounds and interests of the three parties make it difficult to achieve an integration of these knowledges.

We suspect the 'stickier' knowledge components in ERP implementation are organizational requirements and processes, given the need to surface country, sector, industry, and organization specific practices. Hippel (1994) has suggested that where the information is sticky, the optimal strategy is to place the locus of problem solving with the sticky source, in this case, the key users and ISD, rather than the consultants. Users face the demand not only to be competent in their business areas, but also to assimilate the package functionality in some depth. They must now consciously 'get into the ERP software' to evaluate the appropriateness of the new configured system or the alternatives adopted. ISD often have some feel for organization knowledge already, and usually have more IT knowledge to facilitate acquisition of ERP knowledge. Organizations can facilitate the knowledge acquisition process by budgeting for vendors to spend time educating key users and ISD about the system, by shifting the ERP focus training earlier in the implementation process, by planning for detailed data, functionality and output walk-through, and by selecting vendors with significant industry knowledge. Most importantly, users should realize that it is no longer sufficient for them to be passive functional experts as in the traditional system development projects. They have a much bigger role in ERP implementation!

REFERENCES

Davis, G. B. (1988) Commentary on Information Systems: To Buy, Build, or Customize? *Accounting Horizons*, March, 101–103.

Hippel, E. V. (1994) Sticky Information and the Locus of Problem Solving: Implications for Innovation. *Management Science*, **40**(4), 429–439.

Lucas, H. C. Jr., Walton, E. J., and Ginzberg, M. J. (1988) Implementing Packaged Software. *MIS Quarterly*, **12**, 537–549.

Markus, M. L. and Tanis, C. (2000) The Enterprise Systems Experience: From Adoption to Success. In *Framing the Domains of IT Research: Glimpsing the Future Through the Past*, Zmud, R. W. (ed.), Cincinnati, OH: Pinnaflex Educational Resources.

Soh, C., Sia, S. K., Boh, W. F., and Tang, M. (2001) Misalignments in ERP Implementations: A Dialectic Perspective. *Journal of Human-Computer Interaction*, forthcoming.

Soh, C., Sia, S. K., and Tay-Yap, J. (2000) Cultural Fits and Misfits: Is ERP a Universal Solution? *Communications of the ACM*, **43**(4), 47–51.

Volkoff, O. (1999) Enterprise System Implementation: A Process of Individual Metamorphosis. In Proceedings of the Academy of Management '99 Conference, August, Chicago.

17 Implementing Enterprise Resource Packages? Consider Different Organizational and National Cultures!

M. Krumbholz, J. Galliers and Neil A.M. Maiden

17.1 The Impact of Culture on ERP Implementations

A recent Standish Group report on ERP implementation projects reveals that these projects were, on average, 178% over budget, took 2.5 times as long as intended, and delivered only 30% of promised benefit, and a recent survey of 12 recent projects revealed that adapting the implementation to the prevailing cultural style was one important cause of this project underperformance (Densley 1999). These findings support anecdotal evidence for the impact of culture reported in the IT press (for example Warren, 1999). This importance of culture is hardly surprising. A customer who implements an ERP package has to change its business processes to the ERP supplier's best-practice processes (Curran and Ladd, 1998). The change both impacts on the customer's organizational culture (that is the ways that things are done in the organization) and is constrained by it. In Europe, the picture is even more complex because companies also have diverse national cultures which influence this organizational culture and make the successful implementation of multinational ERP implementations difficult, as reported in Gulla and Mollan (1999). Indeed, evidence suggests that ERP implementations in North America have been more effective because of the more complex European organizational and national cultures (Warren, 1999). If more ERP implementations are to deliver their promised benefits within budget, we need to understand how organizational and national culture impact on ERP implementations, and how this understanding will deliver better methods for implementation partners and customers to use.

In contrast to the lack of research in computer science, social and management science researchers have investigated the influence of organizational and national culture on organizational behaviour. Unfortunately this research neither addresses issues that are specific to information systems development, nor does it have a tradition of model-theoretic approaches which are familiar in information systems research to describe and predict problems and their solutions. If we are to implement ERP solutions which recognize organizational and national cultures, we need at least to model culture to describe and to predict its impact on an ERP implementation.

This paper reports the synthesis of a meta-model of culture from social and management science theories of organizational (Schein, 1992) and national (Trompenaars, 1994) culture. It also reports the extension of this modelling approach to existing ERP implementation methods such as ASAP (Curran and Ladd 1998) to enable implementation teams to infer important properties of national and organizational culture in order to provide a theoretically grounded checklist of culture issues that might influence these implementations. Our own studies of current ERP implementations (Krumbholz and Maiden, 2000) have identified three different types of culture-related clash that, this paper predicts, might influence ERP implementations:

P1: the current organizational culture clashes with the planned future culture;
P2: the supplier's culture, implicit in the ERP package, clashes with the customer's organizational culture, and;
P3: the new business processes (configured using the ERP solution) clashes with the existing organizational culture.

The studies reported in this paper will explore these predictions, and we will return to them at the end of the paper.

One consequence of the existence of these three types of culture clash is that, in order to handle culture, the implementation team will need to model: (1) the customer's business processes and solution systems; (2) characteristics of the organizational and national culture that impact on the implementation of these business processes, and; (3) how these facets of organizational and national culture impact on the business and software solutions. Implementation teams will use current ERP modelling approaches (Curran and Ladd, 1998) to model current and future business processes, then extend these models using the meta-model of culture to model important characteristics of the customer's organizational and national culture. It will then analyse these extended models to infer potential problems from the differences between the two business process models using rules that codify the synthesized theories of organizational and national culture. However, to deliver such a method, it is necessary first to synthesize new theories of culture, and to understand better the types of culture-related clashes (P1–P3) that occur during ERP implementations.

The method is outlined in Figure 17.1. This paper reports research that investigates how organizational and national culture causes ERP

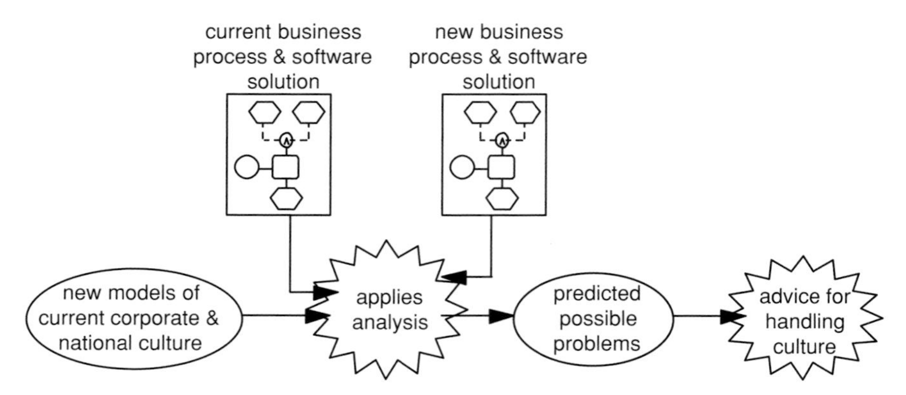

Figure 17.1 An outline of the extended ERP implementation method for culture-sensitive ERP implementation

implementation problems using the meta-model and predictions from social and management science research.

The remainder of this paper is in 5 sections. Section 17.2 summarizes existing social science theories of culture. Section 17.3 presents the first-draft model of culture. Section 17.4 reports empirical studies into the implementation of SAP's R/3 in the Scandinavian subsidiary of a large European pharmaceutical organization. Section 17.5 describes the results from these empirical studies. The paper ends with a discussion of the results relevant to our predictions, implications for the meta-model of culture, and future research plans.

17.2 Social and Management Science Research on Organizational and National Culture

Most social and management science research on culture can be divided into two camps (Dobson, Williams, and Walters, 1993). The first camp claims that culture is something tacit which arises naturally. The other, more common view, is that culture is something explicit which arises from social interaction. Social and management science research also divides research into national cultures, business cultures, and organizational cultures. A good starting point is Schein (1992) who provides the following definition of organizational culture:

> a pattern of basic assumptions – invented, discovered or developed by a given group as it learns to cope with its problems of external adaptation and internal integration – that has worked well enough to be considered valid and therefore to be taught to new members as the correct way to perceive, think and feel in relation to those problems.

He argues that organizational culture can be divided into three layers. In the outer layer there are values which are written down statements about the strategies, missions, and objectives of the organization. In the middle layer there are beliefs, which are the issues that the employees of an organization talk about. In the inner layer there are the 'taken for granted' assumptions which are those aspects of the organizational life which people find it difficult to recall and explain. Schein also describes ten dimensions that he uses to differentiate between organizational cultures in different organizations. These are the observed behavioural regularities of human interaction, the group norms, the espoused organizational values, the formal philosophy, the rules of the game, the climate, the habits of thinking, people's mental models and/or linguistic paradigms, their shared meanings, their embedded skills, and the organization's root metaphors or integration symbols. These dimensions indicate important classes and attributes of culture to model in order to improve our understanding of culture's impact on ERP implementation.

Hofstede (1994) also investigated organizational and national culture. He argues that there are four manifestations of culture, and the differences between national and organizational culture are due to these different manifestations. He also places these manifestations in layers similar to the layers from Schein (1992). Hofstede differentiates between layers that have symbols which represent the most superficial culture often described as practice, layers which have values which represent the deepest manifestations of culture, and intermediate layers which describe heroes and rituals indicative of the organizational culture. He claims that national culture differences reside more in values and less in practices, and organizational culture differences reside more in practices and less in values. Furthermore, he claims that we can detect national and organizational culture differences using a set of dimensions similar to those from Schein (1992). Derived from extensive empirical studies, Hofstede provides four dimensions which differentiate between national cultures: power distance, individualism–collectivism, masculinity–femininity, and uncertainty avoidance (Hofstede, 1994). Likewise, he also detects six dimensions to differentiate between organizational cultures: process versus results oriented, employee versus job oriented, parochial versus professional dependent, open versus closed systems of communication; loose versus tight control; and normative versus pragmatic organizations (Hofstede, 1994). As with Schein's findings, Hofstede's dimensions indicate important elements of culture to model in order to improve our understanding of culture's impact on ERP implementation.

Trompenaars (1994) argues that national culture can be described with three layers similar to those from Hofstede. A central theme of Trompenaar's argument is that people organize themselves in such a way so as to increase the effectiveness of their problem-solving processes, and so have formed different sets of implicit logical assumptions to enable this to happen. Each culture distinguishes itself from others in terms of its solutions to these problems. These problems can also be classified to differentiate between national cultures in a similar way to which Hofstede uses his six dimensions. These classes are how people relate to each other (sub-divided into universalism versus particularism, individualism versus collectivism, neutral versus emotional, specific versus diffuse, achievement versus ascription), people's attitudes to time, and people's attitudes to the environment (Trompenaars, 1994). Again, these classes provide a basis for modelling the critical determinants of organizational and national cultures.

To summarize, the work of these reported and other social science authors reveals four basic conclusions which we might be able to exploit to improve ERP package implementations:

- theories of organizational and national culture have similar definitions of culture and share important concepts that include values, beliefs, and norms (Krumbholz and Maiden, 2001);
- these theories distinguish between the deep manifestations and the superficial characteristics of organizational and of national culture;
- the critical determinants of organizational culture reside more in observable practices, whereas critical determinants of a national culture reside more in the nation's deeper set of values;
- organizational and national culture can be described using multiple dimensions which give us a set of overlapping characteristics with which to describe aspects of culture.

To explore the research predictions reported in this paper, we have applied social science research to develop a first-draft model of organizational and national culture which is applied to describe culture-related problems during ERP package implementation processes.

17.3 A Model of Culture and Its Impact on ERP Implementation

The extended ERP implementation method proposed in section 1 integrates current business processes approaches with models of organizational and national culture to discover potential problems using predictions of possible problems using social science theories of organizational and national culture. The basis of the model is a knowledge meta-schema for modelling the critical characteristics of organizational and national culture. This, in itself, is an innovative advance. Social and management sciences do not have a tradition of conceptual modelling for describing and analysing systems. Indeed, non-computer science disciplines resist conceptual modelling because it is too difficult to capture and describe the knowledge without losing the essential context of the knowledge. However, the common definitions of culture from social science researchers offer exciting opportunities for IT disciplines. We have synthesized and extended these social science theories to model the problems observed in current ERP package implementations.

A Tentative Knowledge Meta-Schema for Culture

The first stage of the research was to design a knowledge meta-schema capable of representing both the deeper manifestations and superficial attributes of organizational and national culture, and the critical causal associations between them. The knowledge meta-schema also incorporates modelling concepts from standard business process models such as SAP's EPC models so that the knowledge meta-schema can be used in methods such as SAP's ASAP method (Curran and Ladd, 1998). For readers unfamiliar with these models, EPC models specify events, processes, and event flows for business processes. The full first-draft knowledge meta-schema is shown in Figure 17.2.

Space does not allow us to give a full definition of the knowledge meta-schema. Instead, first, we provide key definitions drawn from software and business process models common in current ERP implementation methods (for example Rosemann et al., 1999). An agent is a type of object which processes actions (Ncube and Maiden, 1999). Agents perform actions to achieve goals. With respect to culture, agents have beliefs, values, and norms that govern their actions. One instance of an agent can be one individual person, a collection of people, one machine, or a collection of machines. An action is the process of doing something with the intention of achieving a desired goal (Maiden, 1998). Actions are constrained by pre-conditions and post-conditions. *Pre-* and *post-conditions* are conditions which must occur for an action to begin or end. An *event* is a moment in time when something happens. In the knowledge meta-schema events start and end actions. An *object* is something which is manipulated for the attainment of a goal. An object can be a physical object (for example a radio), an infological object (for example information about an incident) or an object with both physical and infological

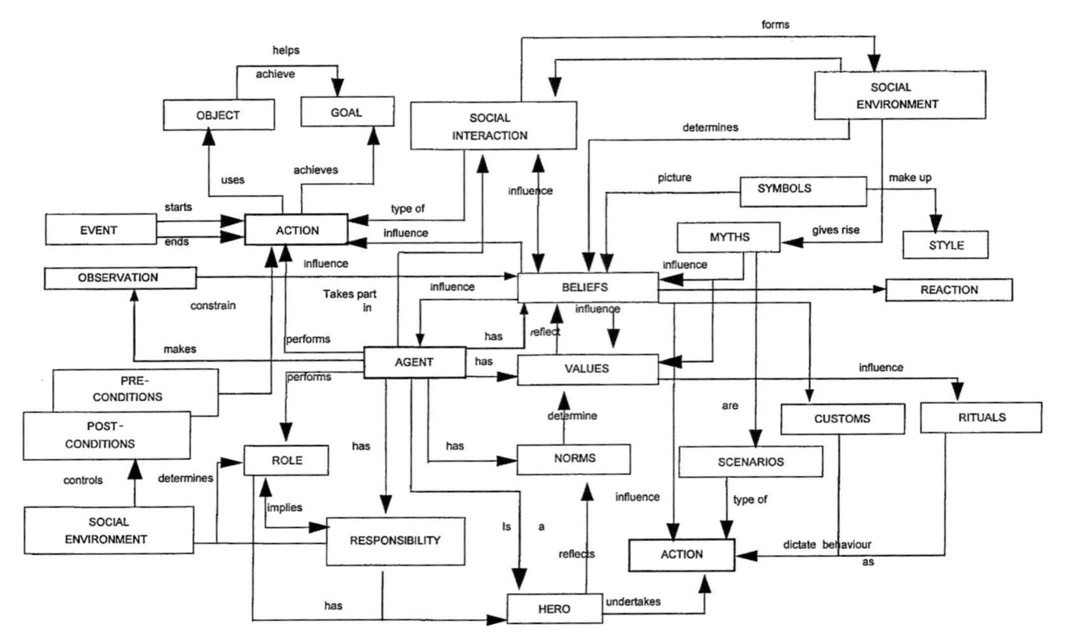

Figure 17.2 The knowledge meta-schema that describes elements of organizational and culture which impact on ERP implementation

elements (for example an incident report). A *goal* is a high-level objective that the system should meet. Goals are achieved by actions performed by agents manipulating objects (Darimont and Van Laamsweerde, 1997). All concepts except for a goal are common in most current ERP business models, and represent many of the more observable indicators of organizational culture according to Hofstede (1994). Goals were specified in the meta-schema to provide a suitable new abstraction of intent missing from current ERP modelling approaches (Mylopoulos, 1999).

The rest of the knowledge meta-schema is drawn from social science research into culture summarized in the previous section, from philosophy, and from research into artificial intelligence models of rational agenthood. These concepts are also critical determinants of organizational and national culture according to the theories reported in the previous section.

A *role* defines the obligations of an agent. An agent can fulfil one or more role type. *Responsibilities* define the liabilities of the agents associated with the role they perform. Agents are responsible for fulfilling a role and initiating, controlling, or undertaking actions related to their role (Maiden, 1998). A *hero* is a human agent admired by other agents in the organization (Deal and Kennedy, 1982). Heroes are agents that, through their role and responsibilities, undertake actions which reflect the organizational beliefs, norms, and values. A *social interaction* describes an interaction between agents. It is a specialization of an action. Social interaction can influence agents and their beliefs.

A *belief* is a mental state held by an agent about a particular proposition. The properties of beliefs can be derived from their representation as a modal operator with a possible-worlds semantics (Hintikka, 1962; Galliers, 1989; Krumbholz et al., 2000). Beliefs influence agents; they are dispositions to act (Engel, 1984; Quine, 1970). A *value* is a special kind of belief; it is a belief that that is relatively less 'vulnerable to removal' (Levi, 1984), that is is more persistent or entrenched (Galliers, 1992; Gardenfors, 1988; Harman, 1986). Values underpin other beliefs about the worth or importance of something. They thus provide for spectrums of belief about good versus evil or normal versus abnormal, for example (Hofstede, 1991; Schein, 1985). A *norm* within a social grouping G is a kind of behaviour A, that all the members of G perform in a certain context C. In addition, all the members of G mutually believe they should perform A in C (Bach and Harnish, 1979). In other words, a norm is a standard expectation of normal behaviour including what is right or wrong in a particular context (Hofstede, 1991; Trompenaars, 1994).

A *scenario* is a sequence of events which start and end actions which describe current or future business processes and/or ERP software use (Maiden, 1998). Scenarios can describe norms, and can be embellished with contextual information to create stories. A *myth* is a scenario, either factual or invented, which encapsulates the organization's and/or agent's beliefs, norms, and values (Johnson, 1992). A *custom* is an established behaviour expressed as actions, codes, or rules of behaviour (Deal and Kennedy, 1982). Customs are created and influenced by norms, values, and beliefs, and influence behaviour in the form of actions and events. A *ritual* is a repeated action or scenario that expresses the goals and values of the organization and dictates behaviour in the form of actions (Hofstede, 1994; Deal and Kennedy, 1982). *Symbols* are objects explicit to people outside the organization such as buildings and logos which are manifestations of the organization's hidden assumptions, beliefs, norms, and values (Hofstede 1994). Finally, the *style* is the way people are dealing with other people within an organization, such as the way they talk to each other. Symbols form the style of the organization (Peters and Waterman, 1982).

More Predictions About Culture

Krumbholz and Maiden (2000) report that the knowledge meta-schema can be applied to detect and model critical determinants of culture that influence problems with the implementation of ERP packages. However, for the proposed ERP implementation method to be effective, the model will also need to explain and predict consequences of organizational and national culture on ERP implementations. Our model also applies results from existing social science research into the impact of culture on change to make predictions about ERP implementations that inform 1 prototypical theoretical prediction.

Hofstede (1994) claims that national culture differences reside more in values and less in practices, and that organizational culture differences reside more in practices and less in values. As our model assumes that a customer's national culture is manifest through its organizational culture, so one further prototype prediction is:

P4: critical determinants of organizational culture that reside more in a customer's observable practices have causal associations with problems that arise during ERP package implementations.

The next section reports the empirical method applied to a large, multinational SAP package implementation that was undertaken to explore these predictions.

17.4 The Empirical Method

Empirical studies were undertaken on-site at a large multinational European supplier of pharmaceuticals and laboratory equipment. The company was implementing the same German ERP package in its different national subsidiaries. This paper reports results from visits to the Scandinavian subsidiary. The Scandinavian subsidiary had sales and administrative offices outside Stockholm and the warehouse in a small town 150 km west of Stockholm.

The Data Gathering Approach

Two bi-lingual (Swedish/English) academic researchers visited the Swedish site for two days in December 1999. The research team gathered data from the IT development managers, sales and distribution developers, materials-management developers, sales managers, sales and distribution key users, the warehouse manager and warehouse key users. A questionnaire was sent to each participant in their native language requesting their name, background, position, length of time in the company and department, nature of their work, and typical work routines. Ten questionnaires were returned.

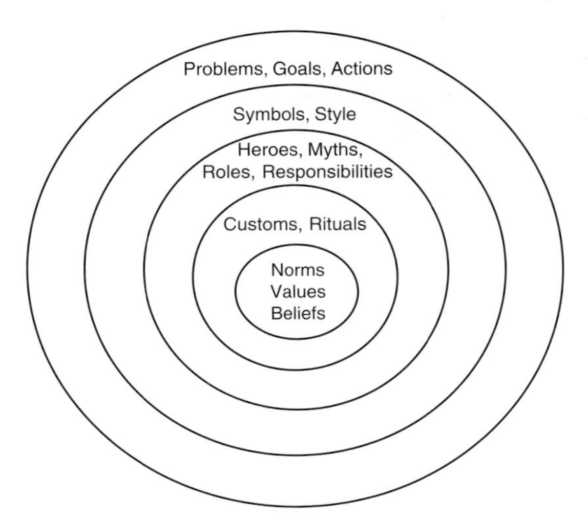

Figure 17.3 A synthesis of the layers of the elements of culture taken from social science research to provide a basis for the questioning method

On-site data gathering was in two phases. The first phase elicited surface manifestations of culture. The second phase used a variation of the laddering technique from knowledge engineering (Rugg et al., 1992) to elicit deeper determinants of culture which were possible causes of these surface manifestations and problems reported in the first phase. The characteristics of culture found on the outer layers of the social science models are the observable manifestations of culture. Characteristics of culture found in the inner layers are tacit, more important determinants of national and, to a lesser degree, organizational culture. Our questions were designed to elicit the observable manifestations of culture described on the outer layers, then use precise verbal probes to elicit tacit rationale for these manifestations. This use of external manifestations of culture to elicit deeper underlying causes was the principal reason for two phases rather than one. The specific questioning approach was based on a synthesis of these layers of culture from social science research shown in Figure 17.3. The questioning method was derived from a synthesis of culture dimensions undertaken by the authors based on dimensions from Hofstede, Schein, and Trompenaars reported in Section 17.2. These dimensions enable us to describe and compare critical manifestations of organizational and national culture. The application of a 'neutral' method from social science research counters claims that the studies 'found what they were looking for' in terms of the posited knowledge meta-schema.

First Phase Data Gathering

The first phase is divided into two stages: the first stage involved making observations on the subsidiary's 'way of doing things'; in the second stage each stakeholder was interviewed individually for 30 minutes in his or her own language. The questions were derived from reports of major problems with ERP implementations and previous studies of culture impact on ERP implementations (Krumbholz and Maiden, 2000). Examples of questions were: 'How was the ERP implementation decided (Communicated throughout the company and mutually decided, or announced)?', and 'How aware would you consider yourself to be of the way the company is doing business (Its business processes, way of doing things etc.)?' All data elicited through the interviews were transcribed to provide a data corpus for analysis. The observations provide a useful general feeling about the culture of an organization and information about the implicit values, beliefs, and norms, which are discussed below as elicited from the data gathering.

Second Phase Data Gathering

The aim of the second phase data gathering is to explore the underlying interviewees' norms, beliefs, and values about the organizational and national culture so as to be able to predict possible problems that could arise during and after ERP implementations. In this stage the organizational, national, and departmental culture was elicited, which together with the data elicited from the other two stages, provided a valuable corpus of information about the culture and possible problems that could arise. The second stage of data gathering provided us with information about implicit characteristics of the client's organizational and national culture. This stage used a statement questionnaire (Appendix 1), which was based on a synthesis of research findings about organizational and national culture (for example Hofstede, 1991; Trompenaars, 1994) and available methods for eliciting information about culture (Deal and Kennedy, 1982). The interviewees received the questionnaire at the end of the first stage of data gathering and have to rank with a scale form A to G (A: strongly agree, G: strongly disagree). Examples of statements are: 'the management is more concerned with employees getting the work done than with the employees as persons' and 'deadlines are loose and flexible'. This stage of data gathering was divided into two phases: the first one examined the responses of each interviewee; in the second stage we returned to all interviewees and asked further 'why' questions, in order to understand the reason why s/he responded the way s/he did to each statement. This is a very important factor, as all the stages of the data gathering approach are linked and interrelated.

17.5 Results

The method was effective for eliciting a large corpus of data about the ERP package implementation and the subsidiary's organizational and national culture. It also facilitated our understanding of the problems that arose because of the ERP implementation by modelling them in the context of the subsidiary's culture (Krumbholz and Maiden, 2001). The Scandinavian subsidiary encountered several problems during the implementation of the ERP package. This section examines a specific problem with the warehousing system by modelling it with the knowledge meta-schema to make explicit otherwise tacit knowledge about relevant organizational and national culture. Data describing this problem were modelled, then the interview transcripts with the relevant personnel were analysed. The aim was to produce a model which identifies stakeholders' norms, values, beliefs, and other critical determinants that indicate the organizational and national culture associated causally with this specific problem.

First of all an overview of the SAP implementation is presented to give the reader the necessary background information. Secondly, the national and organizational culture of the subsidiary is described and analysed using data from the data gathering approaches. Lastly, having set the context for the SAP implementation and analysed its culture, we discuss a specific problem that it faced with the warehouse module.

Overview of the SAP Implementation in the Swedish Subsidiary

The Swedish subsidiary had been using SAP version 4.5 for six months when the study was carried out. The previous system was a terminal-based bespoke system that had been used for ten years. During this period amendments were made to the system continuously, which made the system quick to use and popular amongst the employees. It was nevertheless old-fashioned and the potential for further developments was limited.

For one year before the implementation consultants from a Swedish implementation partner worked at the company site configuring the system and training staff to use it. The involvement of the consultant firm declined after the system went live, but after six month's of use there were still full-time consultants configuring the system. There was widespread discontentment with the consultants' level of support and the lack of knowledge of the business. To overcome these problems an independent expert on SAP implementations was hired to deal with user questions and to support their needs. This expert was highly appreciated, and the company, to a large extent, still depends on his knowledge.

Initially the training was discussed and decided on by a group of representatives from each SAP module. They came up with suggestions that were presented to the personnel in the different departments. The IT staff attended SAP-led courses, but the general opinion was that these courses were too general and could not be applied to the company's specific conditions. The goal was to have in-house training, and key users were trained to support their colleagues. Six months after the implementation the majority of the training was done in-house, and some courses were run by the independent implementation expert. The overall view was that the training could be more extensive, but there were large differences between different departments and users that had to be catered for.

The personnel in the warehouse were unable to use the system when it went live and during the first month after the implementation mainly tried to survive and teach the users how to interact with the system. An additional factor was the recent move of the warehouse to a new location, which resulted in many employees resigning. Consequently, a lot of time and effort was spent in training the new personnel about the old system and developing new routines in the warehouse. Later, they had to learn the new SAP system again, and develop new routines all over again to be consistent with it. Several sub-systems were built to interface with the SAP system, for example for creating and printing labels that indicate how the materials should be handled.

In this section we have briefly discussed the SAP implementation in the Scandinavian subsidiary, and in particular the situation before and after the implementation. In the next section we analyse the national and organizational culture of the subsidiary from the data collected from the two-stage data gathering approach discussed in Section 4.

The Culture of the Swedish Subsidiary

The culture of the Swedish subsidiary, gathered from the two-stage questionnaire described in the previous section, is divided for simplicity reasons into national culture and organizational culture (Hofstede, 1994). The description is based on the observations made, the responses to the first stage questionnaire, and to the culture assessment questionnaire that was distributed in the second stage, all of which are based on the synthesis of the theories and methods of the key researchers in the area of culture. As already mentioned in Section 17.2, there are several dimensions by different authors that help us identify the culture of an organization. Together with our own observations, we classified the culture of the Swedish subsidiary both in terms of its national and its organizational culture, described in more detail below. However, these results of national and organizational culture are often intertwined. It is a very common phenomenon that the national culture of the company of interest is affecting its organizational culture, and therefore certain values and beliefs are coming across from the national culture to the organizational one. The same holds in the other sense, so that the organizational culture is sometimes so strong (whether the one of the subsidiary, or of the parent company) that the national culture elements are not exemplified as strong in the organizational culture.

Organizational culture

The Swedish subsidiary is situated in a new office building just outside Stockholm. The overall impression of the offices was simple and clean with no unnecessary clutter. Employees had their own offices or shared them with one more person. In general, dress codes and the ways in which people interact with each other was informal and relaxed. Guests are treated the same way. There are no visitor passes and guests can access almost every room. When personnel interact it is mainly with other employees of the same department. The company does not organize social events. Those events that do take place from time to time are between employees with common interests and hobbies, for example art excursions, theatre, etc.

The two-stage data gathering approach provided us with interesting data about what the employees perceived the organizational culture of the subsidiary to be. First of all, one aspect of the organizational culture of the Swedish subsidiary that became apparent was that most of the employees feel that they are appreciated and that they as individuals are more important than the actual job they do. This statement is supported by the responses the employees gave in the second stage, as 70% of them agreed with it (Appendix 2, Figure A17.1). Secondly, 90% agreed with the statement that people are treated fairly regardless of their educational background, age, gender, and race (Appendix 2, Figure A17.2). All employees thought that costs are a major concern for the company, since 100% agreed with this statement (Appendix 2, Figure A17.3). The opinions of the employees of the Swedish subsidiary were dispersed regarding to whether the company is a place that makes newly hired people feel at home and adapt to the company easily, as half of them (50%) believed it is, whereas 40% felt neutral about this statement (Appendix 2, Figure A17.4).

National culture

In this section we discuss the national culture of the subsidiary, by testing whether what the key authors of culture claim the Swedish culture to be, is actually supported from the data we gathered. There are many dimensions based on which we can analyse and categorize more extensively the national culture of the Swedish subsidiary, however for the purposes of the overall analysis of its culture we decided to focus on the most representative ones for the Swedish subsidiary.

The first hypothesis is based on Hofstede's (1994) findings, that Sweden has low power distance values. From our data gathering approach, we found that in the Swedish subsidiary there is no real hierarchy between the employees. For example the IT development manager clearly stated that, 'We don't have any real hierarchy in our company, in Sweden hierarchies are flat.' Furthermore, according to the IT developer there is freedom of speech no matter what the job role and responsibilities as he argues that, 'I just talk to my boss if I want to suggest or change something.' However, the data we received from the second stage questionnaire do not strongly support that the Swedish subsidiary has low power distance values. Around 42% of the employees felt that there are low power distance values, and the rest either felt neutral or disagreed with it (Appendix 3, Figure A17.5). Nevertheless, the data we gathered prove that the Swedish subsidiary has low power distance values, but not as low as Hofstede (1994) presents Swedish organizations to have. So, our hypothesis based on Hofstede's findings is neither supported nor disproved by our findings and therefore no conclusive outcomes can be gleaned.

The second hypothesis is based on the findings of Hofstede (1994) and Trompenaars (1994) that people in Sweden prefer to act as individuals rather than as members of a group. From our first stage data gathering on the national culture of the Swedish subsidiary, it was found that overall most of the employees believed they worked as a group and preferred to do so, especially now with the implementation of the SAP system. As the IT developer stated, 'We get projects individually, but we work a lot together on other projects, especially now with the SAP system, but even if we work separately on these individual projects, we talk to each other, ask for advice and help.' Furthermore, this is supported from the findings of the second stage questionnaire from which we found that employees within the Swedish subsidiary prefer to act as members of a group (around 40%) rather than as individuals (around 8%) (Appendix 3, Figure A17.5). The rest of them felt neutral about it. Based on the findings of the key authors, we expected to find the employees of the Swedish subsidiary preferring to work more individually, without any major kind of co-operation, however our findings do not support this. Consequently, we can argue that we reject this hypothesis, as our data do not support the findings of the key authors. A reason for this might be the influence the organizational culture of the parent company might have on the Swedish subsidiary. Maybe the organizational culture of the parent company is so strong that it might have imposed itself on the culture of the subsidiary and the employees might have been influenced to or prefer to work as groups and not as individuals. Also, another reason for that might be that the studies Hofstede and Trompenaars have done might be out of date as the national culture of Sweden might be changing.

Lastly, the third hypothesis is based on the findings of Hofstede (1994), who claims that the Swedish culture is feminine, that values like quality of life and solidarity are prevailing, and people tend to resolve conflicts with negotiation and compromises rather than fights. Furthermore, managers tend to adopt consensus, understanding, etc. From our data gathering it was found that the Swedish subsidiary is a company where the quality of life of the employees is very important; caring and solidarity are values that prevail within the Swedish subsidiary. As the IT developer stated, 'when I was pregnant and was not feeling well, I could stay at home and not feel guilty at all, rather the opposite. I was encouraged to do it.' Moreover, in relation to a competition free environment, as another IT developer stated, 'even if I do get promoted, I would get the job of my boss, and would have more responsibilities, but not more power, so why want something like this?'. Our findings from the second stage questionnaire support the hypothesis that the Swedish culture in the subsidiary has mostly feminine values, as almost 50% of the employees agree with this (Appendix 3, Figure A17.5). Consequently, we can argue that we accept the hypothesis that the Swedish subsidiary has mainly feminine values.

Job Satisfaction Problem in the Warehouse

The cultural value elicited from stakeholders was job satisfaction. An important cultural norm was that warehouse personnel work 'out there' in the warehouse and perform physical tasks, such as moving products. The organization valued employee satisfaction in their work, however the IT materials manager believed that 'especially in the beginning the warehouse staff did not want to sit by the desk, they wanted to be out there in the warehouse'. This reflected a problem that warehouse staff were spending more time using the system and less time 'out there' in the warehouse. The warehousemen were accustomed to their job responsibilities mainly entailing physical tasks and

Table 17.1 Summary of problems with the new warehousing system

Location	Surface problem features	Underlying problem	Cultural norms and values
Sweden	Lack of practical system use before implementation	Lack of training	Employes are trained to use the new system
Sweden	Warehouse staff not working in the warehouse enough	job satisfaction	Warehouse staff should perform physical task
Sweden	System over-acknowledges ware-house operations	Poor fit with ERP package	The computer system should empower the organisation

that this was also what was expected from them. The customs and rituals of a warehouseman before the SAP implementation reflected the expectations of the subsidiary's warehousemen and with the introduction of SAP's R/3 system this has changed. The myth that was circulated in the warehouse was that R/3 would change the nature of the work for the better, leading them to believe that the R/3 system would reduce administrative work and make physical tasks more organized. However, the reality did not reflect the myth and all warehousemen realized that they have to use the system regularly and whoever manages to do both physical and administrative tasks and still be happy is regarded as a hero.

Problems with the Implementation of the Warehouse System

Table 17.1 summarizes our findings from the development of the warehouse system models, similar to the one described in the previous section. Stakeholders in the Swedish warehouse articulated one essential value, that the new system shall meet customer requirements. They also revealed two other critical values, that personnel are well-trained and that personnel are satisfied in their work. Second-phase questionnaires provide evidence to support these findings. For example, Scandinavian warehouse staff were dissatisfied with the volume of training received. Evidence from the second-phase questionnaire also supports this: 5/10 Swedish respondents disagreed with the statement 'there are numerous training and career developments programmes within the company'. Follow-up interviews elicited views from Scandinavian respondents who claimed that: 'I would like to be able to get training through the lab, but there is a time problem', 'I think we should get more training', and 'There was not enough time to get enough education'.

Table 17.2 Summary of problems with the new sales/marketing system

Location	Surface problem features	Underlying problem	Cultural norms and values
Sweden	Unable to enter short delivery times into R/3 system	Inflexible delivery times	Important to have delivery times that are as short as possible
Sweden	Extra work needed to enter data into R/3 system	Increases the work load	Staff are productive and effective at work
Sweden	Unable to obtain holistic, integrated view of the data	Poor system navigation	Expect to obtain the benefits of integration
Sweden	Unable to handle customer enquiries adequately	Poor screen layout	Importance of good customer relations

Modelling Sales/Marketing Problems

Scandinavian stakeholders also identified important problems with the implemented sales function. Relevant stakeholder interview transcripts were again analysed to produce model fragments of stakeholder norms, and values associated with these problems. Table 17.2 summarizes the critical determinants of culture from the model describing problems with the new sales and marketing system. It reveals that the Swedish subsidiary identified good customer service as the single overriding cultural value that was associated with problems in the Sales and Distribution module implementation.

17.6 Conclusions and Discussion

This paper reports studies into the implementation of the German ERP package, SAP's R/3 in the Scandinavian subsidiary of a multinational pharmaceutical and laboratory equipment supplier. The empirical method captured a large corpus of data about the implementation approach and about the problems with the implementation in the subsidiary. It applied a knowledge meta-schema first reported in Krumbholz and Maiden (2000) to analyse data in the corpus both to drive the capture of data and to determine the critical stakeholder values, and norms indicative of the subsidiary's organizational and national culture. In this respect the studies were successful in the development of a large number of model fragments describing critical determinants of culture. This paper focuses on one model related to a specific problem with the implementation of the warehousing system. The model is an important contribution to the increasing body of knowledge about ERP implementations.

To investigate how culture impacts on ERP package implementations, the paper set four predictions from our model of culture outlined in section 3.

Prediction P1: the current organizational culture clashes with the planned future culture.

Studies of the warehouse and sales operations revealed no evidence to support this prediction. Results indicate that the Scandinavian stakeholders did not believe that the organization intentionally undertook an undesirable change in its organizational culture.

Prediction P2: the supplier's culture, implicit in the ERP package, clashes with the customer's organizational culture

The results reveal evidence, in the form of stakeholders' claims about the SAP package implementation problems, that indicate that the supplier's culture, tacit in the ERP package's solution, clashed with the customer's organizational culture. The paper reports three values: (1) warehouse staff should perform physical tasks; (2) computer systems empower the organization; and (3) warehouse staff process orders in a flexible manner, that were not supported by the SAP package. However, implementation of the SAP package led to greater administrative work and inflexible processes in the warehouse that the IT development manager identified as a more 'German' way of doing things.

Prediction P3: the new business processes (configured using the ERP solution) clashes with the existing organizational culture

Whereas P2 reports problems that arose from the immediate implementation of the SAP package, there is little evidence to support prediction P3, that new business processes clashed with the existing organizational culture. Although Scandinavian sales operations are less specific to local conditions, stakeholders did not report problems specific to the new embedded processes: the new organizational culture appears to have been accepted by most stakeholders.

Prediction P4: critical determinants of organizational culture that reside more in a customer's observable practices have causal associations with problems that arise during ERP package implementations

The model fragment reported in this paper provides evidence for this prediction. The model shown in Figure 17.4 describes phenomena elicited verbally from stakeholders in response to focused questions about implementation problems, and causal associations to observable actions and other observations (added to improve the knowledge meta-schema) that describe problems

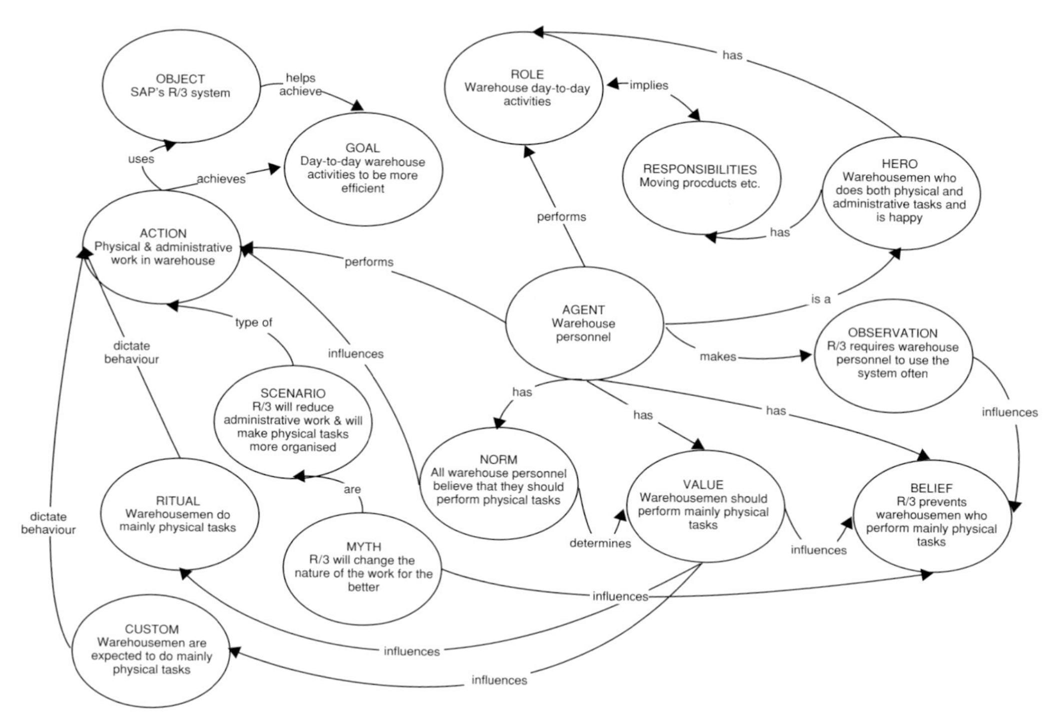

Figure 17.4 Job satisfaction problem with the warehouse module

with the ERP package implementation. More model fragments that provide evidence to support this prediction are reported in Coulianos (2000) and Krumbholz and Maiden (2001) and Krumbholz et al. (2000).

More generally, the results summarized in Tables 17.1 and 17.2 show that the diverse range of ERP implementation problems were associated causally to a smaller number of clashes with core customer values. This has important implications for our extended ERP implementation method: eliciting and analysing core customer values for their fit with the ERP package can give greater leverage when predicting and handling implementation problems shown graphically in Figure 17.3.

Future Work

These results have important implications for the extended ERP package implementation method, and in particular for the need to elicit, model, and analyse the ERP supplier's culture, tacit in the ERP package's solution. The authors are currently planning studies to elicit and model the organizational culture of the German supplier reported in this paper to investigate hypotheses about the impact of the supplier's culture on implementation success. We look forward to reporting our findings in the near future.

Results also have implications for our model of culture posited in Section 17.3. Evidence of culture was elicited from different people in the Scandinavian subsidiary, hence the influence of different personalities has to be considered when modelling a homogeneous organizational culture, and in particular the influence of leaders and founders in the organization. Schein (1983) argues that: 'founders usually have a major impact on how the group defines and solves external problem of surviving and growing, and how it will internally organise itself and integrate its own efforts'. This suggests that founders and leaders in the subsidiary (for example the IT development managers) might have influenced the organizational and national culture and their impact on implementation. However, these studies have not produced any evidence about the nature of this influence, for example whether or not the IT development managers either nullified or re-inforced different facets of organizational and/or national culture. Further studies that take into account stakeholder personalities and influences are needed to answer this question.

ACKNOWLEDGEMENTS

The authors wish to thank SAP and its customer for the support in undertaking the studies reported in this paper.

REFERENCES

Bach, K. and Harnish, R. M. (1979) *Linguistic Communication and Speech Acts*. Cambridge, MA: MIT Press.

Coulianos, N. (2000) The Impact of Culture on the Implementation of ERP Systems such as SAP's R/3. Technical Report, Centre for HCI Design, City University, June 2000.

Curran, T. A. and Ladd, A. (1998) *SAP R/3 Business Blueprint*. Prentice-Hall.

Darimont, R. and Van Lamsweerde, A. (1997) Formal Refinement Patterns for Goal-Driven Requirements Elaboration. *Proceedings 4th ACM Symposium Foundations of Software Engineering*, ACM Press, pp. 179–190.

Deal, T. and Kennedy, A. (1982) *Corporate Cultures: The Rites and Rituals of Corporate Life.* London: Penguin Books.

Densley, B. (1999) The Magnificent Seven: Getting the Biggest Bang from the ERP Buck. *Proceedings 1st International Workshop* EMRPS99, Eder, J., Maiden, N., and Missikoff, M. (eds), Istituto de Analisi dei Sistemi ed Informatica, CNR Roma, 59–65.

Dobson, P., Williams, A., and Walters, M. (1993) *Changing Culture: New Organisational Approaches.* 2nd edn, London: Institute of Personnel Management.

Engel, P. (1984) Functionalism, Belief and Content. In *The Mind and the Machine – Philosophical Aspects of Artificial Intelligence*, Torrance, S. (ed.). Chichester: Ellis Horwood Ltd.

Gabriel, R. (1995) *Patterns of Software.* Oxford University Press.

Galliers, J. (1989) A Theoretical Framework for Computer Models of Cooperative Dialogue, Acknowledging Multi-agent Conflict. Technical Report No. 172, Computer Laboratory, University of Cambridge.

Galliers, J. (1992) Autonomous Belief Revision and Communication. In *Belief Revision*, Gardenfors, P. (ed.), Cambridge: Cambridge University Press.

Gardenfors, P. (1988) *Knowledge in Flux. Modelling the Dynamics of Epistemic States.* Cambridge, MA: MIT Press.

Gulla, J. A. and Mollan, R. (1999) Implementing SAP R/3 in a Multi-Cultural Organisation. *Proceedings 1st International Workshop* EMRPS99, Eder, J., Maiden, N., and Missikoff, M. (eds), Istituto de Analisi dei Sistemi ed Informatica, CNR Roma, 127–134.

Harman, G. (1986) *Change in View – Principles in Reasoning.* Bradford Book, Cambridge, MA: MIT Press.

Hintikka, J. (1962) *Knowledge and Belief.* New York: Cornell University Press.

Levi, I. (1984) Truth, Fallibility and the Growth of Knowledge. In *Decisions and Revisions*, Cambridge: Cambridge University Press.

Hofstede, G. (1984) *Culture's Consequences: International Differences in Work-related Values.* Sage

Hofstede, G. (1991) *Cultures and Organizations.* McGraw-Hill.

Hofstede, G. (1994) Cultures and Organisations, Intercultural Co-operation and Its Importance for Survival. Software of the Mind. *Author of Culture's Consequences*, London: Harper Collins.

Johnson, G. (1992) Managing Strategic Change: Strategy, Culture and Action. *Long Range Planning*, **25**, 28–36.

Krumbholz, M. and Maiden, N. A. M. (2000) How Culture might Impact on the Implementation of Enterprise Resource Planning Packages. *Proceedings CAiSE00 (Computer-Aided Information System Engineering)*, Springer-Verlag LNCS 1789, 279–293.

Krumbholz, M., Galliers, J., Coulianos, N. and Maiden, N. A. M. (2000) Implementing Enterprise Resource Planning Packages in different Corporate and National Cultures. *Journal of Information Technology*, **15**, 267–279.

Krumbholz, M. and Maiden, N. A. M. (2001) The Implementation of Enterprise Resource Planning Packages in different Organisational and National Cultures. *Information Systems Journal*, **26**(3), 185–204.

Maiden, N. A .M. (1998) SAVRE: Scenarios for Acquiring and Validating Requirements. *Journal of Automated Software Engineering*, **5**, 419–446.

Mylopoulos, J. M. (1999) Goal-Oriented Analysis for Software Customisation. *Proceedings 1st International Workshop* EMRPS99, Eder, J., Maiden, N., and Missikoff, M. (eds), Istituto de Analisi dei Sistemi ed Informatica, CNR Roma, 375.
Ncube, C. and Maiden, N. A. M. (1999) Guidance for Parallel Requirements Acquisition and COTS Software Selection. *Proceedings 4th IEEE Symposium on Requirements Engineering.* IEEE Computer Society Press, pp. 133–140.
Peters & Waterman (1982) *In Search of Excellence.* New York: Harper & Raw.
Quine W. V. (1970) *The Web of Belief.* New York: Random House.
Rosemann, M., Frink, D., von Uthmann, C. and Friedrich, M. (1999) Workflow-based ERP: A New Approach for Efficient Order Processing. *Proceedings 1st International Workshop* EMRPS99, Eder, J., Maiden, N., and Missikoff, M. (eds), Istituto de Analisi dei Sistemi ed Informatica, CNR Roma, 239–248.
Rugg, G., Corbridge, C., Major, N. P., Burton, A. M., and Shadbolt, N. R. (1992) A Comparison of Sorting Techniques in Knowledge Elicitation, *Knowledge Acquisition*, **4**(3), 279–291.
Schein, E. H. (1985) *Organizational Culture and Leadership.* San Francisco, CA: Jossey-Bass.
Schein, E. H. (1992) *Organisational Culture and Leadership.* San Francisco: Jossey-Bass Publishers.
Schein, E. H. (1983) The Role of the Founder in the Creation of Organisational Culture. *Organisational Dynamics*
Trompenaars, F. (1994) *Riding the Waves of Culture: Understanding Cultural Diversity in Business.* London: Nicholas Brealey Publishing.
Vernon, M. (1999) ERP Endangered Species? *Computer Weekly*, 4 November, 32.
Warren, L. (1999) ERP Sans Frontiers, *Computer Weekly*, 25 November 1999, 32–33.

Appendix 1

Culture-related questionnaire to elicit information about Company X with respect to the SAP implementation.

The purpose of this questionnaire is to derive information about the culture of Company X and see its effect on the SAP implementation. The results of this questionnaire are going to be analysed together with the 15 minutes interviews, so to be able to elicit culture-related effects on the implementation.

Name:
Position:

Please tick the box that applies best to each statement.
A – Strongly agree
B – Agree
C – Tend to agree
D – Neutral
E – Tend to disagree

F–Disagree G–Strong disagree

**** Responses are handled with strict confidence****

1	The management is more concerned with employees getting the work done, other than with the employees as persons.	A☐	B☐	C☐	D☐	E☐	F☐	G☐
2	Good results are rewarded.	A☐	B☐	C☐	D☐	E☐	F☐	G☐
3	Deadlines are loose and flexible.	A☐	B☐	C☐	D☐	E☐	F☐	G☐
4	There are numerous training and career development programmes within my company.	A☐	B☐	C☐	D☐	E☐	F☐	G☐
5	There are detailed regulation, rules and procedures for most of the things that I do.	A☐	B☐	C☐	D☐	E☐	F☐	G☐
6	Our policies and procedures are formal.	A☐	B☐	C☐	D☐	E☐	F☐	G☐
7	The quality of the physical facilities is satisfying.	A☐	B☐	C☐	D☐	E☐	F☐	G☐
8	The management freely shares information.	A☐	B☐	C☐	D☐	E☐	F☐	G☐
9	Short-term results are more valuable than long-term results.	A☐	B☐	C☐	D☐	E☐	F☐	G☐
10	People are treated fairly regardless of their educational background, age, gender, and race.	A☐	B☐	C☐	D☐	E☐	F☐	G☐
11	The different departments of the company are of equal importance to top management.	A☐	B☐	C☐	D☐	E☐	F☐	G☐
12	Newly hired people feel at home and adapt to the company easily.	A☐	B☐	C☐	D☐	E☐	F☐	G☐
13	Employees communicate a lot and of good quality.	A☐	B☐	C☐	D☐	E☐	F☐	G☐
14	Many social events, where everybody is invited take place within our company.	A☐	B☐	C☐	D☐	E☐	F☐	G☐
15	Costs are a major concern for the company.	A☐	B☐	C☐	D☐	E☐	F☐	G☐
16	Persons that achieve extraordinary results are recognised and rewarded.	A☐	B☐	C☐	D☐	E☐	F☐	G☐
17	I would lie for my company.	A☐	B☐	C☐	D☐	E☐	F☐	G☐
18	I feel comfortable taking risks.	A☐	B☐	C☐	D☐	E☐	F☐	G☐
19	Job roles tend to make employees work more individually and not in groups	A☐	B☐	C☐	D☐	E☐	F☐	G☐

20	Employees are encouraged to take part in the decision making processes of the company. A☐ B☐ C☐ D☐ E☐ F☐ G☐
21	Employees are encouraged to make suggestions. A☐ B☐ C☐ D☐ E☐ F☐ G☐
22	I feel that my work is being appreciated. A☐ B☐ C☐ D☐ E☐ F☐ G☐
23	I feel I that know the mission of the organisation. A☐ B☐ C☐ D☐ E☐ F☐ G☐
24	Company X practices what it advocates. A☐ B☐ C☐ D☐ E☐ F☐ G☐
25	Company X encourages personal development (like concentration seminars, etc.) A☐ B☐ C☐ D☐ E☐ F☐ G☐
26	Company X acts as if its employees are its greatest asset. A☐ B☐ C☐ D☐ E☐ F☐ G☐
27	Company X is a good place to work A☐ B☐ C☐ D☐ E☐ F☐ G☐
28	Employees are happy with the changes the management decides to take place. A☐ B☐ C☐ D☐ E☐ F☐ G☐
29	I am happy with "the way things are done around here". A☐ B☐ C☐ D☐ E☐ F☐ G☐
30	The company has a very structural career path. A☐ B☐ C☐ D☐ E☐ F☐ G☐

Appendix 2

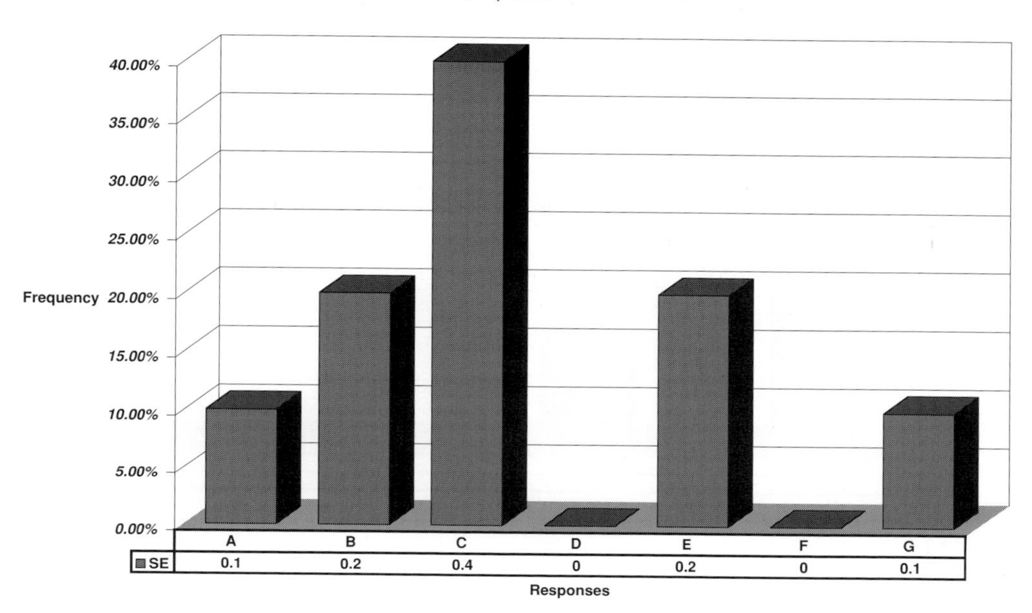

Figure A17.1

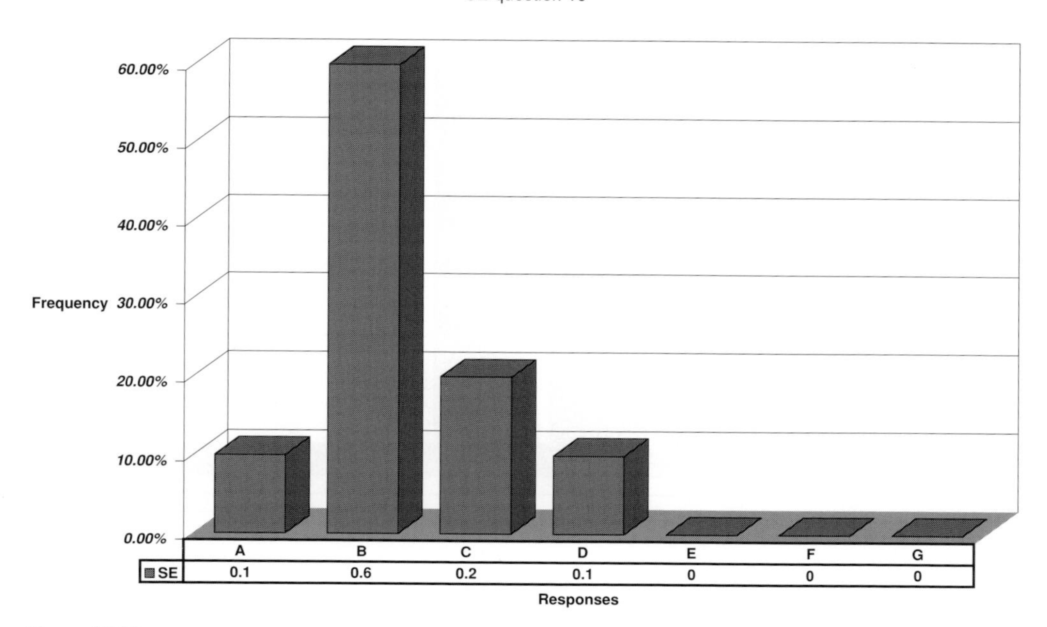

Figure A17.2

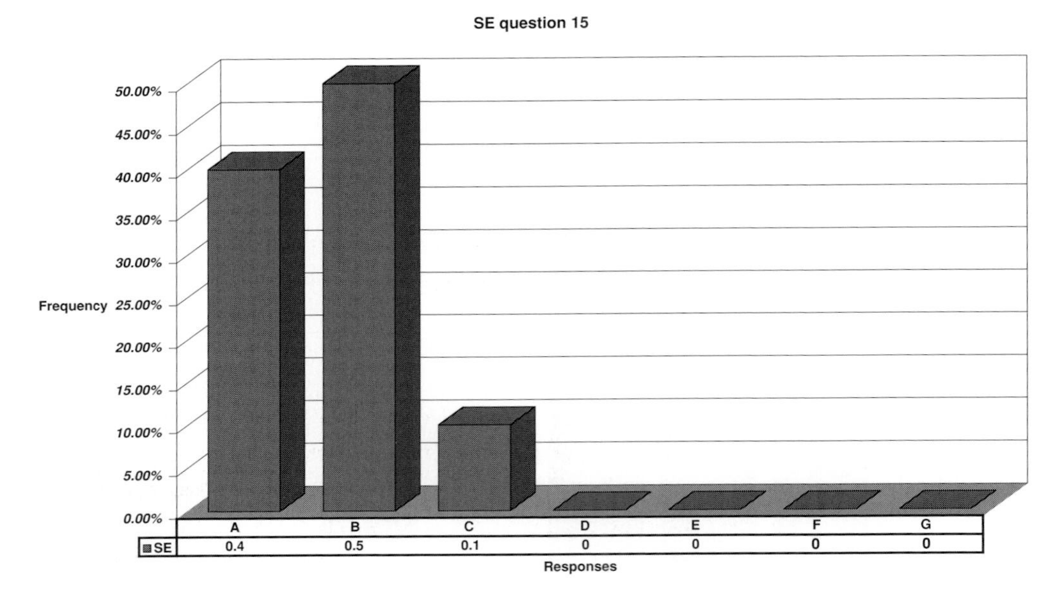

Figure A17.3

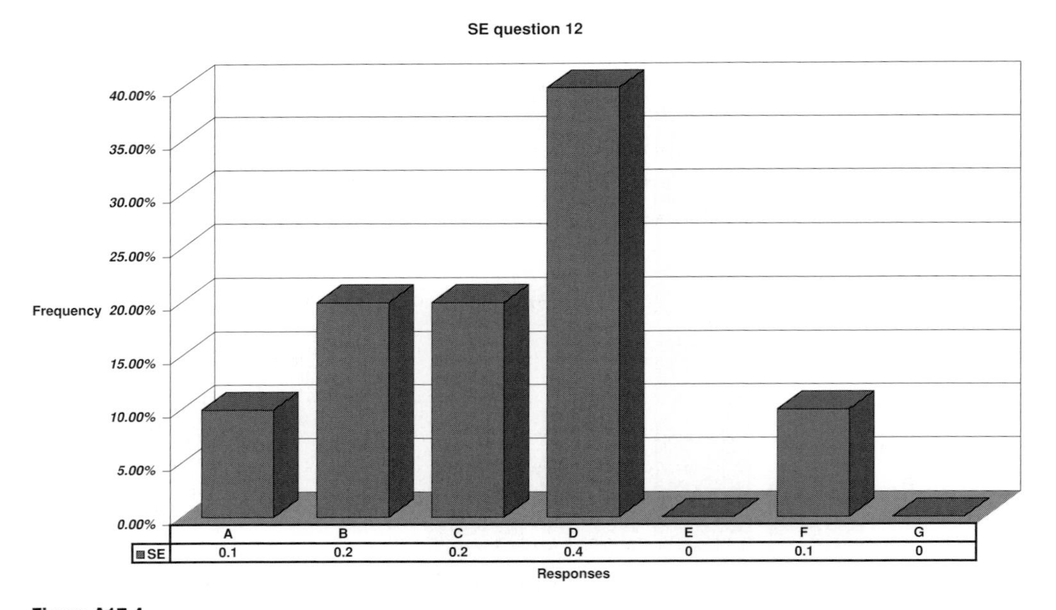

Figure A17.4

Appendix 3

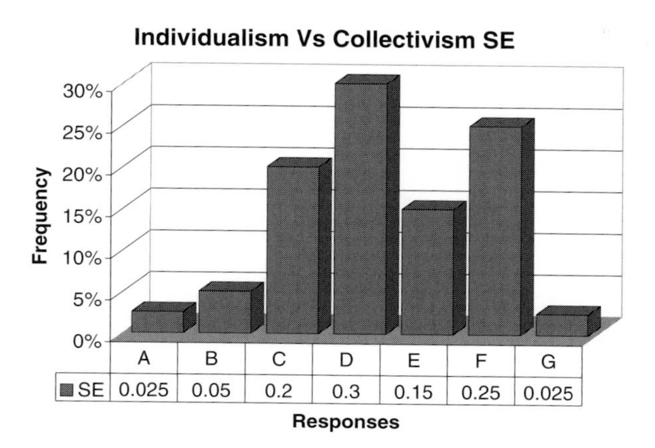

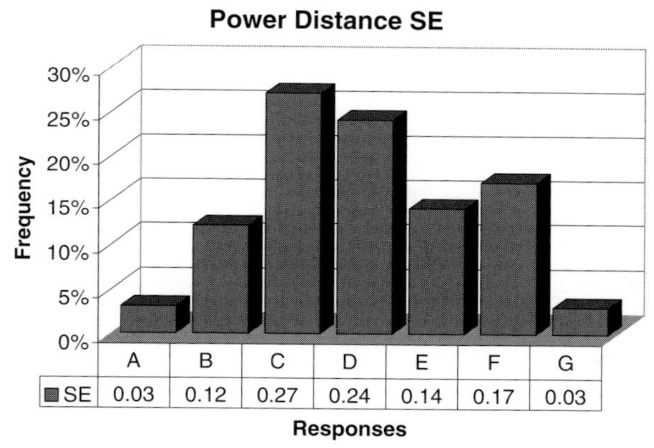

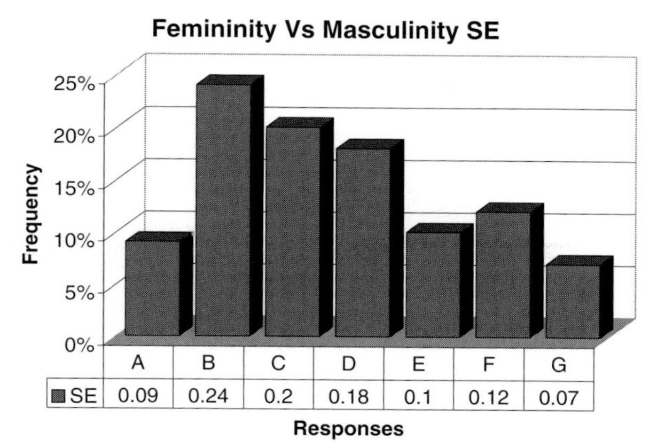

Figure A17.5

Part V

Future Directions

18 Continuity Versus Discontinuity: Weighing the Future of ERP Packages

M. Lynne Markus, David Petrie, and Sheryl Axline

18.1 Introduction

As we look back over the contents of this book, one thing becomes very clear. The Introduction's claim that one of the most significant IT developments in recent years has been the widespread implementation of ERP packages by large and medium companies has been substantially justified. Adopting packaged software allowed many companies to replace their ageing legacy systems in time to avoid Y2K problems and brought a variety of other benefits, including strategic business advantages, improved system architectures, and outsourced software maintenance. With so much going for them, ERP packages seem likely to remain popular for some time to come. In this context, it makes sense to build on the picture painted in the Introduction, and anticipate how ERP packages will evolve in the future.

One view of the future can be constructed from an assessment of the issues faced by today's ERP package adopters and the responses under development by ERP package vendors. In this view, the functionality of ERP packages will expand, and the architecture of ERP packages will evolve, in ways that address many of the 2002–2007 business opportunities and technical challenges. This view of the future assumes a high degree of continuity with today's in-house IT management regime; in particular, it involves a continuation of the current division of labour between ERP package adopting organizations and ERP package vendors (Brehm and Markus, 2000). A clear articulation of this continuity view of the future is offered by Davenport (Davenport, 2000a).

But experts in strategic planning contend that it is often not wise to rely solely on views of the future that are extensions of the past (Markus, 1996; Schwartz, 1991). Instead, they argue, there is value in visualizing alternative future scenarios that incorporate discontinuities. Doing so can enable planners to identify and invest in options that preserve one's flexibility if unexpected situations unfold.

In the scenario planning tradition, this chapter offers a plausible discontinuity view of the future of ERP packages. Our claim is not that our scenario *will* transpire, but only that it *could*. However, if our scenario did occur, it would have major implications for both ERP package adopters and ERP package vendors. Companies that had developed the knowledge, capabilities, and skills suited to this alternative future would prosper relative to those that had not. Therefore, companies would be well advised to assess their ERP strategies in light of the discontinuity view.

In the next section, we summarize the continuity view of ERP packages' future. Then, we lay out our alternative – discontinuity – view. Finally, we discuss the implications of our analysis for ERP package adopters and vendors.

18.2 The Continuity View

Technologies evolve, in part, through responses to the problems people experience in using them (Rosenberg, 1982; Stinchcombe, 1990). This empirical generalization suggests a strategy for forecasting the direction of technology change. While many companies have achieved considerable benefits from their investments in ERP packages, they have not always had an easy time doing so. In the face of problems using ERP packages, some organizations devise their own solutions; others rely on vendors or services providers. As users' difficulties become known to vendors – sometimes through the efforts of user groups, the media, consultants, and market research firms – vendors attempt to address them, often pre-announcing their intentions in order to forestall further criticism and to prevent customer defections to competing products. Thus, it is often possible to sense the direction of package evolution long before users experience relief.

Some Common Problems ERP Package Users Face

Among the most serious problems associated with ERP packages reported in trade and academic literature are the following: limited functionality, lack of decision support, lack of extended enterprise support, implementation and upgrade difficulties, and high total cost of ownership. These issues are dealt with in more detail in many chapters of this book (for example Chapters 1, 6, 12) but it is useful to draw these points together here to review where ERP 'state-of-the-art' has got to, and to pursue our argument further.

Limited functionality

ERP packages differ from traditional software packages most obviously in that they provide a wide range of functionality within a common architecture. Thus, they subsume many of the transaction processing applications that a typical company would have in its applications portfolio: accounting applications, sales order entry programs, inventory management and production scheduling applications, distribution programs, etc.[1] These packages are said to be integrated, because the applications share a common database, and transaction data can flow seamlessly from one 'module' to the next, without rekeying or software interfaces (Davenport, 1998), thereby eliminating problems that plague companies with unintegrated, legacy applications systems.

Because their functionality is so broad, ERP package vendors have tended to sell them as 'complete' solutions to a company's information processing needs. However, the experience of many ERP package adopters has been otherwise. First, although vendors are increasingly delivering 'industry specific' versions of their systems, many organizations have found both that unusual business processes are not well supported by the software and that it is not always possible to change business processes to conform to package features (Brehm and Markus, 2000; Markus and Tanis, 2000 – see also Chapters 6 and 16).

Companies have used a wide variety of approaches for dealing with the lack of appropriate functionality in ERP packages, including:

- leaving some processes un-automated;
- adopting manual workarounds;
- adopting specialized 'bolt-on' packages designed by independent software vendors to work with a particular ERP system;
- integrating multiple enterprise packages in a best-of-breed solution;
- integrating the ERP package with the organization's legacy systems;
- building new custom modules to work with the ERP system;
- modifying ERP package code.

The various 'integration' strategies listed above have spawned a whole new industry of 'EAI' (enterprise application integration) software tool vendors and services providers (see the book's Introduction – Discussion and Figures 1 and 2 there).

Two particular non-industry-specific areas in which many companies found early ERP packages deficient were decision support and support for relationships with customers and suppliers. These areas are discussed next.

[1] ERP systems originated in the manufacturing sector. Today, ERP systems are being developed for retail, financial services, distribution, education and other non-manufacturing industries.

Lack of decision support

In fairness to ERP package vendors, ERP packages were not originally intended to fulfill companies' needs for business reporting; they originated as integrated collections of transaction processing systems. Technologically, decision support and transaction processing are quite different things: a system optimized to do one well does not do the other well, and vice versa.

The companies that adopt ERP packages often have extensive needs for business reporting. Since the late 1960s, a high proportion of business information processing activity has been directed at decision support versus routine transaction processing. And one of the prime motivations for companies to adopt integrated packages was to acquire the integration of data they need for sophisticated decision support. It is small wonder, then, that one of the loudest complaints about ERP packages was lack of adequate decision support (Bashein and Markus, 2000).

Some companies quickly solved this problem on their own. For instance, Microsoft loaded data from its SAP R/3 financials software into a database, developed a custom reporting system and some preformatted reports, and made the software, the reports, and data for *ad hoc* analyses available via the company intranet (Bashein, Markus, and Finley, 1997). Other companies struggled with the vendors' facilities for creating operational reports or simply rekeyed the data into Excel spreadsheets (Koh, Soh, and Markus, 2000).

Lack of extended enterprise support

More recently, enthusiasm for e-commerce led many companies to demand functionality to support their purchasing and marketing transactions and decisions. Early ERP packages did not support such supply chain management functions as 'advanced planning and optimizing (APO)', 'collaborative planning, forecasting, and replenishment (CPFR)', or 'vendor managed inventory (VMI)'. Nor did they support such marketing activities as 'call center operations', 'customer relationship management (CRM)', or 'e-commerce storefronts'.

Some companies made do for these needs with their own legacy systems or new custom software, which had to be custom integrated with ERP packages. Others turned to independent software vendors (and later to their ERP vendors, see below) for programs they hoped could be quickly integrated with their ERP systems. In some cases, the vendors of these extended enterprise packages offered to host the packages (as 'application service providers' or ASPs) and to do the integration themselves: this solution often gets companies up and running on the new capabilities faster than with in-house implementation and integration, but this market has been slow to develop, not least due to customer lack of awareness of, and confidence in, the options (Kern, Lacity et al., 2001; Kern, Lacity, and Willcocks, 2002; see also Chapter 4).

Implementation and upgrade difficulties

Because ERP packages encompass so much functionality (though not so much as adopters may want), they are very complex, which makes it difficult to configure them to the organization's structures and processes. Although they are presented to ERP adopting organizations in 'modules' of functionality, the modules are not entirely independent: the way the sales and distribution module is configured, for example, has implications for the functioning of the accounting module. Because of tight internal integration, ERP packages are often referred to pejoratively as 'monolithic' (Sprott, 2000).

Many organizations rely on experienced implementation consultants to guide them through the difficulties of configuration. Those who go it alone often learn by trial and error: Revel Asia, for example, had to reconfigure its financial module, when it later implemented an operations module (Koh, Soh, and Markus, 2000).

Problems such as those at Revel Asia translate into lengthy and expensive implementations. Adding to implementation difficulties is the need, discussed above, to integrate ERP systems with bolt-ons, legacy systems, other ERP systems, or new custom software. And the integrations themselves can cause problems downstream when companies want to avail themselves of new versions and releases of the vendors' ERP software (Brehm and Markus, 2000). In some cases, the integrations do not work with the new package versions and releases, forcing the company to re-program the integrations or do without with the additional functionality (Markus and Tanis, 2000).

High total cost of ownership

Difficulty in upgrading to later releases or versions is one reason that companies have experienced higher than hoped-for total cost of ERP package ownership (Kremers and Dissel, 2000; Ohlson, 2000). But there are other important reasons as well. With the client-server versions of ERP packages, adopters have had to maintain ERP clients on individual users' PCs, resulting in expensive license fees and a sizeable in-house support burden: *someone* has to load the software on all those machines. (A few companies have faced unexpectedly large expenses for upgrading users' desktop machines to run ERP clients.)

Further burdens have come in the form of training and data entry costs (Wheatley, 2000). Because ERP packages are designed to serve so many different types of organizational needs, vendors would have a nearly impossible task to optimize an interface for each job type. Instead, screens were designed according to program logic; consequently, even a simple task might involve the need to click through multiple screens. With even modest delays in response time, this situation could result in slow responses to customer inquiries or the need to add more staff; some companies found the delays such that they could not do online order entry. Further, the large number of screens involved in some jobs, coupled with the integrations that could propagate data entry errors throughout the business, resulted in substantially increased training costs.

This litany of problems hardly exhausts the difficulties that ERP adopting organizations encountered while implementing and living with ERP packages. But it covers the main ones, and it provides sufficient background to explain a large part of ERP package vendors' development agendas.

Vendor Solutions to Users' Problems With ERP Packages

Among the major items on the close-in[2] development agendas of most ERP package vendors,[3] are providing data warehouses and decision support, developing or acquiring new industry-specific and extended enterprise functionality, componentizing software, web-enabling software and providing mobile access, and externally hosting software (see also this book's Introduction).

Data warehouses and decision support

ERP vendors' first response to customers' reporting needs was to claim that the operational report writing facilities in ERP packages were all that was required. In the face of continuing complaints, however, they rapidly developed data warehousing and decision support capabilities (King, 2000).

New industry-specific and extended enterprise functionality

As customers' attention shifted from core ERP package functionality to industry-specific applications and extended enterprise capabilities, ERP package vendors scrambled to provide this capability themselves. In some cases, they developed their own new software modules. In other cases, they acquired independent software vendors and absorbed their packages or developed marketing alliances with independent software vendors. Today, ERP package

[2] Recent past, current, or short-term future.

[3] Another major challenge, of course, is just staying in business in face of intense competition and industry consolidation; Baan.

vendors offer a range of extended enterprise capabilities as well as a variety of versions tailored to particular industry segments.

ERP package vendors further claim that ERP systems are the necessary foundation for e-commerce success, because their systems provide internal systems integration, assumed to be a precondition of external integration. For example, e-commerce gurus Kalakota and Robinson (1999) outline a 'Roadmap to Market Leadership' that involves a progression from internal enterprise integration to external integration in the supply chain and the larger business community:

> The business world's steady embrace of enterprise applications was arguably the most important development in the corporate use of information in the 1990s. As companies race toward the information economy, their structures are increasingly made up of interlocking business applications. Isolated, stand-alone applications are history. E-business is about how to integrate an intricate set of applications so they work together like a well-oiled machine to manage, organize, route, and transform information. (Kalakota and Robinson, 1999, p. 82).

Kalakota and Robinson further point out that achieving the business goal of creating a richer customer experience requires integrating e-commerce web sites with back-office systems like the ERP packages that fulfill the customers' orders (Kalakota and Robinson, 1999, p. 85). Norris et al. (2000) claim that:

> the companies best positioned to succeed in the near future are those that can balance existing ERP-based infrastructure and capabilities, with exciting new e-business innovations.

Componentization

ERP package vendors could not mistake their customers' calls for breaking up the 'monolithic' ERP package software (without, of course, destroying integration capabilities!). The vendors' response was to 'componentize' the packages – that is, re-develop them using object development methods, component interface protocols such as COM and CORBA, integration standards such as XML, and semantic agreements such as those provided by CommerceOne (Sprott, 2000).

ERP package adopters hope that the vendors' componentization efforts will allow them to upgrade one ERP package module without upgrading all others at the same time and to effortlessly combine modules from various sources, including different ERP vendors, independent software vendors, and their own legacy systems. As Davenport (2000a) points out, it may not be in ERP package vendors' financial interests to provide easy access to other vendors' software: most likely, then, componentization will result in only modest relief for companies' integration needs and upgrading difficulties.

Web-enablement and mobile access

The widespread adoption of internet standards allowed ERP package vendors to promise substantial improvements in adopting companies' total cost of ownership. Web-enablement means that individual users almost anywhere in the world can access ERP processes and data without requiring a local ERP client or the technical support this entails. Because of web-enablement, one company reported to us that implementing ERP in seven countries was 'no big deal'. By 2001, ERP vendors were advertising that their systems could be accessed through hand-held organizers, like the Palm. Of course, acquiring the capabilities of web-enablement and mobile access requires ERP adopters to upgrade to the most recent versions of ERP software.

Application hosting

Perhaps the ultimate solution to implementation and upgrading difficulties and the costs of technical support is application hosting, enabled by the internet, in which the ERP vendor (or another service provider) runs the software for an adopter, pricing this service on a usage basis. While application hosting has found considerable acceptance in the area of CRM, it has yet to become established in the ERP environment. One reason identified by Kern, Lacity, and Willcocks (2002) was that many ERP adopters had already developed in-house operation and support capability (whereas they had not yet done so for CRM), and so were reluctant to outsource it without demonstrated benefits. Another reason those authors identifed was the immaturity of application hosting business models and pricing schemes. For instance, the hosters would prefer to offer the same functionality to all adopters (since it lowers their support costs), whereas customers prefer tailored solutions. Finally, ERP software is often seen as 'mission critical'[4] software, whereas CRM at least initially is not.

In short, a review of the major problems experienced by ERP package adopters explains the key items on the ERP package vendors' development agendas. Taken together, these converging trends suggest a future for ERP packages that is a modest extension of the situation today: many companies will continue to use ERP packages, finding them greater in functionality, somewhat easier to integrate with other capabilities (especially those provided

[4] This is the title of Davenport's book on ERP systems (Davenport, 2000b).

by the ERP vendor), generally easier to upgrade, and less costly to support. Their ERP systems will be the foundation for companies' extended enterprise (e-commerce) initiatives (Norris et al., 2000). Often, companies will continue to do the operating, integrating, and upgrading of ERP systems themselves, though others may begin to make greater use of applications hosting .

This is the continuity view of the future, which represents a modest extension of present (2003) capabilities and needs. In the next section, we present another vision of the future – one that is also plausible, but is sharply discontinuous with the present, because it is predicted on somewhat different current trends.

18.3 The Discontinuity View

One thing that gets lost in the continuity view is the nature of the connections between a company and its 'extended enterprise partners'. Today, many companies have electronic links with some of their suppliers and customers via a technology known as electronic data interchange (EDI). At its most basic, EDI is a unique, dyadic electronic connection (involving custom software and telecommunications linkages) between two firms. It requires that both firms agree on the types of communications between them (for example, purchase orders, shipment notices, etc.), the format of the communications (that is, the syntax and semantics of messages, so that the communication can be understood), and the technology of the communications (for example, telecommunications' protocols). Many developments facilitate the bi-lateral agreements involved in EDI. Industry and cross-industry standards exist for different types of transactions (for example, purchase orders) in different industries; value added networks reduce the burden of telecommunicating with different business partners.

By and large, however, EDI remains a technology of dyads rather than of supply chains or business communities. Powerful companies like retail chains (Bouchard and Markus, 1996) and shipping lines (Damsgaard and Truex, 1999) set their own rules of interaction that their smaller or weaker partners are forced to follow, meaning that they may have to create *multiple* custom integrations to perform the same type of transaction with their different business partners. Since many small companies lack IT resources, they often do not integrate their 'electronic' connections to business partners with their own internal systems: they receive EDI messages via email or fax, print them out and manually reenter them into internal systems (if any). Naturally, this process obviates for the small companies many of the benefits claimed for electronic integration. And, naturally, too, EDI has never been as widely nor as enthusiastically adopted as its proponents expected.

Even the largest organizations that depend heavily on EDI suffer from its dyadic nature. Large companies may have many different EDI connections with the same business partner – and each one of these integrations needs to be maintained. In the terms used earlier, EDI has a high total cost of ownership.

Internet technology promises some relief for the costs of EDI: using the public internet or a virtual private network based on internet protocols substantially cuts telecommunications costs over dial-up lines and value added networks. But it does not, by itself, make any difference to EDI's dyadic nature: a supplier to several manufacturers, for example, must access (ideally electronically interface with) each manufacturer's extranet to accumulate orders.

What *can* make a difference in the dyadic nature of EDI, however, is the emergence of third parties that mediate exchanges between buyers and suppliers. In some industries, wholesalers or distributors have long performed this function: a buyer may need to connect to only one wholesaler to 'communicate' with multiple suppliers. But in large areas of business, intermediaries have been relatively uncommon – at least until the recent explosion of new 'business-to-business' e-commerce ventures.

Enter the Electronic Marketplaces

In the last few years, entrepreneurial firms have sprung up to aggregate supply, aggregate demand, or match buyers and sellers, often in areas where intermediaries did not exist before (Kaplan and Sawhney, 2000). These ventures have been called by various names: hubs, marketplaces, exchanges, vortals (a contraction of 'vertical portal') to name but a few. Originally, many of the marketplaces were technology companies or new 'dot-com' ventures, but consortia of established industry participants also exist (Chircu and Kauffman, 2000): examples include cooperative buying or selling exchanges in PC parts, airline tickets, automobile inputs, and pharmaceuticals and hospital supplies. More recently, 'private trading exchanges' have started to gain favor (Deloitte Consulting and Deloitte & Touche, 2000).

Even in 2002, the future of many of these exchanges were still definitely uncertain: the airline tickets exchange and the automobile parts exchange have attracted antitrust attention (Copeland, 2000; Meehan and Sullivan, 2000). Some companies decline to participate in marketplaces (Ansberry, 2000), fearing lower switching costs, greater price competition, or loss of perceived competitive advantage. Electronic marketplaces may level the playing field in an industry, allowing small or inefficient competitors to gain at the expense of larger or better-performing companies. Consequently, the strongest companies may not be willing to join.

Wal-Mart Stores Inc., for instance, decided not to join fellow retailers on the WorldWide Retail Exchange, because it did not see any reason to link up and help its competitors. Jay Allen, a Wal-Mart spokesman, says the retailer put its own system in place in 1991 and has more than 9,000 vendors participating. 'We've put a lot of effort and time and resources to develop that', he says. 'Why share all that?' (Ansberry, 2000).

But other strong companies disagree:

> Chevron launched the Petrocosm Marketplace in March. Developed with Mountain View, Calif.-based ecommerce developer Ariba, Petrocosm is an industry-owned online marketplace where companies can buy and sell just about anything having to do with oil and gas – from drilling pipe to engineering designs.
>
> Interestingly, Petrocosm is not meant to be a Chevron-only marketplace; the company owns a founding shareholders' stake of 20 percent in the venture, but the site overall will be divided between several 'anchor tenants' from all parts of the industry. It may seem odd to launch a venture that can benefit competitors. Texaco has joined Petrocosm as a partner.
>
> But Chevron's Paul thinks it makes sense. 'The real economic benefits are going to come when you bring lots of suppliers and lots of customers to the marketplace', he says. Moreover, Chevron hopes to benefit like any investor, should Petrocosm's business boom. (Gantenbeim, 2000)

Despite uncertainty about their futures and speed of utilization, electronic marketplaces, whether dot-coms, consortia or private, must be reckoned a significant trend.

As 'hubs' in networks of buyers and sellers, marketplaces can reduce the number of pair-wise connections between buyers and sellers. Instead of one (or more) unique electronic connections between each actual pair of transacting firms, each firm now needs only one standard connection to the marketplace[5] to be able to reach, not only one's current partners, but potential future partners as well. Consider this example:

> To get purchase orders to our suppliers, we [Schlumberger's Oilfield Services Division] use MarketSite, Commerce One's Internet marketplace for business-to-business transactions. MarketSite lets us connect with hundreds of suppliers using a single, open system. It replaces the proprietary electronic data interchange systems we used to have to

[5] Or possibly one connection to each of several electronic marketplaces.

maintain. Unlike EDI, which required a series of expensive, one-to-one connections with individual suppliers, a Web marketplace is a low-cost, many-to-many system. (E-Procurement at Schlumberger, 2000, p. 22)

A further, though unstated advantage, is that the same MarketSite system would allow Schlumberger easily to connect with suppliers with which it does not currently do business, thus potentially expanding the market for Schlumberger in a relatively low-cost way.

Enter the Collaboration Facilitators

Naturally, the picture portrayed above is an oversimplification. Damsgaard and Truex (1999) make it clear that standards represent the least common denominator of communications between firms; they must be reinterpreted by each interacting pair. Thus, many people argue that electronic marketplaces are only appropriate for the most commodity-like products or services. But this is where another interesting development comes in: the emergence of third-party firms that provide the service of easing complex coordinations among firms, where these coordinations were formerly done on a direct, firm-to-firm (pair-wise) way. Consider FreeMarkets OnLine. This company is often admired as electronic provider of reverse auctions: companies post their requirements for goods or services and suppliers bid to fulfill them. But FreeMarkets OnLine is not a java applet. It is an organization of people who work hard with companies that may never have done business together before:

The search for circuit boards for the United Technologies Corp., a big electronics manufacturer, begins the way most supplier searches begin: by scouring lists of more than 2,500 factories in printed catalogs and electronic registries, and calling dozens of knowledgeable sources. At Pittsburgh-based FreeMarkets OnLine Inc., men and women with the primitive-sounding titles of 'market maker' and 'market making engineer' cull that list down to 1,000 factories, based on considerations like the plant location, and then whittle that number down by about two-thirds after reading reports on production capability and listening to feedback from customers. After an extensive written survey, another cut takes the number down to 100, and these are examined with an eye toward their long-term business performance, processes and the capability of their management teams. The 50 most promising suppliers are invited to play a brand-new game that could win them a major new customer, and, coincidentally, forever change the way they do business. (Jahnke, 1998)

In other words, FreeMarkets OnLine makes a bigger market for customers like United Technologies not (just) by providing an electronic marketplace but also by the human tasks of finding and grooming potential suppliers. They also work hard with the buyer to craft an appropriate request for quotation. For example, owing to a lack of integrated internal systems, a buyer may define its needs for a particular type of goods differently in various locations, creating problems when it needs to aggregate these needs into a common order.

Another area in which third-party facilitation is catching on is that of supply chain management. For some time now, companies have begun outsourcing to logistics firms the management of their warehouses and transportation needs (Rao et al., 1998). But things have gone one step farther with CPFR (collaborative planning, forecasting, and replenishment). It has become quite clear that companies can achieve better supply chain performance (for example, reduced stockouts, shorter delivery times, etc.) if they share information about sales, production, inventory, lead-times, etc. with customers, suppliers, and logistics partners (Wouters, Sharman, and Wortmann, 1999).

But obstacles to information sharing and coordination exist (Wouters et al., 1999). First, in many industries, companies distrust each other and are reluctant to share information. Furthermore, for effective supply chain performance, it is not possible to do collaborative planning on a pair-wise basis: one needs to know the demands of the customers and the constraints imposed by suppliers' suppliers, and some of this information is a function of their business with your competitors. In such an environment, the only hope for collaborative planning (and it may be slim)[6] is for a trusted third party,[7] to whom all others would confide their confidential needs and production details, to optimize the production and shipment requirements for the entire supply chain. Given their existing role as intermediaries in transportation and logistics planning, companies like UPS and FedEx are aggressively developing services in collaborative planning, often in partnership with leading ERP package vendors.

It is much too soon to speculate on the future of such ventures, and many may well be sources of disappointment by the time this future view is published. As with electronic marketplaces, there are many barriers to their success. At the same time, however, the energy behind the new collaboration initiatives and the examples of facilitators like FreeMarkets OnLine suggest that collaboration facilitation is a significant trend. The market research firm Gartner Group put it like this:

[6] Several recently attempts at collaborative supply chain optimization have not been as successful as hoped. A more successful effort, called the SLIM project, was discussed by (Wouters, Sharman, and Wortmann, 1999).

[7] Since the transportation-logistics company is usually referred to as the third party between buyers and sellers as in third party logistics (3PL), multi-party collaboration with a trusted intermediary is often called 4PL.

> C-commerce [collaborative commerce] is a form of e-business – the most advanced form. E-business has become synonymous with conducting business over the Internet. It includes a broad set of sales, marketing and service activities that, until now, have focused on connecting an enterprise with suppliers and customers. The c-commerce vision includes interenterprise Internet connection but goes a step further by enabling *multiple enterprises* to work interactively online to find ways to save money, make money and solve business problems – often by dynamically restructuring their relationships. . . . C-commerce applications will replace static, Web-enabled supply chain and value chain applications as the dominant application model by 2004. (Collaborative Commerce, 2000, emphasis added)

Both electronic marketplaces and collaboration facilitators offer participants at least two major benefits. First, they provide business integration benefits such as market expansion or supply chain optimization that cannot be achieved *solely by electronic integration between pairs of companies*. Second, they reduce technology integration costs by reducing the number of electronic connections that need to be set up and maintained: single connections with the hub or facilitator take the place of multiple unique connections with different trading partners. Together, these two advantages suggest that the obstacles to the adoption of significantly different business practices may eventually be overcome, at least in some instances. However, if the trend can be spotted, the speed of implementation is much more difficult to anticipate.

18.4 Implications for the Future of ERP Packages

Let us assume then that the future (2003–2007) may hold many more intermediated business interactions than does the present. What might this mean for the future of ERP packages? There are at least two distinct possibilities. The first is that ERP will remain essential for all parties involved in collaborative commerce. The second is that only the 'hub' organizations will retain what we now call ERP functionality.

ERP – Essential Functionality for All Participants

The first possibility is entirely consistent with the continuity view described above. Indeed, since the ERP vendors have thrown themselves wholeheartedly into collaboration ventures, this is the future they are banking on. The future looks like this: The hubs or facilitators will act as switches, passing electronic transactions between the interacting parties. Each party to the transaction would have its own ERP system, but instead of direct connections as with EDI, the transactions would travel via the intermediary (who would also, of course, add various kinds of value such as facilitation or market making).

Since the ERP package vendors are determined to be major players in collaboration facilitation, and since, as mentioned earlier under the continuity view, it is not necessarily in their best interests to provide connectivity to other ERP systems, one imagines (Davenport, 2000a) that the whole thing would be designed to work best (or only) at least initially for companies that all used the same ERP system. Naturally, this view of the future is highly favorable to the interests of ERP vendors and it does not portend much change for in-house IT departments: they would continue to do pretty much what they were doing in 2001–3.

ERP – Only Needed at the Hub of Collaborative Commerce

But there is a plausible alternative: the intermediaries themselves might become the information processors for the participants, or in today's jargon, hosters of ERP functionality for a trading community or supply chain. As discussed below, the intermediaries may outsource information processing capabilities to other providers – a possibility only enhanced by the 2000–2 economic recession in developed economies, where further outsourcing to cut costs became a renewed imperative (McCartney, 2001). Note that this scenario is different from typical ASP arrangements, discussed under the continuity view. There, a service provider hosts either a company's unique configuration of an ERP package or a company's unique database running on a shared, standard ERP configuration. In the discontinuity scenario, not only is the ERP package shared, but the database is, too.

There are two main arguments in support of this radical view of the future. First, as explained below, ERP packages, even with extended enterprise capabilities, do not have the business functionality needed for collaborative commerce and cannot provide it. Second, for all parties to maintain ERP capability involves unnecessary costs in redundancy and the possibility of errors.

By 2001 there was growing recognition that ERP systems lacked the functionality required for collaborative multi-party commerce. For example, an earlier Delphi study of supply chain executives in European multinationals found that only three or four of 12 key trends in supply chain management were supported by ERP systems (Akkermans et al., 1999). Among the key trends believed not to be supported by ERP packages are: further integration of activities between suppliers and customers across the entire supply chain, changes in who drives supply chain coordination, supply chains consisting of several enterprises, full exchange of information with all the players in the chain, and further outsourcing of activities such as physical distribution, finance, and administration.

According to (Akkermans et al., 1999) the major reasons for ERP packages' failure to support these emerging needs are that ERP packages are monolithic – a factor likely to be addressed by componentization – and that ERP systems have developed along a different trajectory than did supply chain management concepts – a factor that bodes less well for the ability of ERP packages to adapt. Wouters, Sharman, and Wortmann (1999) suggest that the problem lies in the fundamental logic of ERP packages, which is much harder to address with the items currently on the ERP package development agenda. In traditional supply chain relationships, *pairs* of companies undertake all different phases of the sales and fulfillment cycle. In collaborative commerce, by contrast, the different activities are functionally decomposed and allocated to the party in a multi-party supply chain that is best able to do them (e.g., at lowest total cost or with best total delivery time).

ERP packages and their extended enterprise modules have been designed for traditional relationships between pairs of companies. They have not been designed to support three-way and *n*-way interactions.[8] New ERP modules and ERP package componentization, by themselves, will not address this fundamental difference in business logic between traditional supply chain relationships and multi-party collaborative commerce. Further, it is difficult to envision successful coordination of complex collaborative arrangements with multiple decentralized ERP systems.

The second argument in favor of the alternative view of the ERP package future – that collaborative information processing capability need be done only at the hub – concerns 'systems integration' costs – the costs of redundancy and the potential for errors. When two companies 'trade', they execute a complex transaction involving flows of goods, money, and information. From an abstract perspective, the transaction belongs to neither party alone; it belongs to both. (The only part of the transaction truly unique to each party is its tax implications.) With ERP systems and EDI, both parties process and store information about the transaction separately, incurring redundant costs for systems integration and data processing and risking the possibility of errors. A

[8] A major part of the difficulty has to do with differences in the way product data is represented in the information systems of different parties in an extended supply chain (Wouters, Sharman, and Wortmann, 1999).

shared information processing system and database would decrease the total cost of information processing for the interacting firms.[9]

In essence, the alternative future sketched out here is one in which organizations will devolve information processing activities to collaboration intermediaries. The technology market research firm, Forrester, describes a similar vision of the future under the provocative title 'the Death of IT', meaning the decline of in-house IT operations, not of information technology itself (Cameron, Shevlin, and Hardisty, 2000). Their argument is summarized as follows: To support complex, fast-changing business processes that span multiple companies, firms will disperse technology management across an exT (external technology) environment. (Cameron, Shevlin, and Hardisty, 2000)

Forrester's argument is predicated on three business trends – the emergence of electronic marketplaces, the disaggregation of companies (Davenport, 2000a), and the dynamic reconfiguration of business processes (Wouters, Sharman, and Wortmann, 1999). Together, these trends mean that companies will require the ability to integrate and disintegrate their business processes and related systems capabilities frequently and rapidly (Davenport, 2000a; Werbach, 2000). In-house IT organizations, Forrester argues, will not be able to meet these needs with traditional IT processing tools (for example, ERP packages, integration technologies, etc.). Consequently, the ownership and operation of most information processing assets will shift outside the boundaries of the individual e-commerce participants to 'exT' (external technology) service providers.

Our enhancement to Forrester's argument is that the external technology services providers will probably not operate as today's outsourcers do: today, individual companies contract with one or more technology service provider to address their own individual needs. In the future, we expect that individual companies will contract with collaboration facilitators who will coordinate information processing services for the community of collaborating members. The collaboration facilitator may do the information processing in house or may contract information processing out to an ASP or to a consortium of ASPs, each providing a specializing information processing function (for example, application hosting, data management, telecommunications, etc.). As an example of how such an arrangement might work, consider the example of Biztro.

Designed specifically for small businesses, the Biztro system allows a manager or owner to log on to one website to take care of payroll, benefits, human resources, and procurement. Biztro licenses those services and its technology

[9] This argument assumes that telecommunications costs are not a significant factor. Most analysts expect dramatic declines in telecommunications costs in the foreseeable future.

to bigger service providers – such as phone companies, ISPs, and banks – which in turn offer it to their small-business customers. With its far-reaching plan to gain critical mass by scooping up small businesses from the customer lists of established companies, Biztro ambitiously hopes its web-based backbone will become the standard 'operating system' for a new generation of business applications, all running on the web and connected to hundreds of other suppliers and service providers (Donahue, 2000).

In 2001, Microsoft positioned itself to enter this business (AMR Research, 2001).

> With the acquisition of Great Plains Software complete . . . Microsoft could dominate the low- and mid-tier enterprise application market. . . . Microsoft could assemble a one-stop-shopping suite for the SME market with a couple of delivery options. *Users could either purchase the components pre-integrated, or plug into a Microsoft-hosted service center.* This would position Microsoft as a virtual standard, which would make internal [systems] integration unnecessary at the SME level and business-to-business integration would become less complex. (AMR Research, 2001)

Objections and Answers

At least two objections to this model of future information processing are likely: First, how will organizations gain access to the data they need to make management decisions? And, second, how will organizations integrate across multiple business processes and/or trading communities? Both questions make it clear that collaborative commerce would not eliminate 'systems integration' issues. However, their locus would likely shift from in-house IT operations to external service providers.

Even today, companies cannot satisfy their needs for decision support entirely with their own enterprise data, no matter how well integrated via ERP or other means (Gray and Watson, 1998). Effective decision support requires external data, as well as internal. Companies frequently purchase external data from information providers and custom integrate external data with their own internal data. However, a new type of 'exT' provider is springing up to manage data integration for companies on an outsourced basis. One can imagine such companies providing the 'glue' across multiple collaborations and external data sources.

Interestingly, a major hurdle in e-commerce is the lack of standardization in product data across trading partners (Wouters, Sharman, and Wortmann, 1999; Norris et al., 2000). An additional hurdle concerns lack of commonality of business processes (Sawy, 2000). While standards are particularly important in a network of trading relationships, they are often resisted and they can be difficult to implement in the absence of central hierarchical control (Akkermans and Horst, 2000). In the case of EDI, powerful firms often forced their weaker partners to adopt multiple standards (Damsgaard and Truex, 1999). Electronic marketplaces may, however, be the catalysts of more mutually beneficial standards within industry groups.[10]

In short, the emergence of electronic marketplaces and collaboration facilitators suggests a future for ERP packages that is radically different from the continuity view. In the discontinuity future, the capability we now know as ERP, much revamped and optimized for collaborative, rather than dyadic, commerce, is provided on a centralized basis to a community of trading partners. The hub either performs information processing and integration along with other value-added services or outsources information processing and integration to applications service providers. The individual trading partners rely for their decision support needs on specialized data integrators who work with various hubs and external data providers to supply an integrated decision support environment.

18.5 Conclusion

In this final chapter, we have presented two alternative futures for ERP packages. The continuity view represents an extension of existing trends visible in the difficulties companies have had in using ERP systems and in the major steps ERP vendors are taking to reduce those problems. The discontinuity view extrapolates from the business trends of electronic marketplaces and supply community facilitators. Both futures are possible. Both may coexist – in different industries or in different types of firms.

ERP vendors consider the trends underlying the discontinuity future to be very important. Vendors such as SAP, Oracle, and Peoplesoft are investing in technology to support collaboration (Fox, 2000)[11]. Clearly, they are hoping that their core ERP systems will retain a large market. They may even hope that collaboration technologies will draw more ERP package adopters into their folds. Whether or not these hopes are founded, ERP vendors are planning to play an important role in the exchanges between companies.

[10] Although integration is likely to remain a problem across industry groups.

[11] For instance, 'SAP has also been building Internet marketplaces–most notably for chemical companies BASF, Henkel, Degussa-Huls, and Metallgesellschaft, and for food giants Nestle and Danone' (Fox, 2000).

The two futures have very different implications for in-house IT management. The continuity future implies business as usual. The skills most useful in today's in-house IT environment – skills at integrating systems and data – remain most useful in the future. The discontinuity future implies a radical change in in-house IT management. The key skills for the future involve contracting and coordinating with external service providers and understanding needs for new types of services.

Scenario planners (Schwartz, 1991) argue that envisioning alternative futures, however unlikely they may appear in a statistical sense, helps firms develop the strategies they need to do well, no matter what occurs. In the spirit of scenario planning, we suggest that many in-house IT departments should consider how they would approach a discontinuous future. What skills are needed? What projects would provide the flexibility for several different courses of action? What indicators should be tracked to monitor the emerging future? Asking and answering such questions will go a long way toward making the future a better place to be.

ACKNOWLEDGEMENTS

We gratefully acknowledge the help of many individuals, including Lars Brehm, Dana Edberg, Mumtaz Shamsudeen, Graham J. Sharman, and Cornelis Tanis.

REFERENCES

Akkermans, H. A., Bogerd, P., Yucesan, E., and Wassenhove, L. N. V. (1999) The Impact of ERP on Supply Chain Management: Exploratory Findings From a European Delphi Study. Working paper available from enver.yucesan@insead.fr.

Akkermans, H. A. and Horst, H. v. d. (2000) Managing IT Infrastructure Standardisation in the Networked Manufacturing Firm. Working paper available from H.A.Akkermans@tm.tue.nl.

AMR Research (2001) Microsoft Assembling Small and Midsize Enterprise Application Portfolio. 17 April. http://www.amrresearch.com/ftm/010424emsstory3.asp.

Ansberry, C. (2000) Online Supply Networks Boom, But Some Major Hurdles Loom. *Wall Street Journal Interactive Edition*, 17 April. http://www.wsj.com.

Bashein, B. J. and Markus, M. L. (2000) *Data Warehouses: More Than Just Mining*. Morristown, NJ: Financial Executives Research Foundation.

Bashein, B. J., Markus, M. L., and Finley, J. B. (1997) *Safety Nets: Secrets of Effective Information Technology Controls.* Morristown, NJ: Financial Executives Research Foundation Inc.

Bouchard, L. and Markus, M. L. (1996) Managing One's Business Partners: The Selling of EDI. In *Impression Management and Information Technology*, Beard, J. W. (ed.), Westport, CT: Quorum Books, pp. 65–91.

Brehm, L., & Markus, M. L. (2000) The Divided Software Life Cycle of ERP Packages. *Proceedings of the 1st Global Information Technology Management (GITM) World Conference*, 11–13 June Tennessee: Memphis, pp. 43–46.

Cameron, B., Shevlin, R., & Hardisty, A. (2000) *The Death of IT.* Forrester. http://www.forrester.com.

Chircu, A. M. and Kauffman, R. J. (2000) Reintermediation Strategies in Business-to-Business Electronic Commerce. *International Journal of Electronic Commerce* (Summer), available at http://ebusiness.mit.edu/.

Collaborative Commerce (2000) *Gartner Insight, 2*, electronic communication, access date 31 May.

Copeland, L. (2000, May 8) Auto Exchange Hits Potholes. *Computerworld.* http://www.computerworld.com.

Damsgaard, J., & Truex, D. (1999) Binary Trading Relations and the Limits of EDI Standards: The Procrustean Bed of Standards. Working paper Available from dtruex@gsu.edu.

Davenport, T. H. (1998) Putting the Enterprise Into The Enterprise System. *Harvard Business Review*, July–August, 121–131.

Davenport, T. H. (2000a) Chapter 9: The Future of ES-Enabled Organizations. *Mission Critical: Realizing the Value of Enterprise Systems.* Boston, MA: Harvard Business School Press, pp. 255–297.

Davenport, T. H. (2000b) *Mission Critical: Realizing the Value of Enterprise Systems.* Boston, MA: Harvard Business School Press.

Deloitte Consulting & Deloitte & Touche. (2000) The Future of B2B: A New Genesis. Available at http://www.netmarketmakers.com.

Donahue, S. (2000) Order Out: If the newly launched Biztro network catches on, every small business's back office just got a lot bigger – and cheaper. *Business 2.0*, 1 May. http://www.business2.com/content/magazine/indepth/2000/05/01/10756.

E-Procurement at Schlumberger (2000) *Harvard Business Review*, **78**(3), 21–22.

Fox, J. (2000, June 12) Lumbering Toward B2B. German Software Giant SAP Wants to Become the Power in B2B. Fortune, Vol. 141. http://www.fortune.com/fortune/2000/06/12/btb.html.

Gantenbeim, D. (2000, June 1) Gassed Up, Ready to Go: For one petroleum giant, the Internet means nothing short of survival. *Business 2.0.* http://www.business2.com/content/magazine/indepth/2000/06/01/11002.

Gray, P., & Watson, H. (1998) *Decision Support in the Data Warehouse.* Upper Saddle River, NJ: Prentice Hall.

Jahnke, A. (1998) How Bazaar. *CIO Web Business Magazine*, 8 August. http//www.cio.com.

Kalakota, R., & Robinson, M. (1999) *E-Business: Roadmap for Success.* Reading, MA: Addison-Wesley.

Kaplan, S., & Sawhney, M. (2000) E-Hubs: The New B2B Marketplaces. *Harvard Business Review*, **78**(3), 97–103.

Kern, T, Lacity, M., and Willcocks, L. (2002) *Netsourcing: Renting Business Applications and Services over Networks.* New York: Prentice Hall.

Kern, T., Lacity, M., Willcocks, L., Zuiderwijk, R., and Teunissen, W. (2001) *ASP Market-Space Report 2001*. CMG, Netherlands.

King, J. (2000, April 26) R/3 Users Look Beyond SAP for Business-to-business Help. *Computerworld*. http://www.computerworld.com.

Koh, C., Soh, C., and Markus, M. L. (2000) A Process Theory Approach to ERP Implementation and Impacts: The Case of Revel Asia. *Journal of Information Technology Cases and Applications*, **2**(1), 4–23.

Kremers, M. and Dissel, H. v. (2000) ERP System Migrations. *Communications of the ACM*, **43**(4 April), 53–56.

McCartney, N. (2001) A Plethora of New Outsourcing Deals. *Financial Times FT Telecoms Special Report*, 21 November, p. 11.

Markus, M. L. (1996) The Futures of IT Management. *The Data Base for Advances in Information Systems*, **27**(4 Fall), 68–84.

Markus, M. L. and Tanis, C. (2000) The Enterprise Systems Experience – From Adoption to Success. In *Framing the Domains of IT Research: Glimpsing the Future Through the Past*, R. W. Zmud (ed.), Cincinnati, OH: Pinnaflex Educational Resources.

Meehan, M. and Sullivan, B. (2000) Feds Scrutinize Airline's Ticket Site. *Computerworld*, 22 May. http://www.computerworld.com.

Norris, G., Hurley, J., Hartley, K., Dunleavy, J., and Balls, J. (2000) *E-Business and ERP: Transforming The Enterprise*. Chichester: Wiley.

Ohlson, K. (2000) Study: R/3 Users Face High Costs For Upgrades. *Computerworld*. http://www.computerworld.com.

Rao, B., Navoth, Z., Wuebker, R., and Horwitch, M. (1998) The FDX Group: Building the Electronic Commerce Backbone for the Future. http://www.ite.poly.edu/people/brao/fedex_case.htm.

Rosenberg, N. (1982) *Inside the Black Box: Technology and Economics*. Cambridge: Cambridge University Press.

Sawy, O. (2000). *Re-designing Enterprise Processes For E-Business*. Boston: McGraw Hill.

Schwartz, P. (1991). *The Art of the Long View*. New York, NY: Doubleday.

Sprott, D. (2000) Componentizing the Enterprise Applications Packages. *Communications of the ACM*, **43**(4 April), 63–69.

Stinchcombe, A. L. (1990) *Information and Organizations*. Berkeley, CA: University of California Press.

Werbach, K. (2000) Syndication: The Emerging Model for Business in the Internet Era. *Harvard Business Review*, **78**(3), 84–93.

Wheatley, M. (2000, June 1) ERP Training Stinks. *CIO*. http://www.cio.com.

Wouters, M. J. F., Sharman, G. J., and Wortmann, H. C. (1999). Reconstructing the Sales and Fulfillment Cycle to Create Supply Chain Differentiation. *International Journal of Logistics Management*, **10**(2), 83–98.

Index